Measurement While Drilling (MWD) Signal Analysis, Optimization and Design

Scrivener Publishing
100 Cummings Center, Suite 541J
Beverly, MA 01915-6106

Publishers at Scrivener
Martin Scrivener (martin@scrivenerpublishing.com)
Phillip Carmical (pcarmical@scrivenerpublishing.com)

Measurement While Drilling (MWD) Signal Analysis, Optimization and Design

by

Wilson C. Chin, Ph.D., M.I.T.

Stratamagnetic Software, LLC, Houston, Texas

Yinao Su, Limin Sheng, Lin Li, Hailong Bian and Rong Shi

China National Petroleum Corporation (CNPC), Beijing, China

Scrivener
Publishing

WILEY

Co-published by John Wiley & Sons, Inc. Hoboken, New Jersey, and Scrivener Publishing LLC, Salem, Massachusetts.
Published simultaneously in Canada.

For general information on our other products and services or for technical support, please contact our Customer Care Department within the United States at (800) 762-2974, outside the United States at (317) 572-3993 or fax (317) 572-4002.

Wiley also publishes its books in a variety of electronic formats. Some content that appears in print may not be available in electronic formats. For more information about Wiley products, visit our web site at www.wiley.com.

For more information about Scrivener products please visit www.scrivenerpublishing.com.

Cover design by Kris Hackerott

Library of Congress Cataloging-in-Publication Data:

ISBN 978-1-118-83168-7

Printed in the United States of America

10 9 8 7 6 5 4 3 2 1

Contents

Opening Message   **xiii**

Preface   **xv**

Acknowledgements   **xix**

1 Stories from the Field, Fundamental Questions and Solutions   **1**

1.1 Mysteries, Clues and Possibilities   1

1.2 Paper No. AADE-11-NTCE-74, "High-Data-Rate Measurement-While-Drilling System for Very Deep Wells," updated   10

    1.2.1 Abstract   10

    1.2.2 Introduction   10

    1.2.3 MWD telemetry basis   12

    1.2.4 New telemetry approach   13

    1.2.5 New technology elements   15

        1.2.5.1 Downhole source and signal optimization   15

        1.2.5.2 Surface signal processing and noise removal   18

        1.2.5.3 Pressure, torque and erosion computer modeling   19

        1.2.5.4 Wind tunnel analysis: studying new approaches   22

        1.2.5.5 Example test results   41

    1.2.6 Conclusions   44

    1.2.7 Acknowledgements   45

    1.2.8 References   45

1.3 References   46

2 Harmonic Analysis: Six-Segment Downhole Acoustic Waveguide   **47**

2.1 MWD Fundamentals   48
2.2 MWD Telemetry Concepts Re-examined   49
  2.2.1 Conventional pulser ideas explained   49
  2.2.2 Acoustics at higher data rates   50
  2.2.3 High-data-rate continuous wave telemetry   52
  2.2.4 Drillbit as a reflector   53
  2.2.5 Source modeling subtleties and errors   54
  2.2.6 Flow loop and field test subtleties   56
  2.2.7 Wind tunnel testing comments   58
2.3 Downhole Wave Propagation Subtleties   58
  2.3.1 Three distinct physical problems   59
  2.3.2 Downhole source problem   60
2.4 Six-Segment Downhole Waveguide Model   62
  2.4.1 Nomenclature   64
  2.4.2 Mathematical formulation   66
    2.4.2.1 Dipole source, drill collar modeling   66
    2.4.2.2 Harmonic analysis   68
    2.4.2.3 Governing partial differential equations   69
    2.4.2.4 Matching conditions at impedance junctions   71
    2.4.2.5 Matrix formulation   72
    2.4.2.6 Matrix inversion   74
    2.4.2.7 Final data analysis   75
2.5 An Example: Optimizing Pulser Signal Strength   77
  2.5.1 Problem definition and results   77
  2.5.2 User interface   80
  2.5.3 Constructive interference at high frequencies   81
2.6 Additional Engineering Conclusions   83
2.7 References   85

3 Harmonic Analysis: Elementary Pipe and Collar Models   **86**
3.1 Constant area drillpipe wave models   86
  3.1.1 Case (a), infinite system, both directions   87
  3.1.2 Case (b), drillbit as a solid reflector   88
  3.1.3 Case (c), drillbit as open-ended reflector   88
  3.1.4 Case (d), "finite-finite" waveguide of length 2L   89
  3.1.5 Physical Interpretation   89

3.2	Variable area collar-pipe wave models	92
	3.2.1 Mathematical formulation	92
	3.2.2 Example calculations	94
3.3	References	96

4 Transient Constant Area Surface and Downhole Wave Models   **97**

4.1 Method 4-1. Upgoing wave reflection at solid boundary,
single transducer deconvolution using delay equation,
no mud pump noise   99
    4.1.1 Physical problem   99
    4.1.2 Theory   100
    4.1.3 Run 1. Wide signal – low data rate   101
    4.1.4 Run 2. Narrow pulse width – high data rate   103
    4.1.5 Run 3. Phase-shift keying or PSK   104
    4.1.6 Runs 4, 5. Phase-shift keying or PSK, very high
        data rate   107

4.2 Method 4-2. Upgoing wave reflection at solid boundary,
single transducer deconvolution using delay equation,
with mud pump noise   108
    4.2.1 Physical Problem   108
    4.2.2 Software note   109
    4.2.3 Theory   109
    4.2.4 Run 1. 12 Hz PSK, plus pump noise with S/N = 0.25   110
    4.2.5 Run 2. 24 Hz PSK, plus pump noise with S/N = 0.25   111

4.3 Method 4-3. Directional filtering – *difference* equation
method requiring two transducers   112
    4.3.1 Physical problem   112
    4.3.2 Theory   113
    4.3.3 Run 1. Single narrow pulse, S/N = 1, approximately   114
    4.3.4 Run 2. Very noisy environment   116
    4.3.5 Run 3. Very, very noisy environment   117
    4.3.6 Run 4. Very, very, very noisy environment   118
    4.3.7 Run 5. Non-periodic background noise   119

4.4 Method 4-4. Directional filtering – *differential* equation
method requiring two transducers   120
    4.4.1 Physical problem   120
    4.4.2 Theory   121
    4.4.3 Run 1. Validation analysis   122

4.4.4 Run 2. A very, very noisy example 124
4.4.5 Note on multiple-transducer methods 125
4.5 Method 4-5. Downhole reflection and deconvolution at the bit, waves created by MWD dipole source, bit assumed as perfect solid reflector 126
4.5.1 Software note 126
4.5.2 Physical problem 127
4.5.3 On solid and open reflectors 127
4.5.4 Theory 128
4.5.5 Run 1. Long, low data rate pulse 130
4.5.6 Run 2. Higher data rate, faster valve action 130
4.5.7 Run 3. PSK example, 12 Hz frequency 131
4.5.8 Run 4. 24 Hz, Coarse sampling time 132
4.6 Method 4-6. Downhole reflection and deconvolution at the bit, waves created by MWD dipole source, bit assumed as perfect open end or zero acoustic pressure reflector 133
4.6.1 Software note 133
4.6.2 Physical problem 133
4.6.3 Theory 134
4.6.4 Run 1. Low data rate run 135
4.6.5 Run 2. Higher data rate 136
4.6.6 Run 3. Phase-shift-keying, 12 Hz carrier wave 137
4.6.7 Run 4. Phase-shift-keying, 24 Hz carrier wave 137
4.6.8 Run 5. Phase-shift-keying, 48 Hz carrier 138
4.7 References 139

5 Transient Variable Area Downhole Inverse Models 140
5.1 Method 5-1. Problems with acoustic impedance mismatch due to collar-drillpipe area discontinuity, with drillbit assumed as open-end reflector 142
5.1.1 Physical problem 142
5.1.2 Theory 143
5.1.3 Run 1. Phase-shift-keying, 12 Hz carrier wave 147
5.1.4 Run 2. Phase-shift-keying, 24 Hz carrier wave 147
5.1.5 Run 3. Phase-shift-keying, 96 Hz carrier wave 148
5.1.6 Run 4. Short rectangular pulse with rounded edges 149

5.2   Method 5-2. Problems with collar-drillpipe area
      discontinuity, with drillbit assumed as closed end, solid
      drillbit reflector   150
      5.2.1   Theory   150
      5.2.2   Run 1. Phase-shift-keying, 12 Hz carrier wave   150
      5.2.3   Run 2. Phase-shift-keying, 24 Hz carrier wave   151
      5.2.4   Run 3. Phase-shift-keying, 96 Hz carrier wave   151
      5.2.5   Run 4. Short rectangular pulse with rounded edges   151
5.3   References   152

6   Signal Processor Design and Additional Noise Models   **153**
6.1   Desurger Distortion   154
      6.1.1   Low-frequency positive pulsers   156
      6.1.2   Higher frequency mud sirens   157
6.2   Downhole Drilling Noise   160
      6.2.1   Positive displacement motors   161
      6.2.2   Turbodrill motors   162
      6.2.3   Drillstring vibrations   162
6.3   Attenuation Mechanisms   164
      6.3.1   Newtonian model   164
      6.3.2   Non-Newtonian fluids   165
6.4   Drillpipe Attenuation and Mudpump Reflection   167
      6.4.1   Low-data-rate physics   168
      6.4.2   High data rate effects   169
6.5   Applications to Negative Pulser Design in Fluid Flows and
      to Elastic Wave Telemetry Analysis in Drillpipe Systems   170
6.6   LMS Adaptive and Savitzky-Golay Smoothing Filters   172
6.7   Low Pass Butterworth, Low Pass FFT and Notch Filters   174
6.8   Typical Frequency Spectra and MWD Signal Strength
      Properties   175
6.9   References   176

7   Mud Siren Torque and Erosion Analysis   **177**
7.1   The Physical Problem   177
      7.1.1   Stable-closed designs   179
      7.1.2   Previous solutions   179
      7.1.3   Stable-opened designs   181

	7.1.4	Torque and its importance	182
	7.1.5	Numerical modeling	183
7.2	Mathematical Approach		183
	7.2.1	Inviscid aerodynamic model	185
	7.2.2	Simplified boundary conditions	186
7.3	Mud Siren Formulation		188
	7.3.1	Differential equation	188
	7.3.2	Pressure integral	189
	7.3.3	Upstream and annular boundary condition	190
	7.3.4	Radial variations	192
	7.3.5	Downstream flow deflection	193
	7.3.6	Lobe tangency conditions	194
	7.3.7	Numerical solution	194
	7.3.8	Interpreting torque computations	195
	7.3.9	Streamline tracing	196
7.4	Typical Computed Results and Practical Applications		198
	7.4.1	Detailed engineering design suite	198
7.5	Conclusions		204
	7.5.1	Software reference	204
7.6	References		205

8 Downhole Turbine Design and Short Wind Tunnel Testing   206
8.1	Turbine Design Issues	206
8.2	Why Wind Tunnels Work	208
8.3	Turbine Model Development	211
8.4	Software Reference	215
8.5	Erosion and Power Evaluation	219
8.6	Simplified Testing	221
8.7	References	223

9 Siren Design and Evaluation in Mud Flow Loops and Wind Tunnels   224
9.1	Early Wind Tunnel and Modern Test Facilities		225
	9.1.1	Basic ideas	226
	9.1.2	Three types of wind tunnels	227
	9.1.3	Background, early short wind tunnel	228
	9.1.4	Modern short and long wind tunnel system	229
	9.1.5	Frequently asked questions	233

9.2   Short wind tunnel design   236
    9.2.1   Siren torque testing in short wind tunnel   240
    9.2.2   Siren static torque testing procedure   243
    9.2.3   Erosion considerations   246
9.3   Intermediate Wind Tunnel for Signal Strength
    Measurement   248
    9.3.1   Analytical acoustic model   249
    9.3.2   Single transducer test using speaker source   251
    9.3.3   Siren Δp procedure using single and differential
         transducers   252
    9.3.4   Intermediate wind tunnel test procedure   254
    9.3.5   Predicting mud flow Δp's from wind tunnel data   257
9.4   Long Wind Tunnel for Telemetry Modeling   259
    9.4.1   Early construction approach - basic ideas   259
    9.4.2   Evaluating new telemetry concepts   264
9.5   Water and Mud Flow Loop Testing   264
    9.5.1   Real-world flow loops   265
    9.5.2   Solid reflectors   267
    9.5.3   Drillbit nozzles   268
    9.5.4   Erosion testing   269
    9.5.5   Attenuation testing   270
    9.5.6   The way forward   272

10 Advanced System Summary and Modern MWD Developments 273
10.1 Overall Telemetry Summary   274
    10.1.1   Optimal pulser placement for wave interference   274
    10.1.2   Telemetry design using FSK   277
    10.1.3   Sirens in tandem   279
    10.1.4   Attenuation misinterpretation   280
    10.1.5   Surface signal processing   284
    10.1.6   Attenuation, distance and frequency   287
    10.1.7   Ghost signals and echoes   290
10.2 MWD Signal Processing Research in China   291
10.3 MWD Sensor Developments in China   300
    10.3.1   DRGDS Near-bit Geosteering Drilling System   300
        10.3.1.1   Overview   300
        10.3.1.2   DRGDS tool architecture   300
        10.3.1.3   Functions of DRGDS   309

	10.3.2	DRGRT Natural Azi-Gamma Ray Measurement	314
	10.3.3	DRNBLog Geological Log	318
	10.3.4	DRMPR Electromagnetic Wave Resistivity	320
	10.3.5	DRNP Neutron Porosity	321
	10.3.6	DRMWD Positive Mud Pulser	325
	10.3.7	DREMWD Electromagnetic MWD	326
	10.3.8	DRPWD Pressure While Drilling	329
	10.3.9	Automatic Vertical Drilling System – DRVDS-1	332
	10.3.10	Automatic Vertical Drilling System – DRVDS-2	336
10.4	Turbines, Batteries and Closing Remarks		337
	10.4.1	Siren drive	337
	10.4.2	Turbine-alternator system	337
	10.4.3	Batteries	338
	10.4.4	Tool requirements	339
	10.4.5	Design trade-offs	340
10.5	References		341

Cumulative References   342

Index   347

About the Authors   354

Opening Message

Yinao Su, Ph.D., Academician
Chinese Academy of Engineering

In modern oil and gas exploration, drilling offers many engineering challenges. Multiple economical objectives are targeted, among them rapid penetration rates and productive payzones. To achieve this, high-data-rate MWD systems are urgently needed for detailed and accurate downhole characterization, real-time information being central to control, optimization and safety. But the downhole environment is not forgiving: high noise levels, strong signal distortion and interference, together with severe attenuation, impede data transmission rate. To overcome these difficulties, completely new systems oriented designs are required to replace simple fixes to existing tools.

Several approaches are available, e.g., electromagnetic wave, intelligent wired pipe and drillpipe acoustics, each possessing its unique shortcomings. Here we have asked, "Is it possible to improve mud pulse telemetry, the most popular and by far least intrusive operationally?" The answer is, "*Yes!*" We have applied wave propagation principles to hardware development, telemetry design and surface signal processing, treating our challenge from an integrated systems perspective. With research guided by theory and experiment, we have shown that basic transmission rates can be increased significantly, with further improvements possible through data compression.

At China National Petroleum Corporation, through its Drilling Research Institute, new technology, research and innovation aim at responsibly providing society with clean, safe and reliable energy. In this book, we wish to share our experiences with the industry in achieving our goal for "Developing Energy, Creating Harmony." We hope that the methods we have pioneered, described in detail, will contribute to finding oil and gas more safely and efficiently.

Beijing, China

Preface

The physical theories behind Measurement-While-Drilling design should be rich in scientific challenges, engineering principles and mathematical elegance. To develop the next generation of high-data-rate tools, these must be understood and applied unfailingly without compromise. But one does not simply peruse the latest petroleum books, state-of-the-art reviews, or the most recent patents to understand their teachings. Most descriptions are just wrong. The science itself does not exist. All simply rehash hearsay and misconceptions that have proliferated for more than three decades – recycled street narratives and folklore about sirens, positive and negative pulsers, and yes, mud attenuation; over-simplified product brochures from oil service companies that monopolize the industry; and, unfortunately, all preach the same complaints about low data rates and industry's failure to address modern logging needs.

The truth is, there have been no substantive developments in MWD telemetry and design over the years. Not one paper has appeared that deals with telemetry in a manner worthy of scientific publication. New tools, more like muscle-machines than intelligent instruments, are designed without regard for acoustic concepts, while signal optimization and surface processing, more often than not based on "hand-waving" arguments, proceed without guidance from wave equation models. True, tools *are* better engineered; mechanical parts erode less, pulser modulation is controlled more reliably, high-powered microprocessors have replaced simple circuit boards, electronic components survive higher temperatures and pressures, and overall reliability is impressive, all of which enables the logging industry to reach deeper targets. However, these are incremental improvements unlikely to change the big picture. And the big picture is bleak: unless conceptual breakthroughs are made, the present low data-rate environment is likely to persist.

Through this rapid progress, several disturbing problems are apparent. The first author, having consulted for established as well as start-up companies over the past ten years, is aware of no comprehensive theory addressing MWD acoustics. There are no university courses developed to educate the next generation of telemetry designers. The one-dimensional wave propagation models that are available are no more sophisticated that organ acoustics formulas from Physics 101. And tight-lipped service companies have been reluctant to publicize their failings, for obvious reasons, a business decision that has stymied progress in an important commercial endeavor. But unless companies are willing to share ideas and experiences, no one will benefit.

All of this is not new to science and certainly not unique to the commercialization of new products. The aerospace industry, decades ago just as subdued and secretive, suffered from similar failings. In that era though, just as the first author completed his Ph.D. from the Massachusetts Institute of Technology in aerospace engineering, companies like Boeing, Lockheed and McDonnell-Douglas, for instance, finally recognized that the best way forward was free dissemination of scientific methods. Engineers openly carried their Fortran decks from one company to the next, published their findings in open journals and debated their ideas with new-found colleagues near and far. Increased employment mobility only increased idea dissemination more rapidly. The rest is history: the Space Shuttle, the Space Station, the 767, 777 and 787. It is in this spirit that the present book is written: intellectual curiosity and honesty and a genuine interest to see MWD data rates improve.

The author, no new-comer to MWD, earned his stripes at Schlumberger and Halliburton, managing MWD telemetry efforts that developed and refined new hardware concepts and signal processing techniques. However, research funding was fragmented and scientific objectives were unclear. Knowing the right questions, it is understood, solves half the problem. But it was not until the new millenium that progress in the formulation and solution of rigorous wave equation models took hold. Numerical models, notorious for artificial dissipation and dispersion, that is, phase error, were abandoned in favor of more challenging exact analytical solutions. Physical principles could, for once, be clearly understood. New methods to model acoustic sources were developed and special studies were initiated to define broad classes of noise together with the requirements for their elimination. New experimental procedures based on acoustics models were designed, as were special "short" and "long wind tunnels" that accommodated subtle physical mechanisms newly identified.

Theories and models, even the most credible, can be incorrect. In the final analysis, well designed experiments are needed to validate or disprove new ideas. In this regard, China National Petroleum Corporation (CNPC) offered to build laboratory facilities, test siren designs, educate staff and evaluate new telemetry methods, and importantly, to share its results and technology openly with the petroleum industry.

A comprehensive project overview was first presented by the authors in "High-Data-Rate Measurement-While-Drilling System for Very Deep Wells," Paper No. AADE-11-NTCE-74, at the American Association of Drilling Engineers' 2011 AADE National Technical Conference and Exhibition, Houston, Texas, April 12-14, 2011. The paper summarized key ideas and results, but given page limits, could not provide details. All of our theoretical and experimental methods are now explained and summarized in this book, with numerous examples, providing useful tools to students and designers alike – our signal processing methods, dealing with signal reflection, distortion and optimization, are formulated, solved, validated and described for the first time. In addition, we offer a new prototype road-map for high-data-rate MWD that has found strong support from knowledgeable industry professionals.

Since publication of the above paper, numerous commercial drivers have made high-data-rate telemetry needs increasingly urgent. In the "old days," conventional well logging data, e.g., resistivity, sonic or positioning, was simply transmitted to the surface for monitoring and evaluation. However, recent trends call for near-bit geosteering and rotary-steerable capabilities, in support of real-time economic and pore and annular pressure measurements. Despite their importance, few industry publications or websites provide "behind the scenes" descriptions of tool and software development processes, offering little to newer engineers eager to understand the technology – an unfortunate circumstance occurring even as the industry's "great crew change" takes place.

To fill this need, China National Petroleum Corporation (CNPC) has encouraged us to document in detail its engineering processes, new tools and well logging sensors, in a comprehensive collection of laboratory and field photographs. Much of this work parallels ongoing developments in the West and sheds considerable insight into the country's efforts to embrace high technology, e.g., stealth fighters, moon missions, fast computers and deep-sea submersibles, and its new-found open-ness in sharing its intellectual property. This book also captures the spirit of MWD engineering in China – we have provided recent paper abstracts and described advanced sensor development activities. It is the authors' hope that the new technologies offered in the following chapters will contribute to the industry's continuing need and increasing demand for real-time data as deeper, higher potential and more dangerous wells are drilled.

Wilson C. Chin, Ph.D., M.I.T.
Houston, Texas
Email: wilsonchin@aol.com
Phone: (832) 483-6899

Acknowledgements

The lead author gratefully acknowledges the insights, experiences and friendships he acquired during his early MWD exposure at Schlumberger, Halliburton and other companies – pleasant memories that much more than compensate for the frustrations and sleepless nights brought upon by the challenges of high data rate telemetry. All of the authors are indebted to China National Petroleum Corporation for its support and encouragement throughout this project, and in particular, for its willingness and desire to share its results and activities with the petroleum engineering and well logging community.

Phillip Carmical, Acquisitions Editor and Publisher, has been extremely supportive of this book project and others in progress. His philosophy, to explain scientific principles the way they must be told, with equations and algorithms, is refreshing in an environment often shrouded in secrecy and commercialism. The authors are optimistic that their story-telling will advance the technology and explain why "black boxes" aren't so mysterious after all.

Finally, the authors thank Xiaoying "Jenny" Zhuang for her hard work and commitment to ably working both sides of the language barrier (the lead author neither speaks nor reads Chinese, while the CNPC team is newly conversant in English). Without her interpretation skills and willingness to learn and understand MWD design issues, our efforts would not have yielded the successes they have and would not have led to friendships and lasting memories. And without Jenny's personal devotion to a cause, this book would never have seen publication – and who knows, low data rates may remain just that.

1
Stories from the Field,
Fundamental Questions and Solutions

This chapter might aptly be entitled "Confessions of a confused, high-tech engineer." And here's why. In 1981, I was Manager, Turbomachinery Design, at Pratt & Whitney Aircraft, United Technologies Corporation, the company that supplied the great majority of the world's commercial jet engines. Prior to that, I had served as Research Aerodynamicist at Boeing, working with pioneers in computational fluid dynamics and advanced wing design. What qualified me for these enviable positions was a Ph.D. from the Massachusetts Institute of Technology in acoustic wave propagation – and I had joined a stodgy M.I.T. from its even stodgier cross-town rival, the California Institute of Technology. These credentials in acoustics and fluid mechanics design made me eminently qualified to advance the state-of-the-art in Measurement-While-Drilling (also known as, "MWD") telemetry – or so I, and other companies, unknowingly thought. At this juncture in my life, the journey through the Oil Patch begins.

1.1   Mysteries, Clues and Possibilities

As a young man, I had dreaded the idea of forever making incremental improvements to aircraft systems, merely as a mainstay to the art of survival and paying the mortgage, sitting at the same desk, in the same building, for decades on end. That possibility, I believed, was a fate worse than death. Thus, in that defining year, I answered a Schlumberger employment advertisement in The New York Times for scientists eager to change the world – the petroleum world, anyway. But unconvinced that any normal company would hire an inexperienced aerospace engineer, and of all things, for a position chartered with high-tech underground endeavors, I was unwilling to give up one of my ten valuable, hard-earned vacation days. Still, the company was stubborn in its pursuit and, for better or worse, kindly accommodated my needs.

Carl Buchholz, the division president at the time, interviewed me that one fateful Saturday. "What do you know about oil?" he bluntly asked, giving me that honest Texan look in the eye. To be truthful, I did not know anything, zilch.

"Nothing, but I've watched Jed Clampett shoot it out of the ground," I confessed (Clampett was the hillbilly in the television sitcom who blasted his rifle into the ground, struck oil and moved to Los Angeles to live in his new mansion in "The Beverly Hillbillies"). Buchholz broke out in uncontrolled laughter. That type of honesty he appreciated. I got the job. And with that, I became Schlumberger's Supervisor, MWD Telemetry, for 2nd generation mud siren and turbine design.

The company's Analysts division, at the time responsible for an ambitious next-generation, high-data-rate MWD design program, had built ultra-modern office and flow loop facilities in southwest Houston. The metal pipe test section was housed in an air-conditioned room where engineers could work in a clean and comfortable environment away from the pulsations of the indoor mudpump that supplied our flow. A small section of the flow loop was accessible in this laboratory with the main plumbing carefully hidden behind a wall – details no self-respecting, white-collar Ph.D. cared for nor admitted an interest to.

My charter was simple. We were transmitting at 3 bits/sec in holes shallow by today's standards with a 12 Hz carrier frequency. Our objective was N bits/sec, where $N \gg 3$ (the value of N is proprietary). The solution seemed straightforward, as company managers and university experts would have it. Simply "crank up the carrier to $(N/3) \times 12$ Hz and run." I did that. But my transducers would measure only confusion, with new pressure oscillations randomly adding to old ones and results depending on mud type, pump speed and time of day. What happened "behind the wall" controlled what we observed but we were too naïve to know. Anecdotal stories told by different field hands about new prototypes were confusing and contradictory. One simply did not know what to believe. Thirty years later, the data rate is still comparable, a bit better under ideal conditions, as it was then. Clearly, there were physical principles that we did not, or perhaps were never meant to, fully comprehend.

Fast-forward to 1992 at Halliburton Energy Services, an eternity later, where I had been hired as Manager, *FasTalk* MWD. Again, mass confusion prevailed. Some field engineers had reported excellent telemetry results in certain holes, while others had reported poor performance under seemingly identical conditions. The company had acquired several small companies during that reign of corporate acquisitions in the oil service industry. It would turn out that "good versus bad" depended, with all other variables constant, on whether the signal valve was a "positive" or a "negative" pulser. No one really distinguished between the two: because the MWD valve was simply viewed as a piston located at the end of the drillpipe, exciting the drilling fluid column residing immediately above, it didn't matter if it was pushing or pulling. Sirens were a different animal; no one, except Schlumberger, it seemed, understood them. But nobody really did.

Additional dependencies on drilling conditions only added to the confusion. Industry consensus at the time held that MWD telemetry characteristics depended on drillbit type and nozzle size and, perhaps, rock properties, to some extent. It also appeared that whether or not the drillbit was off-bottom mattered. Very often, common sense dictated that the drillbit acted as a solid reflector, since nozzle cross-sectional areas were "pretty small" compared to pipe dimensions. Yet, this line of reasoning was contradictory and had its flaws; strong MWD signals by then had been routinely detected in the borehole annulus, where their existence or lack of was used to infer gas influx. It became clear that what the human eye visually perceived as small may not be small from a propagating wave's perspective.

Lack of controlled experiments also pervaded the industry and still does. Whenever any service company design team was lucky enough to find a test well, courtesy of obliging operating company customers, engineering "control" usually meant installing the same pressure transducer in the same position on the standpipe. New tools that were tested in one field situation would perform completely differently in others: standpipe measurements had lives of their own, it seemed, except at very low data rates of 1 bit/sec or less, barring mechanical tool failure, which was often. Details related to surface plumbing, bottomhole assembly, bit-box geometry, drilling motor details and annular dimensions, were not recorded and were routinely ignored. The simple "piston at the end of pipe model" didn't care – and neither did most engineers and design teams.

By the mid-1990s, the fact that higher data rate signaling just might depend on wave propagation dawned upon industry practitioners. This revelation arose in part from wave-equation-based seismics – new then, not quite understood, but successful. I began to view my confusion as a source of inspiration. The changing patterns of crests and troughs I had measured *had* to represent waves – waves whose properties *had* to depend on mud sound speed and flow loop geometry. At Halliburton, I would obtain patents teaching how to optimize signals by taking advantage of wave propagation, e.g., signal strength increase by downhole constructive wave interference (without incurring erosion and power penalties), multiple transducer array signal processing to filter unwanted signals based on direction and not frequency, and others.

Still, the future of mud pulse telemetry was uncertain, confronting an unknowing fate – a technology held hostage by still more uncontrolled experiments and their dangerous implications. At the time, industry experts had concluded that mud pulse telemetry's technology limits had been attained and that no increases in data rate would be forthcoming. At Louisiana State University's ten-thousand-feet flow loop, researchers had carefully increased MWD signal frequencies from 1 to 25 Hz, and measured, to their dismay, continually decreasing pressures at a second faraway receiver location. At approximately 25 Hz, the signal disappeared. Completely. That result was confirmed by yours truly, at the same facility, using a slightly different pulser system. Enough said – the story was over. Our MWD research efforts were terminated in 1995 and I resigned from the company in 1999.

The key revelation would come years later as I watched children play "jump rope" in the park. A first child would hold one end of the rope, while a second would shake the opposing end at a given frequency. Transverse waves on a rope are easy to visualize, but the ideas apply equally to longitudinal waves. The main idea is this. At any given frequency, a standing wave system with nodes and antinodes is created that depends on material properties. If the frequency changes, the nodal pattern changes and moves. If one fixes his attention at one specific location, the peak-to-peak displacement appears to come and go. Node and anti-node positions move: what may be interpreted as attenuation may in fact be amplitude reduction due to destructive wave interference – a temporary effect that is not thermodynamically irreversible loss.

This was exactly the situation in the 10,000 feet LSU flow loop. At one end is a mudpump whose pistons act like solid reflectors, assuming tight pump seals, while at the opposite end, a reservoir serves as an open-end acoustic reflector. Pressure transducers were located at *fixed* positions along the length of the acoustic path. Unlike the jump-rope analogy, the MWD pulser was situated a distance from one of the ends, adding some complexity to the wave field since waves with antisymmetric pressures traveled in both directions from the source. The exact details are unimportant for now. However, the main idea drawn from the jump rope analogy applies: increasing frequency simply changes the standing wave pattern and we (and others) were measuring nothing more than expected movements nodes and antinodes. Attenuation results were buried in the mass of resulting data. This is easy to understand in hindsight. Recent calculations, in fact, show that large attenuation is impossible over the length of the flow loop for the mud systems used.

One crucial difference was suggested above. Whereas, in our jump rope example, excitations originated at the very end of the waveguide (i.e., "at the bit"), the excitations in the LSU flow loop occurred within the acoustic path, introducing subtleties. For example, when a positive pulser or a mud siren closes, a high-pressure signal is created upstream while a low-pressure signal is formed downstream, with both signals propagating away from the valve; the opposite occurs on closure. These long waves travel to the ends of the acoustic channel, reflect accordingly as the end is a solid or open, and travel back and forth through the valve (which never completely closes) to set up a standing wave patterns whose properties depend on mud, length and source.

Had our pulser created disturbance pressure fields that were symmetric with respect to source position, as opposed to being antisymmetric – that is, had we tested a "negative pulser," our results and conclusions would have been completely different. Any theory of wave propagation applicable to MWD telemetry *had* to accommodate end boundary conditions, acoustic impedance matching conditions at area (pipe and collar junctions) or material discontinuities (rubber interfaces in mud motors), and importantly, signal source "dipole" or "monopole" properties. Fortunately, such a general theory is now available for signal prediction and inversion and forms, in part, the subject matter of this book. Using the six-segment-waveguide model in Chapter 2, one can confirm the LSU findings. Importantly, one can show that MWD signals can survive well beyond published 25 Hz limits and explain why the industry's very slow pulsers always create strong signals. In fact, in Chapter 10, the method is used to design a conceptual prototype system capable of transmitting more than 10 bits/sec without data compression in very deep wells.

Engineers even today give overly simplified explanations for MWD signal generation; these physical inadequacies are reflected in models which do not, and cannot, extrapolate the full potential of mud pulse telemetry. We describe some of these fallacies. First, many believe that an "obvious" pressure drop, or "delta-p" (denoted by Δp), created by a valve is essential for MWD signal generation. Very often, this is incorrectly measured in laboratory flow loops using slowly opening and closing pulsers and orifices. This unfortunately measures pressure drops associated with viscous losses about blunt valves – and has nothing to do with the acoustic water hammer pulses associated with high data rate – that is, the "banging" of the mud column that brings it to a near stop. This dynamic element of the testing cannot be ignored or compromised.

But even more troublesome is the Δp explanation itself. Viewed as an essential requirement for MWD signal generation, the concept is completely inapplicable to negative pulsers. For positive pulsers and sirens, the created acoustic pressures are antisymmetric with respect to source location, and a nonzero pressure differential always exists. But while it is true that such pulsers create acoustic Δp's that excite the telemetry channel, Δp's are not necessary for all MWD systems. A negative pulser on opening (or closing) creates acoustic disturbance pressure fields that are symmetric with respect to source position. As such, the corresponding Δp is identically zero; for such systems, it is the (nonzero) discontinuity in axial velocity across the source position that is directly correlated to the signal. The formulation differences between acoustic "dipoles" (that is, positive pulsers and mud sirens) and "monopoles" (negative pulsers) are carefully distinguished in this book. Because negative pulsers can damage or even fracture underground formations, and are therefore a liability in deepwater applications, we will not focus on their design in this book.

Competent engineering requires one to distinguish between length scales that are relevant and those that are not. As will become clear in Chapter 2, and as suggested in our discussion on drillbit geometry, the ratio between nozzle and drillpipe diameters is one such measure that is mostly irrelevant to long wave acoustics. Another meaningless measure is the ratio of the pulser-to-drillbit distance to drillpipe length. The extreme smallness of this dimensionless number is often used to justify, for modeling purposes, the placement of the pulser at the bottom end of an idealized drilling channel. In effect, this reduces the formulation to a simple "piston at the end of a pipe" model which can be solved by most graduate engineers. But as we will demonstrate, this simplification amounts to "throwing out the baby with the bath water."

And why is this? Piston models are unable to deal with source properties: they cannot distinguish between created pressures that are antisymmetric with respect to source position and those that are symmetric. Thus they predict like physics for both dipoles and monopoles. What's worse, the possibility that upgoing waves can interact constructively with those that travel downward and then reflect up cannot be addressed – this potential application is extremely important to signal enhancement by constructive wave interference, which is achievable by tailoring the telemetry scheme to take advantage of phase properties associated with the mud sound speed and bottomhole assembly.

Moreover, the simple piston model precludes signal propagation up the borehole annulus, which as discussed, has proven to be useful in gas influx detection while drilling. When the complete waveguide – to include the annulus and bit-box as essential elements – is treated as an integrated system, as will be done in Chapter 2, it becomes clear that our simple description of the drillbit as a solid or an open reflector – offered only for illustrative purposes – is too simplified. By extending our formulation to allow pulsers to reside *within* the drill collar and not simply *at* the drillbit, we will demonstrate a wealth of physical phenomena and engineering advantages previously unknown.

The subject of surface signal processing and reflection cancellation is similarly shrouded in mystery. An early patent for "dual transducer, differential detection" draws analogies with electric circuit theories, however, using methods with sinusoidal $e^{i\omega t}$ dependencies. But why time periodicity is relevant at all in systems employing randomly occurring phase shifts is never explained. Rules-of-thumb related to quarter-wavelength interactions, appropriate only to steady-state waves (which do not convey information) used in the patent, prevail to this day for transient situations. They can't possibly work and they don't.

Just as troubling are more recent company patents on multiple transducer surface signal processing which sound more like accounting recipes than scientific algorithms, e.g., "subtract this, delay that, add to the shifted value," when, in fact, formal methods based on the wave equation (derived later in this book) yield more direct, rigorous and generalizable results. We take our cues directly from wave-equation-based seismic processing where all propagation details, including those related to source properties, are treated in their full generality. With this approach, new multiple transducer position and multiple time level reflection cancellation schemes can be inferred straightforwardly from finite difference discretizations of a basic solution to the wave equation.

As if all of this were not bad enough, we take as our final example, the infamous "case of the missing signal," the mystery which had stymied many of the best minds one too many times – a situation in which MWD tools of all kinds refused to yield discernible standpipe pressures despite their near-perfect mechanical condition. It turned out that, of all things, operators were using inexpensive centrifugal (as opposed to positive displacement) pumps. This illustration offers the strongest, most compelling evidence supporting the wave nature underlying MWD signals. Pistons on positive displacement mud pumps function as solid reflectors, which double the upgoing signal at the piston face; centrifugal pumps with open ends, to the contrary, enforce "zero acoustic pressure" constraints which destroy signals. An understanding of basic acoustics would have reduced frustration levels greatly and saved significantly on time and money.

Despite the problems raised, there are reasons for optimism in terms of understanding the physics and modeling it precisely. At high data rates, acoustic wavelengths λ are short but not too short. For example, from $\lambda = c/f$ where c is mud sound speed and f is excitation frequency in Hertz, a siren in a 3,000-5,000 ft/sec - 12 Hz environment would have a λ of about 300-500 ft. At 100 Hz, the wavelength reduces to 30-50 ft, which still greatly exceeds a typical drillpipe diameter. It is "long" acoustically. Thus, one-dimensionality applies to downhole signal generation and three-dimensional complications do not arise. More importantly, the waves are still long even in wind tunnels. Hence, signal creation and acoustic-hydraulic interactions at the pulser can be studied experimentally in convenient laboratory environments.

For those who have forgotten, one-dimensional acoustics is taught in high school and amply illustrated with organ pipe examples. Classical mathematics books give the general solution "f(x + ct) + g(x – ct)" showing that any solution is the sum of left and right-going waves; books on sound discuss impedance mismatches and conservation laws applicable at such junctions. Basic frequency-dependent attenuation laws have been available for over a hundred years. In this sense, the field is well developed. But in other respects the field offers fertile ground for nurturing new and practically useful ideas.

These new ideas include, for example, (1) formal derivations for receiver array reflection and noise cancellation based on the wave equation, (2) model development for elastic distortions of MWD signal at desurgers, (3) constructive and destructive wave interference in waveguides with multiple telescoping sections, (4) downhole signal optimization by constructive wave interference, (5) reflection deconvolution of multiple echoes created within the downhole MWD drill collar, and so on. All of these topics are addressed in this book. In fact, forward models are developed which create transient pressure signals when complicated waveguide geometries and telemetering schemes are specified, and complementary inverse models are constructed that extract position-encoded signals from massively reverberant fields under high-data-rate conditions, with mathematical consistency between the two demonstrated in numerous examples.

While innovative use of physical principles is emphasized for downhole telemetry design and signal processing, testing and evaluation of hardware and tool concepts are equally important, but often viewed as extremely time-consuming, labor-intensive and, simply, expensive. This need not be – and is not – the case. In "Flow Distribution in a Tricone Jet Bit Determined from Hot-Wire Anemometry Measurements," SPE Paper No. 14216, by A.A. Gavignet, L.J. Bradbury and F.P. Quetier, presented at the 1985 SPE Annual Technical Conference and Exhibition in Las Vegas, and in "Flow Distribution in a Roller Jet Bit Determined from Hot-Wire Anemometry Measurements," by A.A. Gavignet, L.J. Bradbury and F.P. Quetier, SPE Drilling Engineering, March 1987, pp. 19-26, the investigators, following ideas suggested by the lead author, who had by then routinely used *wind tunnels* to study sirens and turbines, showed how more detailed flow properties can be obtained using aerospace measurement methods in *air*. The scientific justification offered was the "highly turbulent nature of the flow." This counter-intuitive (but correct) approach to modeling mud provides a strategically important alternative to traditional testing that can reduce the cost of developing new MWD systems. Wind tunnel use in the petroleum industry was, by no means, new at the time. For instance, Norton, Heideman and Mallard (1983), with Exxon Production Research Company, and others, had published studies employing wind tunnel use in offshore platform design, extrapolating air-based results dimensionlessly to water flows using standard Strouhal and Reynolds number normalizations.

Additional reasons for wind tunnel usage are suggested by some simpler, but deeper arguments, than those in Gavignet *et al.* For static measurements (e.g., those for stall torque, power determination, erosion trends and streamline pattern) wind tunnels apply also to laminar flows. From basic fluid mechanics, for two flows to be alike dynamically, their Reynolds numbers need to be similar. This dimensionless parameter is given by $Re = \rho UL/\mu = UL/\nu$ where ρ and μ are density and viscosity, U is the speed of the oncoming flow, and L is a characteristic length (ν is the kinematic viscosity μ/ρ). It can be shown that if both U's and both L's are identical in an experiment (which is actually ideal and doable since full-scale testing of plastic or wood mockups at full speed is inexpensive and straightforward for downhole tools) then dynamic similarity is achieved when both kinematic viscosities match. In fluid-dynamics, even a ten-fold difference is "close" for modeling purposes. Reference to physical tables shows that this is remarkably the case – the kinematic viscosities for mud and air are very close and justify wind tunnel usage!

Additionally, a common normalization given in turbomachinery books can be used to reduce static and dynamic torque properties for various flow rates and densities to a single dimensionless performance curve – simply plot torque (normalized by a dynamic head) against the velocity swirl or "tip speed" ratio. This also motivates intelligent test matrix design: by judiciously choosing widely separated test points, *everything* there is to know about torque can be inferred – there is no need to perform hundreds of tests for different flow rates, rotation speeds and mud weights. Taken together, the two recipes just discussed allow simple and rigorous characterization of siren and turbine properties over the entire operating envelope with a minimum of labor, time and expense!

Short wind tunnels, envisioned for torque and erosion objectives, are easily and inexpensively designed and built within days. What is generally not known is the justifiable use of intermediate length (100-200 feet) and very long wind tunnels (say, 2,000 feet) in evaluating telemetry concepts, for instance, acoustic Δp signal strength, wave interference effects, surface signal processing schemes, downhole wave-based signal optimization methods, and so on. There are two reasons supporting such applications. For one, acoustic waves, even in wind tunnels, are still "long" in the classical sense. The sound speed in air is approximately 1,000 ft/sec. For a very high 100 Hz carrier wave, the wavelength λ = 1,000/100 ft = 10 ft still greatly exceeds the diameter of a typical drillpipe, say, six inches. Second, it can be shown that if ω is circular frequency, μ is viscosity, ρ is mass density, c is sound speed and R pipe radius, then the pressure P corresponding to an initial signal P_0 can be determined from $P = P_0 e^{-\alpha x}$ where x is the distance traveled by the wave and α is the attenuation rate given by $\alpha = (Rc)^{-1} \sqrt{\{(\mu\omega)/(2\rho)\}} = (Rc)^{-1} \sqrt{\{(\nu\omega)/2\}}$. The kinematic viscosity ν again appears, although fortuitously, but its presence indicates that signal tests can be cleverly designed to mimic attenuation using air as the working fluid! Thus, baseline MWD designs can be evaluated in air-conditioned offices and labs using short and long wind tunnels, deferring expensive hardware considerations related to mechanical reliability, vibrations, dynamic seals, corrosion, and so on, to the tail end of the design process.

The subject matter of this monograph represents years of both mental satisfaction and endless frustration, that is, continuing "love-hate" conflicts in confronting imposing challenges. These chapters summarize key ideas and highlight new theoretical results, physical insights, and testing and evaluation strategies that were developed in thinking "outside the box." But the endeavor would not come full circle until the suggestions were put to real tests in real engineering design and field testing programs.

Under the leadership of Dr. Yinao Su, Director of CNPC's Downhole Control Institute, comprehensive wind tunnel facilities were developed, and procedures, algorithms and theories were tested. The recent work described in "High-Data-Rate Measurement-While-Drilling System for Very Deep Wells," Paper No. AADE-11-NTCE-74 presented at the American Association of Drilling Engineers' 2011 National Technical Conference and Exhibition in Houston, summarizes findings aimed at an MWD system architecture that provides at least 10 bits/sec (without data compression) in very deep wells with lengths up to 30,000 ft. An updated version concludes the present chapter, providing an overview of current MWD project results and objectives. We emphasize that all of the theoretical and experimental methods in this book are available to the industry. The authors hope that, by openly identifying and discussing problems, solutions and strategies, petroleum exploration can be made more efficient with greater emphasis on safety, while reducing economic and exploration risk and educating the next generation of engineers.

1.2 Paper No. AADE-11-NTCE-74 – "High-Data-Rate MWD System for Very Deep Wells"
Significantly expanded with photographs and detailed annotations . . .

1.2.1 Abstract.

Measurement-While-Drilling systems presently employing mud pulse telemetry transmit no faster than one or two bits/sec from deep wells containing highly attenuative mud. The reasons – "positive pulsers" create strong signals but large axial flow forces impede fast reciprocation, while "mud sirens" provide high data rates but are lacking in signal strength. China National Petroleum Corporation research in MWD telemetry focuses on improved formation evaluation and drilling safety in deep exploration wells. A high-data-rate system providing 10 bits/sec and operable up to 30,000 ft is described, which creates strong source signals by using downhole constructive wave interference in two novel ways. First, telemetry schemes, frequencies and pulser locations in the MWD drill collar are selected for positive wave phasing, and second, sirens-in-series are used to create additive signals without incurring power and erosion penalties. Also, the positions normally occupied by pulsers and turbines are reversed. A systems design approach is undertaken, e.g., strong source signals are augmented with new multiple-transducer surface signal processing methods to remove mudpump noise and signal reflections at both pump and desurger, and mud, bottomhole assembly and drill pipe properties, to the extent possible in practice, are controlled to reduce attenuation. Special scaling methods developed to extrapolate wind tunnel results to real muds flowing at any downhole speed are also given. We also describe the results of detailed acoustic modeling in realistic drilling telemetry channels, and introduce by way of photographs, CNPC's "short wind tunnel" for signal strength, torque, erosion and jamming testing, "very long wind tunnel" (over 1,000 feet) for telemetry evaluation, new siren concept prototype hardware and also typical acoustic test results. Movies demonstrating new test capabilities will be shown.

1.2.2 Introduction.

The petroleum industry has long acknowledged the need for high-data-rate Measurement-While-Drilling (MWD) mud pulse telemetry in oil and gas exploration. This need is driven by several demand factors: high density logging data collected by more and more sensors, drilling safety for modern managed pressure drilling and real-time decision making, and management of economic risk by enabling more accurate formation evaluation information.

Yet, despite three decades of industry experience, data rates are no better than they were at the inception of mud pulse technology. To be sure, major strides in reliability and other incremental improvements have been made. But siren data rates are still low in deeper wells and positive pulser rates also perform at low levels. Recent claims for data rates exceeding tens of bits/sec are usually offered without detailed basis or description, e.g., the types of mud used and the corresponding hole depths are rarely quoted.

From a business perspective, there is little incentive for existing oil service companies to improve the technology. They monopolize the logging industry, maintain millions of dollars in tool inventory, and understandably prefer the status quo. Then again, high data rates are not easily achieved. Quadrupling a 3 bits/sec signal under a 12 Hz carrier wave, as we will find, involves much more than running a 48 Hz carrier with all else unchanged. Moreover, there exist valid theoretical considerations (via Joukowski's classic formula) that limit the ultimate signal possible from sirens. Very clever mechanical designs for positive pulsers have been proposed and tested in the past. Some offer extremely strong signals, although they are not agile enough for high data rates. But unfortunately, the lack of complementary telemetry schemes and surface signal processing methods renders them hostage to strong reverberations and signal distortions at desurgers.

Figure 1.1a. Prototype single-siren tool (assembled).

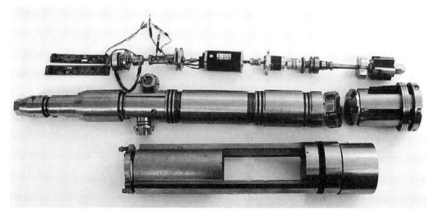

Figure 1.1b. Prototype single-siren tool (disassembled).

One would surmise that good "back of the envelope" planning, from a systems engineering perspective underscoring the importance of both downhole and surface components, is all that is needed, at least in a first pass. Acoustic modeling in itself, while not trivial, is after all a well-developed science in many engineering applications. For example, highly refined theoretical and numerical models are available for industrial ultrasonics, telephonic voice filtering, medical imaging, underwater sonar for submarine detection, sonic boom analysis for aircraft signature minimization, and so on, several dealing with complicated three-dimensional, short-wave interactions in anisotropic media.

By contrast, MWD mud pulse telemetry can be completely described by a single partial differential equation, in particular, the classical wave equation for long wave acoustics. This is the same equation used, in elementary calculus and physics, to model simple organ pipe resonances and is subject of numerous researches reaching back to the 1700s. Why few MWD designers use wave equation models analytically, or experimentally, by means of wind tunnel analogies implied by the identical forms of the underlying equations, is easily answered: there are no physical analogies that have motivated scientists to even consider models that bear any resemblance to high-data-rate MWD operation. For instance, while it has been possible to model Darcy flows in reservoirs using temperature analogies on flat plates or electrical properties in resistor networks, such approaches have not been possible for the problem at hand.

1.2.3   MWD telemetry basics.

Why is mud pulse telemetry so difficult to model?   In all industry publications, signal propagation is studied as a piston-driven "high blockage" system where the efficiency is large for positive pulsers and smaller for sirens. The source is located at the very end of the telemetry channel (near the drillbit) because the source-to-bit distance (tens of feet) is considered to be negligible when compared to a typical wavelength (hundreds of feet).

For low frequencies, this assumption is justified.   However, the mathematical models developed cannot be used for high-data-rate evaluation, even for the crudest estimates. In practice, a rapidly oscillating positive pulser or rotating siren will create pressure disturbances as drilling mud passes through it that are antisymmetric with respect to source position. For instance, as the valve closes, high pressures are created at the upstream side, while low pressures having identical magnitudes are found on the downstream side. The opposite occurs when the pulser valve opens.

The literature describes only the upgoing signal. However, the equally strong downgoing signal present at the now shorter wavelengths will "reflect at the drillbit" (we will expand on this later) with or without a sign change – and travel through the pulser to add to upgoing waves that are created later in time. Thus, the effect is a "ghost signal" or "shadow" that haunts the intended upgoing signal. But unlike a shadow that simply follows its owner, the use of "phase-shift-keying" (PSK) introduces a certain random element that complicates signal processing: depending on phase, the upgoing and downgoing signals can constructively or destructively interfere. Modeling of such interactions is not difficult in principle since the linearity of the governing equation permits simple superposition methods. However, it is now important to model the source itself:

it must create antisymmetric pressure signals and, at the same time, allow up and downgoing waves to transparently pass through it and interfere. It is also necessary to emphasize that wave refraction and reflection methods for very high frequencies (associated with very short wavelengths) are inapplicable. The solution, it turns out, lies in the use of mathematical forcing functions, an application well developed in earthquake engineering and nuclear test detection where long seismic waves created by local anomalies travel in multiple directions around the globe only to return and interfere with newer waves.

Wave propagation subtleties are also found at the surface at the standpipe. We have noted that (at least) two sets of signals can be created downhole for a single position-modulated valve action (multiple signals and MWD drill collar reverberations are actually found when area mismatches with the drill pipe are large). These travel to the surface past the standpipe transducers. They reflect not only at the mudpump, but at the desurgers. For high-frequency, low amplitude signals (e.g., those due to existing sirens), desurgers serve their intended purpose as the internal bladders "do not have enough time" to distort signals. On the other hand, for low-frequency, high amplitude signals (e.g., positive and negative pulsers), the effects can be disastrous: a simple square wave can stretch and literally become unrecognizable.

Thus, robust signal processing methods are important. However, most of the schemes in the patent literature amount to no more than crude "common sense" recipes that are actually dangerous if implemented. These often suggest "subtracting this, delaying that, adding the two" to create a type of stacked waveform that improves signal-to-noise ratio. The danger lies not in the philosophy but in the lack of scientific rigor: true filtering schemes must be designed around the wave equation and its reflection properties, but few MWD schemes ever are. Moreover, existing practices demonstrate a lack of understanding with respect to basic wave reflection properties. For example, the mud pump is generally viewed with fear and respect because it is a source of significant noise. It turns out that, with properly designed multiple-transducer signal processing methods, piston induced pressure oscillations can be almost completely removed even if the exact form of their signatures is not known. In addition, theory indicates that a MWD signal will double near a piston interface, which leads to a doubling of the signal-to-noise ratio. Placing transducers near pump pistons works: this has been verified experimentally and suggests improved strategies for surface transducer placement.

1.2.4 New telemetry approach.

A nagging question confronts all designers of high-data-rate mud pulse systems. If sirens are to be the signal generator of choice (say, if lowered torques enable faster direction reversals), how does one overcome their inherently weaker signal producing properties? The Joukowski formula "p = ρUc" provides an exact solution from one-dimensional acoustics stating that the pressure induced by an end-mounted piston is equal to the product of fluid density ρ, impact velocity U and sound speed c. It closely describes the acoustic performance of the positive pulser. And because the positive pulser brings the mud column to an almost complete stop – in a way that mud sirens cannot – the Joukowski formula therefore provides the upper limit for siren performance at least as presently implemented.

This understanding prompts us to look for alternatives, both downhole and uphole. We first address downhole physics near the source. We have observed that up and downgoing waves are created at the siren, and that reflection of the latter at the drillbit and their subsequent interaction with "originally upgoing" waves can lead to "random" constructive or destructive wave interference that depends on the information being logged. This is certainly the case with presently used phase-shift-keying which position-modulates "at random" the siren rotor. However, if the rotor is turned at a constant frequency, random wave cancellations are removed. The uncertainty posed by reflections of phase-shifted signals, whose properties depend on nozzle size, wavelength, annular geometry, logging data, and so on, are eliminated in the following sense: a sinusoidal position modulation always creates a similar sinusoidal upgoing pressure wave without "kinks" and possible sign changes. In fact, depending on the location of the source within the MWD drill collar, the geometry of the bottomhole assembly, the transmission frequency and the mud sound speed, the basic wave amplitude can be optimized or de-optimized and controlled with relative ease. Pump and desurger reflections at the surface, of course, still require surface signal processing.

Information in the form of digital 0's and 1's can therefore be transmitted by changes in frequency, that is, through "frequency-shift-keying" (FSK). But, unlike conventional FSK, we select our high frequencies by using only those values that optimize wave amplitude by constructive interference. Neighboring low-amplitude waves need not be obtained by complete valve slowdown, as in conventional PSK. If, say, 60 Hz yields a locally high FSK amplitude, it is possible (and, in fact, we will show) that 50 Hz may yield very low amplitudes, thus fulfilling the basic premise behind FSK. The closeness in frequencies implies that mechanical inertia is not a limiting factor in high data rate telemetry because complete stoppage is unnecessary, so that power, torque and electronic control problems are minimal and not a concern. Eliminating complete stoppage also supports data rate increases because the additional time available permits more frequency cycles. In fact, using a frequency sequence like "60-50-70-80" would support more than 0's and 1's, suggesting "0, 1, 2 and 3" encoding.

In order to make constructive interference work, the time delay between the downgoing waves and their reflections, with the newer upgoing waves, must be minimized. This is accomplished by placing the siren as close to the drillbit as possible, with the downhole turbine now positioned at the top of the MWD drill collar. This orientation is disdained by conventional designers because "the turbine may block the signal." However, this concern is unfounded and disproved in all field experiments. This is obvious in retrospect. The "see through area" for turbines is about 50% of the cross-section. If signals can pass through siren rotor-stator combinations with much lower percentages, as they have time and again, they will have little difficulty with turbines.

1.2.5 New technology elements.

The above discussion introduces the physical ideas that guided our research. An early prototype single-siren tool designed for downhole testing is shown assembled and disassembled in Figures 1.1a and 1.1b. Multiple siren tools have been evaluated. To further refine our approach and understanding of the scientific issues, math models and test facilities were developed to fine-tune engineering details and to obtain "numbers" for actual design hardware and software. We now summarize the technology.

1.2.5.1 Downhole source and signal optimization.

As a focal point for discussion, consider the hypothetical MWD drill collar shown in Figure 1.2a. Here, physical dimensions are fixed while siren frequency and position are flexible. Up and downgoing signals (with antisymmetric pressures about the source) will propagate away from the pulser, reflect at the pipe-collar intersection, not to mention the interactions that involve complicated wave transfer through the drillbit and in the borehole annulus.

A six-segment acoustic waveguide math model was formulated and solved, with the following flow elements: drillpipe (satisfying radiation conditions), MWD drill collar, mud motor or other logging sub, bit box, annulus about the drill collar, and finally, annulus about the drillpipe (also satisfying radiation conditions). The "mud motor" in Figure 1.2a could well represent a resistivity-at-bit sub. At locations with internal impedance changes, continuity of pressure and mass was invoked. The siren source was modeled as a point dipole using a displacement formulation so that created pressures are antisymmetric. Numerical methods introduce artificial viscosities with unrealistic attenuation and also strong phase errors to traveling waves. Thus, the coupled complex wave equations for all six sections were solved analytically, that is, exactly in closed form, to provide uncompromised results.

Calculated results were interesting. Figure 1.2b displays the actual signal that travels up the drillpipe (after all complicated waveguide interferences are accounted for) as functions of transmission frequency and source position from the bottom. Here, "Δp" represents the true signal strength due to siren flow, i.e., the differential pressure we later measure in the short wind tunnel. For low frequencies less than 2 Hz, the red zones indicate that optimal wave amplitudes are always found whatever the source location. But at the 12 Hz used in present siren designs, source positioning is crucial: the wrong location can mean poor signal generation and, as can be seen, even "good locations" are bad.

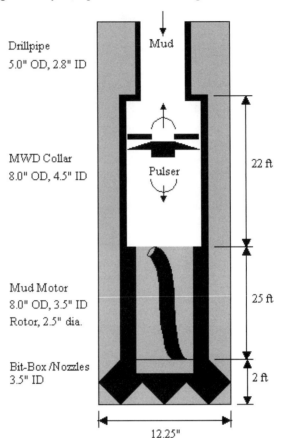

Figure 1.2a. Example MWD collar used for siren frequency and source placement optimization analysis.

These calculations are repeated for upper limits of 50 Hz and 100 Hz in Figures 1.2c and 1.2d. In these diagrams, red means optimal frequency-position pairs for hardware design and signal strength entering the drillpipe. Our objective is p/Δp >> 1 (Δp is separately optimized in hardware and wind tunnel analysis). That present drilling telemetry channels support much higher data rates than siren operations now suggest, e.g., carrier waves exceeding 50 Hz, is confirmed by independent research at www.prescoinc.com/science/drilling.htm (see Figure 10.5). In our designs, we select the frequencies and siren positions, or for sirens-in-tandem, in such a way that high amplitudes are achieved naturally without power or erosion penalties (mud siren signal amplitudes are typically increased by decreasing rotor-stator gap, which leads to higher resistive torques and local sand-convecting flow velocities).

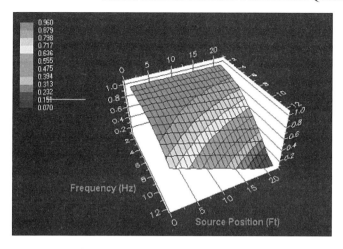

Figure 1.2b. Drillpipe p/Δp to 12 Hz.

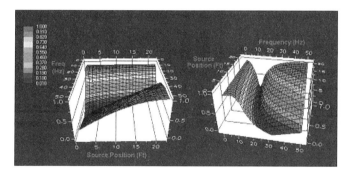

Figure 1.2c. Drillpipe p/Δp to 50 Hz.

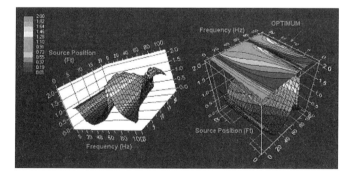

Figure 1.2d. Drillpipe p/Δp to 100 Hz.

1.2.5.2 Surface signal processing and noise removal.

Downhole signal optimization, of course, has its limits. To complement efforts at the source, surface signal processing and noise removal algorithms must be developed that are robust. Our approach is based on rigorous mathematics from first principles. The classic wave equation states that all "solutions (measured at some point "P" along the standpipe) are superpositions of upgoing "f" and downgoing "g" waves. A differential equation for "f" is constructed. It is then finite differenced in space and time as if a numerical solution were sought. However, it is not. The Δz and Δt in the discretized result are re-interpreted as sensor spacing (in a multiple transducer array) and time step, whose pressure parameters are easily stored in surface data acquisition systems. The solution for the derivative of the signal was given in U.S. Patent 5,969,638 or Chin (1999). At the time, it was erroneously believed that telemetered data could be retrieved from spatial derivatives but this proved difficult. In recent work, the method was corrected by adding a robust integrator that handles abrupt waveform changes. The successful recovery of "red" results to match "black" inputs, using the seemingly unrelated green and blue transducer inputs, is shown in Figures 1.3a and 1.3b. Mudpump generated noise can be almost completely removed. Experimental validations are given later.

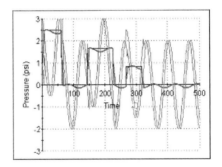

Figure 1.3a. Three step pulse recovery from noisy environment.

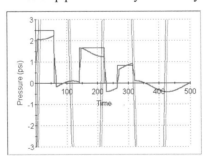

Figure 1.3b. Three step pulse recovery (*very* noisy environment).

1.2.5.3   Pressure, torque and erosion computer modeling.

The mud siren, conceptualized in Figure 1.4a, is installed in its own MWD drill collar and consists of two parts, a stationary stator and a rotor that rotates relative to the stator. The rotor periodically blocks the oncoming mud flow as the siren valve opens and closes. Bi-directional pressure pulses are created during rotation. At the very minimum, the cross-sectional flow area is half-blocked by the open siren; at worst, the drill collar is almost completely blocked, leaving a narrow gap (necessary for water hammer pressure signal creation) between stator and rotor faces for fluid passage. This implies high erosion by the sand-laden mud and careful aerodynamic tailoring is needed. Because there are at least a dozen geometric design parameters, testing is expensive and time-consuming. Thus, the computational method in Chin (2004), which solves the three-dimensional Laplace equation for the velocity potential in detail, is used to search for optimal designs. Computed results, displayed for various degrees of valve closure, are shown in Figures 1.4b and 1.4c. Other results include "resistive torque vs angle of closure" important to the design of fast-action rotors. Results are validated and refined by "short wind tunnel" analyses described later.

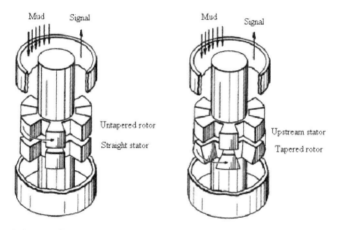

Figure 1.4a. Early 1980s "stable closed' siren (left) and improved 1990s "stable-opened" downstream rotor design.

While apparently simple in design, unanticipated flow effects are to be found. The upstream rotor design used in early designs produces numerous operational hazards, the least of them being stoppage of data transmission. When rock debris or sudden jarring occurs, the rotor is known to stop at a closed azimuthal position that completely blocks mud flow. This results in severe tool erosion, extremely high pressures that affect well control, not to mention surface safety issues associated with high pressure buildup at the mudpump. Early solutions addressed the symptoms and not the cause, e.g., mechanical springs or pressure relief valves that unload the locked rotor, strong permanent magnets that bias special steel assemblies to open positions (thus compromising direction and inclination measurements and requiring the use of nonmagnetic drill collars), and so on. It can be shown that "stable closed" tendencies are a natural aerodynamic consequence of upstream rotor configurations – the rotors tend to close even in clean water. Numerous unsuccessful tests addressing this problem were performed in the 1970s: operational failures associated with jamming valves were catastrophic.

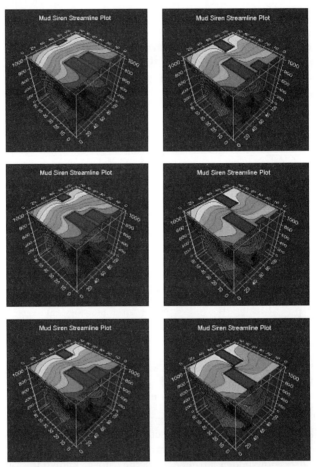

Figure 1.4b. Streamline traces for erosion analysis.

U. S. Patent 4,785,300 or Chin and Trevino (1988) solved the problem by placing the rotor downstream as indicated in Figure 1.4a. The rotor, now "stable open," is augmented with special tapered sides. Torques required to turn, stop or speed up the rotor are much lower than those associated with upstream rotors. From a telemetry standpoint, this means faster position modulation requiring less torque and power, or, much higher data rate. Of course, mechanical considerations are a small part of the problem. Downhole signal enhancement and surface noise removal are equally important, as noted earlier. In our research, all are addressed and fine-tuned to work in concert to provide a fully optimized system.

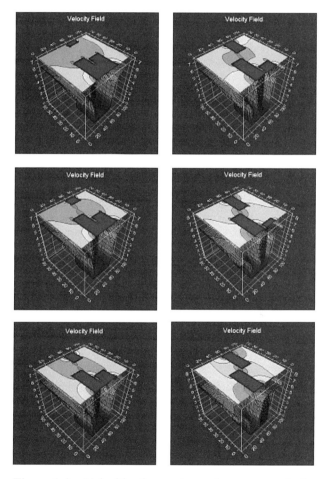

Figure 1.4c. Velocities for erosion and pressure analysis.

1.2.5.4   Wind tunnel analysis: studying new approaches.

While computer models are useful screening tools, they alone are not enough. Gridding effects mask the finest flow details that can be uncovered only through actual testing. The use of wind tunnels in modeling downhole mud flow was first proposed and used by the last author of this paper during his tenure with Schlumberger. Technical details and justification are disclosed, for instance, in Gavignet, Bradbury and Quetier (1987) who used the method to study flows beneath drill bits nozzles. This counter-intuitive (but correct) approach to modeling drilling muds provides a strategically important alternative to traditional testing and reduces the time and cost of developing new MWD systems.

The CNPC MWD wind tunnel test facility consists of two components, a "short flow loop" where principal flow properties and tool characteristics are measured, and a "long loop" (driven by the flow in the short wind tunnel) designed for telemetry concept testing, signal processing and noise removal algorithm evaluation. Field testing procedures and software algorithms for tool properties and surface processing are developed and tested in wind tunnel applications first and then moved effortlessly to the field for evaluation in real mud flows. This provides a degree of efficiency not possible with "mud loop only" approaches.

Our "short wind tunnel," actually housed at a suburban site, is shown in Figure 1.5a. This laboratory location was selected because loud, low-frequency signals are not conducive to office work flow. The created signals are as loud as motorcycle noise and require hearing protection for long duration tests. More remarkable is the fact that internal pipe pressures are several orders of magnitude louder than the waves that escape – in practice, this is further multiplied by the (large) ratio of mud to air density, about 800 in the case of water. Thus, careful and precise acoustic signal measurement is required to accurately extrapolate those to mud conditions. Similarly, torques acting on sirens in air are at least 800 times lower. In fact, air-to-mud torque scaling is simply proportional to the dynamic head "ρU^2" ratio, where U is the oncoming speed. Thus, wind tunnel tests can be run at lower speeds with inexpensive blowers provided a quadratic correction factor is applied for downhole flow extrapolation. The MWD turbine, similarly designed and tested, is not discussed in this paper.

In Figure 1.5a, a powerful (blue) blower with its own power supply pumps more or less constant flow rate air regardless of siren blockage. A sensitive flow meter is used to record average flow rate. Flow straighteners ensure uniform flow into the siren and to remove downstream swirl for accurate differential pressure measurement. The siren test section deserves special comment. The motion of the rotor is governed by its own electrical controller and is able to effect position-modulated motions as required for telemetry testing.

Blower with muffler and water coolant

Independent power supply

Flow meter

Flow straighteners

Siren test section and electric controller

Torque meter

Drive motor

Total length, about 60-70 feet

Observation window to outdoor flowloop for multiple transducer and wave interference testing

Data acquisition for flow rate, torque, multiple and differential pressure transducers

Siren Δp transducer

Connection to outside long wind tunnel

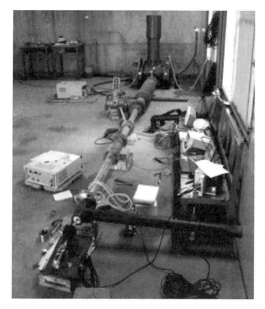

Figure 1.5a.1. Short "hydraulic" wind tunnel, or simply "short wind tunnel."

Figure 1.5a.2. Short wind tunnel, "bird's eye" perspective.

Siren motion is driven electrically as opposed to hydraulically; azimuthal position, torque and Δp signal strength, i.e., the differential pressure between upstream and downstream sides of the siren, are measured and recorded simultaneously. This data is important to the design of control and feedback loops for actual modulation software. At the bottom of Figure 1.5a, a black PVC tube turns to the right into the wall. This emerges outside of the test shed, as shown in Figure 1.5b, into a long flow section more than 1,000 feet in length. Because the waves are acoustically "long," they reflect minimally at bends, even ninety-degree bends. The long wind tunnel wraps itself about a central facility several times before exhausting into open air. This boundary condition is not, of course, correct in practice; we therefore minimize its effect by reducing signal amplitude, so that end reflections are not likely to compromise data quality.

Also shown in Figure 1.5b.1 are "a single transducer" near the test shed (bottom left) and a three-level "multiple transducer array" (bottom right). The former monitors signals that leave the MWD collar, as they are affected by constructive or destructive wave interference, while the latter provides data for echo cancellation and noise removal algorithm evaluation. For the simplest schemes, two transducers are required; three allow redundancies important in the event of data loss or corruption. Additional (recorded) noise associated with real rigsite effects is introduced in the wind tunnel using low frequency woofers.

Numerous concepts were evaluated. Several sirens shown are impractical but were purposely so; a broad range of data was accumulated to enhance our fundamental understanding of rotating flows as they affect signal, torque and erosion. We re-evaluated conventional four-lobe siren designs and developed methods that incrementally improve signal strength and reduce torque. Results reinforced the notion that the technology has reached its performance limits. Radically different methods for signal enhancement and minimization of resistive torque were needed.

As noted earlier, constructive wave interference provides "free" signal amplitude without erosion or power penalties. This is cleverly implemented in two ways. First, FSK with alternating high-low amplitudes is used. High amplitudes are achieved by determining optimal frequencies from three-dimensional color plots such as those in Figures 1.2b,c,d. Design parameters include sound speed, source position and frequency, MWD collar design, and whenever possible, drillpipe inner diameter and mud density. This information is used in the waveguide model of Figure 1.2a and also in a model for non-Newtonian attenuation applicable over the length of the drillpipe.

Figure 1.5b.1. Very long "acoustic" wind tunnel.

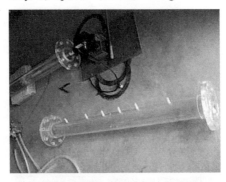

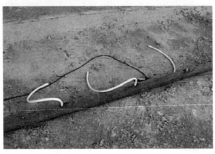

Figure 1.5b.2. Multiple transducer array for surface noise removal
(top, accommodating multiple frequencies; bottom, one configuration used).

Low amplitudes need not be achieved by bringing the rotor to a complete
stop. If a high-amplitude is associated with, say 60 Hz, then a useful low-
amplitude candidate can be found at, say 55 Hz, as suggested by Figures
1.2b,c,d. Thus, FSK can be efficiently achieved while minimizing the effects of
mechanical inertia. Rotor torque reduction, while an objective in wind tunnel
analysis, is useful, but need not be the main design driver in our approach.

In order to make constructive wave interference work, the siren must be
located as close to the most significant bottom reflector, normally the drillbit, as
possible (intervening waveguides, e.g., mud motors, resistivity-at-bit subs, and
so on, support wave transmission). Thus, the siren is placed beneath the turbine
in the MWD collar, in contrast to existing designs. This reduces the time needed
for waves to meet and reinforce. Figure 1.2a shows a "mud motor." This
acoustic element may, in fact, represent a resistivity-at-bit sub. Calculations
show that 10 bits/sec can be accomplished provided this section is
approximately fifteen feet in length or less. Tests confirm that long waves pass
effortlessly through turbines without reflection. Detailed waveguide analyses
suggest that signal gains of 1.5-2.0 are doable using single-siren designs alone.
PSK methods, again, are undesirable because they result in wave cancellations
and ghost signals that hinder signal processing.

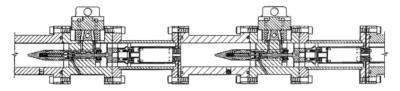

Figure 1.5c. A pair of ganged or tandem mud sirens.

Additional signal enhancement is possible using constructive interference of a different nature: multiple sirens arranged in series or in tandem. If the distance between sirens is small and siren apertures are properly phased, signals will be additive. This idea was first proposed in U.S. Patent 5,583,827 or Chin (1996) and a possible design from that publication is reproduced in Figure 1.5c. This design, incidentally, is not CNPC's preferred embodiment.

Two sirens, for instance, mean twice the signal. If the amplification afforded in the previous paragraphs provides a modest signal gain of 1.5, that is, 50%, the net would be a three-fold signal increase more than enough to overcome attenuation at the higher frequencies used. In principle, any number of sirens can be connected to provide signal increases as needed. Performance is determined by the single transducer in Figure 1.5b.1 (second row, far left) which measures the signal leaving the MWD collar. The extent to which constructive wave interference works is found by comparison with the measured differential pressure Δp taken across the siren (e.g., see Figure 1.5h, middle and right). This Δp depends on siren geometry, flow rate and rotation speed only: it is independent of reflections since waves pass through without interaction (that is, reflected waves do not affect differential pressure measurements providing both sensing ports are close).

Note that Figures 1.2b,c,d suggest that frequencies in the 50-60 Hz range are not unrealistic, a conclusion independently reached at the website www.prescoinc.com/science/drilling.htm (see Figure 10.5). This use of higher frequencies is also supported by test results from actual flow loop tests with real muds. We stress that attenuation measurements are subtle since the effects of acoustic nodes and antinodes (which depend on frequency and flow loop boundary conditions) must be properly accounted for. Almost all existing papers on attenuation fail to even recognize this problem, let alone correct for it.

Our systems approach to high-data-rate design requires an equal focus on surface systems. As implied earlier, signal strength enhancement must be accompanied by using the most sensitive piezoelectric transducers and robust multiple-transducer echo cancellation methods. Figure 1.5b.2 shows a transducer array located far from the test shed. Noise can be introduced by playing back actual field recordings. We have found, to our amusement, that the large firecrackers used at Chinese weddings, e.g., see Figure 1.5o, provide a useful source of low-frequency, plane-wave "pump" noise when all else is unavailable.

Conventional siren designs are built with four lobes cut along radial lines. Rotating sirens with additional lobes would surely increase frequency or data rate, but large lobe numbers are associated with much lower Δp signals. For this reason, they are not used in designs to the authors' knowledge. Because constructive interference now enhances our arsenal of tools against attenuation, we have been able to reassess the use of higher lobe numbers. Downhole and uphole telemetry concepts are easily tested in our wind tunnels.

Wind tunnel usage enables a scale of knowledge accumulation, together with cost, time and labor efficiencies not previously possible. Numerous parameters can be evaluated, first by computational models, and then by testing in air. Design parameters are numerous: lobe number, stator and rotor thicknesses, stator-rotor gap, rotor clearance with the collar housing, rotor taper angles, and so on.

Tests are not limited to signal strength. Torque is important, as is the ability to pass lost circulation material – this is assessed by introducing debris at the upstream end of the short wind tunnel and observing the resulting movement. Erosion tendencies are determined by noting the convergence effects of threads glued to solid surfaces – rapid streamline convergence implies high erosion, e.g., see Figures 1.4b,c.

Two new parameters were included in our test matrix. The bottom left photograph in the top group of Figure 1.5d.1 shows a "curved siren" with swept blades. Research was performed to determine the degree of harmonic generation associate with constant speed rotations. Since the sound generation process is nonlinear, a rotation rate of ω will not only produce pressure signals with ω, but those with frequencies $\pm 2\omega$, $\pm 3\omega$, $\pm 4\omega$ and so on. Higher harmonics are associated with acoustic inefficiencies we wish to eliminate, not to mention added surface signal processing problems – the ability of blade sweep-back in reducing undesirable energy transfer was one objective of the test program. Sweep effects are likely to affect jamming due to lost circulation materials, so that jamming considerations cannot be ignored – they may help by cutting in scissor-like fashion or hinder by obvious reductions in flow area.

The second parameter considered upstream effects. Figure 1.5.d.2 shows conical flow devices which guide inlet flow into the siren. Their effects on torque and signal generation were studied. Convergence and divergence, together with additional grooved helical tracks etched into the hub surface, could possibly affect signal strength, acoustic harmonic distribution and torque, and a variety of hub designs was defined to evaluate possible outcomes.

In order to achieve a given high data rate, one might employ sirens with "few lobes, turning rapidly" or "many lobes, turning slowly." Each design, characterized by different signal strength versus torque relationships, must be assessed in the wind tunnel. Very often, strong signals are accompanied by high resistive torques, an undesirable situation that impedes rapid angular reciprocation. Experiments help to identify optimal designs.

Figure 1.5d.1. Siren concepts tested in wind tunnel (standard size).

Figure 1.5d.2. Evaluation of hub convergence effects.

Figure 1.5e.1. Siren concepts, miniature sirens for slim tools
(the United States quarter shown is similar to the one-yuan Chinese coin).

Figure 1.5e.2. Siren concepts, miniature sirens for slim tools (cont'd).

The sirens shown in Figures 1.5.d.1 and 1.5.d.2 are standard ones, typically several inches in diameter, intended for use in conventional tools. In some applications, the need for ultra-slim MWD devices arises and very small sirens are required. Example designs are shown in Figure 1.5.e.1, where tapers and notches are milled into rotor sides to examine effects on torque and signal. Over twenty were built. Some spontaneously spin due to wind action, drawing kinetic energy from the flow, while others jam due to "stable closed" behavior.

Miniature sirens, mounted on fine bearings, are easily tested even without wind tunnels, and various novel designs have been identified. The candidate shown in Figure 1.5.e.2, interestingly, remains stable-open regardless of wind direction, as demonstrated with the lead author breathing in-and-out as the siren spins and creates signals without interruption (note the presence and absence of condensation in the plastic tube associated with breathing in-and-out). A continuous air-hammer "roar" can be heard, which is indicative of low resistive torque and low turbine power demand – an ingredient for high-data-rate telemetry. A short movie for this experiment is available from the lead author.

In the figures presented here, we provide photographs of actual sirens tested and devices used. Many are self-explanatory and are followed by brief comments. In addition, numerous videos are available for viewing upon request. All provide a useful "engineering feel" for the types of testing conveniently and economically performed in our challenge to find high-data-rate solutions. Improvements to both short and long wind tunnel testing have been and are continuously being made in light of our experiences described below.

Figure 1.5f.1. Flow straighteners for upstream and downstream use.

Comments: Flow straighteners eliminate large-scale vortical structures in the air flow induced near pipe bends and fan blades. Small scale turbulence remains – much as it does in real drillpipe flows, while acoustic plane waves associated with MWD signals pass effortlessly. For turbine testing, flow straighteners are essential to eliminate any azimuthal biases to torque that are inherent in an uncorrected wind.

Figure 1.5f.2. Flow straightener construction.

Comments: Flow straighteners can be inexpensively fabricated from raw PVC stock. Different combinations of diameter and length were tested to evaluate their effectiveness and influence on wave propagation and pressure drop. Generally speaking, twelve inch lengths with inner tube diameters as shown in Figure 1.5f.1 suffice.

Figure 1.5g.1. Digital flowmeter.

Comments: Accurate flow rate measurements are critical to wind tunnel testing, since air-based data must be extrapolated to mud conditions. While mud flow rates can be accurately estimated, that is, calculated from piston displacements and "strokes per minute," those in air are generally inferred from rotating impeller data obtained in different conduits with siren blockages. The consequences can be serious. Suppose torque data from a wind tunnel test at 100 gpm is extrapolated to flow at 500 gpm. Under perfect conditions, the torque would increase quadratically as $(500/100)^2$ or twenty-five-fold based on velocity considerations alone. A 10% error in flow rate, that is, a measurement range of 90-110 gpm, would replace the "25" by "21-31."

Errors in power extrapolation are more severe. For the same problem, the exact power increase varies cubically, that is $(500/100)^3$ or 125; the extrapolated range would vary by factors in the range 94-171. These uncertainties introduce obvious difficulties in engineering design, e.g., seal performance, alternator selection, and so on. It is important to understand that calibrations marked on flowmeter boxes rarely perform as indicated – some flowmeters are designed for heating and air conditioning applications where the conduit assumed standard rectangular cross-sections, while others for circular pipes assume blockage-free flows, meaning no mud sirens!

Thus, wind flow rate measurements must be carefully calibrated. Fortunately, experimental data can be easily checked for physical consistency. For instance, suppose torque data for a siren or turbine configuration are available at two speeds U, say T_a and T_b for U_a and U_b. Since torque is known to vary quadratically with oncoming speed, these measurements must satisfy $(T_a/T_b) = (U_a/U_b)^2$ – if not, either the calibration procedure or the torque measurement itself is incorrect, or both. A separate density ratio correction, of course, applies to both torque and power.

We have emphasized flow rate accuracy for torque analysis, and similar considerations apply to "signal strength versus flow rate." Here the dependence of Δp on U is less obvious. At high speeds, a water (or air) hammer mechanism prevails and the dependence on U is linear; at lower speeds, an orifice effect dominates with a quadratic dependency. Care must be exercised since these trends are used in estimating tool signal generation in field applications – an incorrect result might lead to weak signals observed at the standpipe.

Finally, we offer a note on data integrity. For rotating sirens and turbines, recorded movies for the flowmeter in Figure 1.5g.1 show very steady gauge readings at constant pump settings. This is always achieved since, on the average, sirens and turbines are half-open. This, of course, is not the case with reciprocating positive pulsers that close with very small gaps; for such applications, it is necessary to define other types of time-averaged means. Flow rate is not always important. For example, "stable open versus stable closed" behavior is completely independent of flow rate. A completely satisfactory test is possible by spinning the siren and allowing it to open or close as it slows down. This "roulette wheel" approach is documented in our movies.

Figure 1.5g.2. Safety message – note plastic tape at right.

Comments: The foregoing discussion focuses on "opened versus closed" stability. When a siren jams, arising from "stable closed" tendencies, high wind tunnel pressures build dangerously and may fracture the test section. Figure 1.5g.2 shows "plastic tape repairs" which were needed not infrequently. In our tests, plastic tubing lengths of six feet with 0.25 in wall thicknesses were used. Wind tunnel fracturing is a safety hazard that must be avoided since it may lead to flying debris and plastic shards. Safety goggles and protective clothing are recommended since "pipe blowouts" are possible without warning.

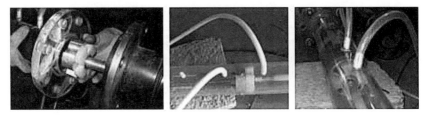

Figure 1. 5h. Siren test sections with differential transducers for signal strength measurement (not influenced by reflections and propagating noise).

Comments: The above figure shows how upstream and downstream ports of a differential pressure transducer are installed for Δp measurement as functions of flow rate and rotation speed. Note that siren signal strength cannot be measured using a single transducer, since the presence of multiple reflections will cause complications and introduce interpretation uncertainty. When the two ports of a differential pressure transducer are closely spaced, however, reflections cancel identically since the instrument "does not have time" to respond to the traveling disturbance (this is obviously untrue of large separations). Just how close is close? This was easily determined by shouting into the tube, or by setting off a short-duration firecracker – if the differential transducer does not respond, then the separation used is "close enough."

Figure 1.5i. Real-time data acquisition and control system.

Comments: Experiments are automated using a Labview user interface, as shown in Figure 1.5i, and standard data acquisition instruments monitor signal strength, rotation speed, torque and local pressures in both short and long wind tunnels. Motor drivers, pump and a "long wind tunnel observation window" are shown in Figures 1.5j – 1.5l without further comment.

Figure 1.5j. Torque, position and rpm counter.

Figure 1.5k. Water-cooled air blower with separate power source.

Figure 1.5l. Test shed window overlooking long wind tunnel.

Figure 1.5m. Single piezoelectric transducer closest to siren
for constructive interference and nonlinear harmonic generation study.

Comments: For high data rate applications, piezoelectric transducers are essential since they provide the required frequency response and data resolution needed for signal processing. In the above, a single transducer close to the test shed monitors the effects of constructive or destructive interference near the pulser – this "efficiency" is compared with the Δp measurement obtained from short wind tunnel analysis. In a sense, Δp characterizes the "brute force" ability of a siren to create sound, while the single-transducer reading measures how cleverly we can amplify it. Far away, transducer arrays collect data for multiple transducer signal processing for echo cancellation and pump noise removal.

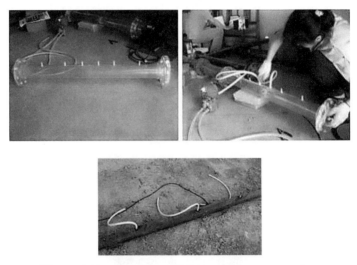

Figure 1.5n. Distant multiple transducer array setup.

Comments: In the top photographs, a short plastic section is tapped for close transducer placements needed for "very, very high" data rates characterized by very short wavelengths. In the lower figure, a similar configuration is used in the long acoustic wind tunnel.

Figure 1.5.o.1. Chinese fireworks.

Figure 1.5o.2. Artificial low frequency "mud pump" noise.

Comments: During one test session, a vendor failed to deliver our "mud pump emulator" on time. Improvising was key to obtaining qualitative data. Chinese firecrackers will produce strong plane waves at approximately once per second. These were constrained in the wind tunnel and ignited as needed.

Figure 1.5p.1. Lead author tests wind effects at the outlet.

Figure 1.5p.2. Volume flow sensor at long wind tunnel outlet characterizes frequency or wavelength-dependent effects.

Comments: It is imperative that long wind tunnels model reality. In actual MWD field application, a downhole pulser creates an upward propagating wave that reflects at a solid mudpump piston (much slower piston movement in progress is irrelevant to this reflection). Pressure waves impinging at solid piston boundaries will reflect with their signs or polarities unchanged. In the setup shown, the far end of the long wind tunnel is *open* to the atmosphere and pressures will reflect with opposite signs and return toward the source. To minimize this effect, signal strengths were controlled so that unrealistic reflections do not affect other transducer measurements (a special wind tunnel has since been designed which exhausts wind differently while allowing impingement at a solid boundary). Interestingly, theory shows that pressure waves impinging at a solid boundary locally double in magnitude. This was verified in experiments, suggesting that pressure transducers mounted near the mud pump could provide improved signal to noise ratios – this idea has been applied to the "one hundred feet hose" amplifier described elsewhere in this book, a significant invention that has been field tested and patented. Figures 1.5q and 1.5r show work sessions captured during wind tunnel testing.

Figure 1.5q. Early prototype tool.

Figure 1.5r. At work in the test shed.

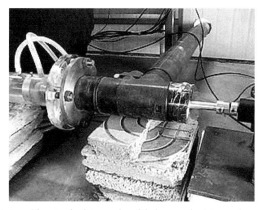

Figure 1.5s. Short wind tunnel, with long wind tunnel connection.

Comments: When long wind tunnel testing is required, the siren is driven by the motor with its shaft penetrating the end of the short section at the junction shown. This leads to a small amount of tolerable air leakage. Note that ninety-degree turns, while requiring higher pressures to move air, do not induce reflections, since the wavelengths are long. This was in fact validated by monitoring reflection arrival times associated with "firecracker" blasts at an inlet. For example, if a signal is created at an open end, and the hollow pipe is 1,000 ft long, the reflection time will be two seconds since the sound speed is 1,000 ft/sec. Any arrivals prior to this would indicate undesirable reflections.

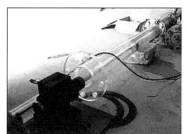 

Figure 1.5t. Short wind tunnel "alone."

Comments: When the "short wind tunnel alone" is required without the "black tube" of Figure 1.5s, e.g., for Δp signal strength, harmonic distribution and torque testing for a given frequncy, or for erosion flow visualization, it is not necessary to connect its long wind tunnel appendage. Thus, air blowers may operate at much lower pressures, although high pressure wind tunnel fracturing is still possible under "stable closed" conditions. At the above right, a six lobe siren is shown closing in the "stable open" position. Its low torque and good stability characteristics make it a front-running candidate for use in "multiple sirens in series" designs.

And now, a few informal photographs showing daily routines, work and living conditions . . .

Figure 1.5u. Wind tunnel design and test planning (Bian Hailong).

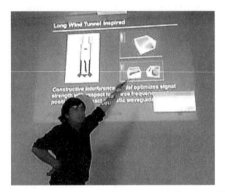

Figure 1.5v. Wave interference ideas explained (Wilson Chin).

Figure 1.5w. Test shed, left; local restaurant, center, where
experimental results and test plans were reviewed and updated daily; our
simple hotel, right, residing next to a pumping unit, in contrast to –

Figure 1.5x. CNPC Headquarters, Beijing.

1.2.5.5 Example test results.

Here we highlight some interesting test results. The first pertains to signal strength as a function of rotation rate with flow speed fixed. Early Schlumberger papers claim that Δp's obtained at high frequencies are independent of frequency, i.e., the siren behaves as an orifice. We believed otherwise. As the rotor turns, it brings oncoming mud to a halt whatever the frequency. However, the water hammer signal must weaken as rotation rate increases because less time is available for fluid stoppage and rebound. Our test setup is summarized in Figure 1.6a.1. The expected monotonic decrease of Δp with increasing frequency is seen, for instance, in Figure 1.6a.2, where we typically test up to 60 Hz as suggested by calculations in Figures 1.2b,c,d. The low Δp's associated with existing "siren alone" approaches reinforced our efforts to seek more innovative signal enhancement methods utilizing ideas from constructive wave interference. Good signal strength alone does not imply a usable siren – a workable design must be low in resistive torque to enable rapid rotational speed changes. Figure 1.6a.3 provides a spreadsheet example of torque data collected as functions of flow and rotation rate. Signal strength, torque and erosion pattern (obtained by flow visualization) are catalogued for each siren prototype.

Figure 1.6a.1. Simultaneous signal strength and torque measurements.

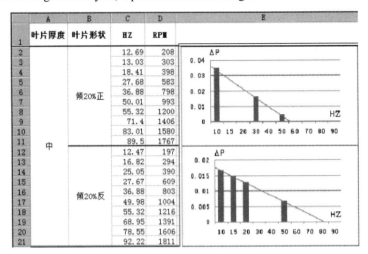

	叶片厚度	叶片形状	HZ	RPM
2			12.69	208
3			13.03	303
4			18.41	398
5			27.68	583
6		倾20%正	36.88	798
7			50.01	993
8			55.32	1200
9			71.4	1406
10			83.01	1580
11	中		89.5	1767
12			12.47	197
13			16.82	294
14			25.05	390
15			27.67	609
16		倾20%反	36.88	803
17			49.98	1004
18			55.32	1216
19			68.95	1391
20			78.55	1606
21			92.22	1811

Figure 1.6a.2. Siren Δp signal strength vs ω at with flow rate fixed.

Group Number	Thickness	Shape	Rotor Vs Stator	Assembly Status	Frequency (Hz)	Percent Open	Flow Rate (m3/h)	Torque (mN m)	GPM	Torque IN-LBF	GPM*	IN-LBF*
1	Medium 中	Slanted 斜	R>S=20%	Positive 20% 正	15	50%	234	114.3	1030	1.0116	1000	747
						100%	273	114.3	1202	1.0116	1000	549
						0%	157	224.5	691	1.9868	1000	3260
					25	50%	394	-28.9	1735	-0.2558	1000	-67
						100%	439	344.9	1933	2.1674	1000	455
						0%	265	255.1	1167	2.2576	1000	1300
					35	50%	526	-55.3	2316	-0.4894	1000	-72
						100%	570	265.3	2510	2.3479	1000	292
						0%	350	316.3	1541	2.7993	1000	924
2	Medium 中	Slanted 斜	R>S=20%	Negative 20% 反	15	50%	239	-42.1	1052	-0.3726	1000	-264
						100%	276	244.9	1215	2.1674	1000	1151
						0%	217	265.3	955	2.3479	1000	2016
					25	50%	345	-68.4	1519	-0.6053	1000	-206
						100%	442	-15.8	1946	-0.1398	1000	-29
						0%	334	244.9	1471	2.1674	1000	786
					35	50%	543	-68.4	2391	-0.6053	1000	-83
						100%	555	265.3	2444	2.3479	1000	308
						0%	359	326.5	1581	2.8895	1000	907
3	Thick 厚	Curved 曲			15	50%	221	234.7	973	2.0771	1000	1720
						100%	250	204.1	1101	1.8063	1000	1169
						0%	205	224.5	903	1.9868	1000	1912
					25	50%	349	244.9	1537	2.1674	1000	720
						100%	404	204.1	1779	1.8063	1000	448
						0%	238	-81.6	1048	-0.7222	1000	-516
					35	50%	397	326.5	1748	2.8895	1000	741
						100%	522	255.1	2298	2.2576	1000	335
						0%	318	-275.5	1400	-2.4382	1000	-975

Three-Lobe Siren

Figure 1.6a.3. Siren torque versus ω at with flow rate fixed.

In Figure 1.6b, pressure data from the near transducer in Figure 1.5m appears at the left, while data from two far transducers in Figure 1.5n are shown at the center and right. At the left, the pure sinusoid shows that high-order harmonics have been completely eliminated by the siren design. The two right figures, which contain additive noise, are almost identical. Multiple transducer signal processing in Figure 1.6c shows how the red signal is successful extracted from the blue and green to match the black upgoing waveform.

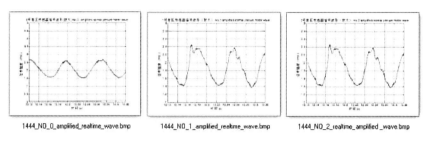

1444_NO_0_amplified_realtime_wave.bmp   1444_NO_1_amplified_realtime_wave.bmp   1444_NO_2_realtime_amplified_wave.bmp

Figure 1.6b. Low-frequency (10 Hz) long wind tunnel data.

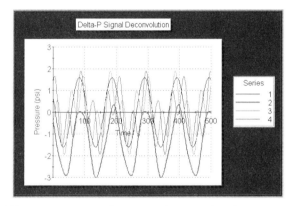

Figure 1.6c. Low-frequency (10 Hz) signal recovery.

Complementary results obtained at 45 Hz are shown in Figures 1.6d,e. The left diagram of Figure 1.6d is clearly not sinusoidal and provides evidence of nonlinear harmonics. Their magnitudes are measured from frequency domain analysis and efforts are made to determine their physical origin. Again, we have successfully extracted the MWD signal from a noisy environment. Our work showed that transducer spacings of 10% of a wavelength sufficed for signal extraction. At the present, not all noise sources are included in our modeling efforts. Vibration and other sound mechanisms will be included in future work.

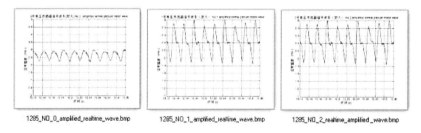

1285_NO_0_amplified_realtime_wave.bmp   1285_NO_1_amplified_realtime_wave.bmp   1285_NO_2_realtime_amplified_wave.bmp

Figure 1.6d. High-frequency (45 Hz) long wind tunnel data.

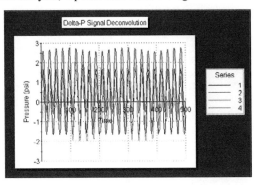

Figure 1.6e. High-frequency (45 Hz) signal recovery.

Our experiences with constructive wave interference "at the drillbit" are also worth noting. In U.S. Patent 5,583,827 or Chin (1996), where the use of downhole constructive interference for signal enhancement was first suggested, the published analytical model mistakenly assumed the bit as a solid reflector. In fact, it is now known that MWD signals are detectable in the annulus, where their absence is used as an indicator of gas influx. The six-segment waveguide model presented in Chapter 2 is used to study typical MWD collars, e.g., see Figure 1.2a. It is more general and does not assume any particular reflection mechanism on an a priori basis. Detailed calculations show that, more often than not, the drillbit acts as an open reflector – attesting to the dangers of "common sense" and visual inspection. This model creates plots similar to Figures 1.2b,c,d. The wave characteristics of siren and positive pulsers from present MWD vendors are consistent with those in Figure 1.2b.

1.2.6 Conclusions.

We have summarized our strategy for high-data-rate mud pulse telemetry and means for developing the technology. Our target objective of 10 bits/sec at 30,000 feet appears to be doable. The signal amplification approach used, together with new surface signal processing techniques, plus the use specially designed tools that are integrated with mud and drillpipe properties, provide a systems oriented process that optimizes data transmission. Needless to say, we have acquired much in our testing program, and we are continually learning from our mistakes and developing new methods to improve the technology. Prototype (metal) tools have been built, using one or more sirens, and are presently being tested for mechanical integrity and telemetry performance; an example is given in Figures 1.1a,b. The top photograph in Figure 1.5b.1 shows the "long wind tunnel" described in this paper, however, it also operates with mud or water using a mud pump and redesigned pulser section (not shown) that is controllable from a test shed at the center of the loop. Real mud laboratory and field tests are in progress and results will be presented at a later date.

1.2.7 Acknowledgments.

The project "Downhole High-Data-Rate Continuous Wave Mud Pulse Telemetry" was sponsored by China National Petroleum Corporation's Technology Management Division under Contract 2008C-2101-1. The authors thank the management of CNPC for permission to publish this research and for its support and interest throughout the effort. We also thank the University of Petroleum (East China), whose College of Electromechanical Engineering provided test facilities for research experiments, and Professor Jun Fang, Professor Xueshi Gao and Dr. Peng Jia, who diligently contributed their expertise during numerous wind tunnel test sessions. Special thanks are also due to Ms. Qiumei Zhang and Ms. Xiaoying Zhuang for their efforts in facilitating the project. We appreciate the support of all of our colleagues, whose contributions ensured the success of the project.

1.2.8 References.

Chin, W.C., "Measurement-While-Drilling System and Method," U.S. Patent No. 5,583,827, Dec. 10, 1996.

Chin, W.C, "Multiple Transducer MWD Surface Signal Processing," U.S. Patent No. 5,969,638, Oct. 19, 1999.

Chin, W.C., "MWD Siren Pulser Fluid Mechanics," *Petrophysics*, Journal of the Society of Petrophysicists and Well Log Analysts (SPWLA), Vol. 45, No. 4, July – August 2004, pp. 363-379.

Chin, W.C., Su, Y., Sheng, L., Li, L., Bian, H. and Shi, R., *MWD Signal Analysis, Optimization and Design*, E&P Press, Houston, 2011.

Chin, W.C. and Trevino, J., "Pressure Pulse Generator," U. S. Patent No. 4,785,300, Nov. 15, 1988.

Gavignet, A.A., Bradbury, L.J. and Quetier, F.P., "Flow Distribution in a Tricone Jet Bit Determined from Hot-Wire Anemometry Measurements," SPE Paper No. 14216, *SPE Annual Technical Conference and Exhibition*, Las Vegas, NV, 1985.

Gavignet, A.A., Bradbury, L.J. and Quetier, F.P., "Flow Distribution in a Roller Jet Bit Determined from Hot-Wire Anemometry Measurements," *SPE Drilling Engineering*, March 1987, pp. 19-26.

1.3 References

Chin, W.C., "Measurement-While-Drilling System and Method," U.S. Patent No. 5,583,827, Dec. 10, 1996.

Chin, W.C, "Multiple Transducer MWD Surface Signal Processing," U.S. Patent No. 5,969,638, Oct. 19, 1999.

Chin, W.C., "MWD Siren Pulser Fluid Mechanics," *Petrophysics*, Journal of the Society of Petrophysicists and Well Log Analysts (SPWLA), Vol. 45, No. 4, July – August 2004, pp. 363-379.

Chin, W.C., Su, Y., Sheng, L., Li, L., Bian, H. and Shi, R., *MWD Signal Analysis, Optimization and Design*, E&P Press, Houston, 2011.

Chin, W.C. and Trevino, J., "Pressure Pulse Generator," U. S. Patent No. 4,785,300, Nov. 15, 1988.

Gavignet, A.A., Bradbury, L.J. and Quetier, F.P., "Flow Distribution in a Tricone Jet Bit Determined from Hot-Wire Anemometry Measurements," SPE Paper No. 14216, *SPE Annual Technical Conference and Exhibition*, Las Vegas, NV, 1985.

Gavignet, A.A., Bradbury, L.J. and Quetier, F.P., "Flow Distribution in a Roller Jet Bit Determined from Hot-Wire Anemometry Measurements," *SPE Drilling Engineering*, March 1987, pp. 19-26.

Oppenheim, A.V. and Schafer, R.W., *Digital Signal Processing*, Prentice-Hall, New Jersey, 1975.

Oppenheim, A.V. and Schafer, R.W., *Discrete-Time Signal Processing*, Prentice-Hall, New Jersey, 1989.

2
Harmonic Analysis:
Six-Segment Downhole Acoustic Waveguide

High-data-rate pulsers in continuous wave MWD telemetry not only create acoustic signals that propagate uphole, but in addition, down-going signals that "reflect at the drillbit," reverse direction and combine with the former to create waves that may constructively or destructively interfere. The ultimate signal that travels up the drillpipe depends on mud sound speed, the position of the pulser in the drill collar and its operating frequency, and importantly, the details of the bottomhole assembly forming the host waveguide. The implications are both good and bad. Destructive interference in signal generation can severely limit data rate and transmission distance, but constructive wave interference, properly applied, can enhance MWD signal strength without the usual power and erosion penalties incurred by purely mechanical methods.

"Reflection at the bit," a phenomenon noted above, actually is more complicated than simple acoustic "open" or "closed" models would have us believe. In reality, signals *do* propagate through nozzles that may be small, and signals *are* detectable in the annulus. Thus, a telemetry model used to study fundamental physics and potential technical capabilities must not disallow transmissions into the annulus; moreover, as explained in Chapter 1, it is additionally important to model the antisymmetric disturbance pressure field about the source, so that downhole constructive and destructive interference processes and reverberant fields within the MWD collar can be properly studied.

Exact solutions for a basic six-segment waveguide model representing the complete acoustic transmission channel are obtained for siren and poppet-type sources and their implications in high-data-rate tool design are discussed. Detailed solutions for a representative bottomhole assembly are provided in which signal generation efficiency is evaluated as a function of source position and frequency. These results suggest the use of new telemetry schemes that specifically focus on the positive aspects of wave propagation in signal creation and also the possibility that transmission rates and distances greatly exceeding the values conventionally accepted can be developed. In Chapter 9, we find that the model is more accurate than long flow loop analysis, since many industry loops are built satisfying practical boundary conditions that are not consistent with actual "radiation conditions" found in the drillpipe and annulus.

47

2.1 MWD Fundamentals

In conventional Measurement-While-Drilling operations, the problems associated with surface reflections, e.g., those at the mudpump, the desurger, the rotary hose, and so on, are well known. Several echo cancellation schemes using delay line models based on single and multiple transducers are available for surface signal processing, these importantly acknowledging the perils and subtleties associated with acoustic wave motions. These problems will be compounded at higher data rates because they are necessarily associated with higher frequencies and much shorter wavelengths. A single pressure transducer installed arbitrarily on the standpipe is not likely to be optimal; careful planning and the use of transducer arrays may be required. Many surface signal processing ideas are appreciated by MWD designers, but perhaps are not as well understood as they should be. As we will find in this book, important and useful models for forward and inverse applications can be developed using rigorous solutions to the one-dimensional acoustic wave equation.

The technical and patent literature has generally focused on the consequences of reflections as they affect surface signal processing. However, as described in the chapter introduction, they are equally prominent downhole in the way that they affect signal creation at higher data rates. For example, consider a mud siren (or other pulser type) transmitting information and employing a phase-shift-keying (PSK) scheme, but at carrier frequencies much higher than those presently used. The "usual" wave will propagate uphole to the surface with encoded data and it is *this* wave that is addressed in surface signal processing. However, a downgoing wave is also created at the source that reflects "at the drillbit" with amplitude and phase changes that depend on mud sound speed, pulser location and the details of the bottomhole assembly and annulus. This reflected wave (still "long" in the sense that its wavelength greatly exceeds a typical diameter) passes through the mud siren (or other pulser) and combines with later created waves that are moving upwards.

The net signal that travels up the drillpipe contains both the "usual" wave and the more complicated reflected one – it is impossible to determine, at the surface using present signal processing methods, which phase-shifts are real and which are not. These "ghost reflections" cannot be eliminated without additional information – these are not problematic if they are not created in the first place, of course, and Chapter 10 suggests a practical solution using an alternative "FSK" telemetry scheme. Because ghost echoes do exist, improvements to existing continuous wave telemetry over the years have been limited. This observation suggests the use of telemetry schemes which are more robust from a downhole source oriented perspective – and perhaps, one that can additionally increase transmission distance as well as data rate – by harnessing the use of constructive wave interference. These applications suggested the development of a math model that can be used in both telemetry job planning and in high-data-rate tool design.

2.2   MWD Telemetry Concepts Re-examined

The Measurement-While-Drilling literature has, for almost four decades, classified telemetry methods in a simple-minded manner according to three well-known categories: "positive pulser," "negative pulser," and "mud siren," the latter synonymously referred to as "continuous wave telemetry" (continuous transmissions are, of course, possible with positive and negative pulsers). These are conceptually illustrated in Figures 2.1, 2.2 and 2.3, respectively, where only the upgoing signals (as in conventional publications) are shown. It turns out that this representation is valid at very low frequencies, as we will demonstrate from our more general model.   For now, we begin by explaining accepted conventional views and their original rationale, and then, re-examine extensions to the physical problem in the high-data-rate context in each instance.

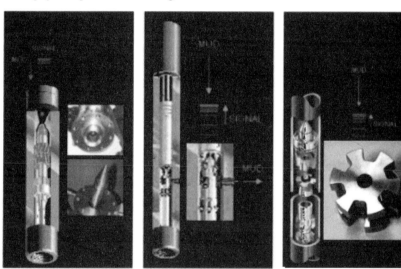

Figures 2.1, 2, 3. Positive, negative and siren pulsers (left to right).

2.2.1 Conventional pulser ideas explained.

Positive pulsers are essentially poppet valves that *slowly* plug and unplug small orifice openings by axial motions. When the orifice closes, the speed of the fluid column upstream reduces significantly while pressure slowly increases – a pressure that returns to ambient conditions when the valve fully opens. For this reason, this signal generator is known as a "positive pulser." Positive versus ambient hydrostatic pressures are used to communicate "0's" and "1's" to the surface. It is clear that such pulsers require high levels of mechanical power as well as the ability to withstand significant levels of erosion.

Negative pulsers can be idealized as "holes" in drill collars that *slowly* open and shut. When the orifice opens, drilling fluid in the collar is diverted into the annulus and pressure decreases – this pressure returns to ambient conditions upon closure. For this reason, the signal generator is known as a "negative pulser." Negative versus ambient pressures are used to communicate "0's" and "1's" to the surface. Such pulsers, in addition to requiring high levels of mechanical power to operate, erode valve components and may damage the formation and lead to well control problems. For this reason, they present liabilities in deepwater applications and are not recommended.

Sirens are high-data-rate devices. While both positive and negative pulsers are lower in data rate, typically offering 1 bit/sec or less, sirens potentially offer 10 bits/sec or more. The mud siren is known as a "continuous wave" pulser with a rotor component that rotates relative to an immobile stator. It resembles a single-stage turbine except that the turbine's thin blades are replaced by thick block-like "lobes" which almost completely block the flow when in a closed position. In industry publications, a positive over-pressure is shown traveling uphole much as it would with positive pulsers – however, the ability of the siren to increase and decrease its rotary speed allows it to send data at much higher speeds than with positive and negative pulsers. The siren makes use of acoustic transmissions, whereas slower positive and negative pulsers often do not.

2.2.2 Acoustics at higher data rates.

Now we re-examine positive pulser operation at higher axial reciprocation speeds. As the valve rapidly closes, it creates an over-pressure (by "banging" into the mud) that travels upstream as before; but at the same time, it creates an under-pressure ("pulling away" from the flowing mud) that travels downstream. When the valve opens, an under-pressure travels uphole while an over-pressure travels downhole. In other words, the pulser creates both positive and negative signals that travel away from the valve at both sides, noting that, at either side of the valve, signals may be positive, negative or zero. While this description is correct at high reciprocation speeds, we will continue, for historical purposes, to refer to high speed poppet valve pulsers as positive pulsers.

We might note that, because the dynamic pressure field (relative to hydrostatic conditions) corresponding to a positive pulser is antisymmetric with respect to the source position, a nonzero pressure difference (or "delta-p," usually denoted by "Δp") generally exists across the valve. We will refer to such antisymmetric pressure fields as "dipole" fields produced by dipole tools. If the valve were positioned in an infinite uniform pipe, "½ Δp" will propagate in one direction while the other "½ Δp" will travel in the opposite direction with an opposite sign by virtual of symmetry. If "transmission efficiency," a new term introduced here, were defined as the pressure transmitted in a single direction relative to the source Δp, the transmission efficiency for an infinite uniform pipe is exactly 0.5. We also emphasize that Δp is an "acoustic delta-p" and *not* the pressure drop associated with viscous pressure losses in the wake of a blunt body. For example, a nearly-closed stationary poppet valve is associated with high pressure losses, but these do not propagate and therefore cannot be used to create traveling sound waves. While viscous pressure drops may be important insofar as mudpump power requirements or tool losses are concerned, they are irrelevant to high data rate MWD signal generation.

When a mud siren closes, the upstream flow slows significantly and the associated over-pressure travels uphole; at the same time, fluid pulls away at its downstream side and an under-pressure travels downhole. The reverse occurs when the siren opens. This, of course, applies whether the siren rotor rotates about its axis or simply oscillates back-and-forth. This physical description from the acoustical perspective is identical to that provided above for rapidly reciprocating positive pulsers, that is, both pulser classes are associated with pressure disturbances that are antisymmetric with respect to source position. Both sirens and positive pulsers represent realizations of dipole tools. In this wave propagation sense, positive pulsers and mud sirens function identically although there are obvious mechanical and practical design differences. As before, the siren signal can also be characterized by a "source strength" we shall term "Δp." In general, the Δp for a positive pulser or a siren valve will depend on geometry, the size of the drill collar, flow rate, mud sound speed and reciprocation or rotation; it can be determined independently in a flow loop, or in a wind tunnel, if measured results are properly rescaled. We importantly emphasize that Δp may be frequency dependent – with all parameters fixed, it may increase or decrease with frequency depending on mechanical design.

Negative pulsers function differently. When the drill collar orifice opens, an under-pressure is locally created, but this under-pressure propagates both uphole and downhole; likewise, when the valve closes, a local over-pressure is created that propagates both uphole and downhole. The pressure difference, if measured at identical distances upstream and downstream of the orifice, identically vanishes because the disturbance pressure field is symmetric with respect to source position. Negative pulsers therefore represent "monopole" tools. For historical reasons, we will continue to refer to such pulsers as negative pulsers. It is clear that a non-zero Δp is *not* necessary for negative pulser operation and, in fact, is never achieved because Δp is always zero (for such pulsers, it is "delta velocity" that is important). As with positive pulsers and sirens, the source strength associated with a negative pulser depends on the acoustic "water hammer" component of the valve motion and not the pressure drop due to viscous losses across the drill collar wall. The pressure symmetries and antisymmetries noted for different pulser types are not speculative and have been observed and recorded in detailed wind tunnel and mud flow loop studies.

2.2.3   High-data-rate continuous wave telemetry.

Sirens have offered relatively high data rates in the past.   But it is important to observe that axially reciprocating positive pulsers and rapidly opening-and-closing negative pulsers can also be used to provide waves for continuous *wave* telemetry.   The rationale for selecting one pulser type over another, in the past, was based on power and erosion considerations and, in this sense, on limitations in mechanical design and performance.   However, if the wave properties of signal production used in continuous wave generation can be harnessed by taking advantage of constructive interference in order to significantly increase source strength without incurring power and erosion penalties, a game-changing means for high-data-rate telemetry and high resolution well logging would become reality.

The importance of wave motions in telemetry physics cannot be deemphasized.   For existing positive and negative pulsers, signal detection and decoding at the surface presently involve the monitoring of slowly changing pressure levels – a relatively elementary endeavor.   At high data rates, new classes of subtle surface signal processing problems arise and we describe these complications next – but again, these same subtle effects offer the potential for significant gain if their physics can be intelligently and robustly harnessed.

Consider the problem of signal generation.   Whether we use positive pulsers, negative pulsers or mud sirens, waves are created that travel both uphole and downhole.   Trade journals (and even scientific papers) normally depict only the upgoing wave, which ideally is correlated with valve displacement or velocity (which is, in turn, driven by assumed strings of "0's" and "1's dictated by logging sensors).   In reality, the down-going wave reflects "at the drillbit" to travel uphole; this wave will add to newly created up-going waves, and the wave that ultimately travels uphole contains the intended signal plus "ghost reflections" of created data from earlier times.

The above paragraph only "scratches the surface" insofar as complications are concerned.   Depending on the data stream transmitted (that is, the detailed motion history of the valve), the geometry of the bottomhole assembly, the mud sound speed, the transmission frequency, the pulser type and the telemetry scheme, the results of constructive and destructive wave interference can lead to good versus bad signals, but more than likely, simply unpredictable signals.   In designing a reliable continuous wave telemetry system, it is important to understand in detail the physics of downhole wave interactions as this understanding is critical to surface signal processing – these can have disastrous operational consequences but they can also be used to advantage.

We have casually noted "reflection at the bit," but we again emphasize that reflections at the drillbit represent only those at one obvious reflector.   In fact, changes in acoustic impedance are found at drillpipe-collar junctions for the downgoing mud flow and at other area changes; they are additionally found, for example, at mud-motor and MWD collar interfaces, as well as at entrances and exits of drillbit nozzles and bit subs or bit boxes and in the annulus. All of these effects must be modeled correctly, but unfortunately, none of the numerous models with which the author is familiar addresses even the simplest reflection problem properly. We will review some fundamental issues before developing a comprehensive acoustic waveguide model encompassing all of these effects.

2.2.4   Drillbit as a reflector.

In many studies, the ratio of total nozzle cross-sectional area to the cross-sectional area of the bit or bit sub is seen to be small and the drillbit is accordingly modeled as a solid reflector. This appears to be reasonable from an engineering perspective but it is completely wrong. This is obvious from the following "thought experiment." It is worthwhile noting that, for a 12 Hz carrier wave in water with a sound speed of 5,000 ft/sec, the wavelength is 5,000/12 or about 500 feet – at lower frequencies, the wavelengths are much longer. Let us consider a positive pulser or a siren, located about 100 feet from the drillbit, in the process of obstructing the oncoming mud flow. A positive signal (that is, an over-pressure) is created that propagates uphole. At the same time, a negative signal is created which travels downhole. If the drillbit is a solid reflector, this negative signal will reflect as a negative signal and travel uphole past the source to combine with the long positive signal already traveling uphole. The net result is a wave, by virtue of destructive interference, with mostly zero disturbance pressure amplitude.

It is known from operational experience, of course, that signals created using positive pulsers and mud sirens *are* measurable at the surface – and, in fact, that they have been observed even in the surface borehole annulus – therefore, the solid reflector model must be wrong. Let us, then, reconsider the downgoing negative wave on this basis. If the drillbit nozzles represent, alternatively, an effective open-ended reflector, negative pressures will reflect as positive pressures – these positive pressures will add to the positive pressures that initially travel uphole and hence increase the possibility of detection. An open-end reflector will therefore double the propagating pressure associated with an infinite uniform pipe – the "transmission efficiency" should approach 1.0, or nearly twice the 0.5 considered previously, only because the end is not entirely opened in an acoustic sense. An open-end model also allows wave transmission up the annulus and is consistent with field observation.

These simple explanations on drillbit reflections are in fact supported by results of our detailed waveguide calculations which assume very general acoustic impedance matching conditions. Again, we note that the transmission efficiency for a uniform drillpipe infinite in both directions is 0.5. For the simple open reflector model above, this increases to 1.0, and interestingly, values exceeding 1.0 are possible, a phenomenon not surprising to designers of telescoping acoustic waveguides with internal area changes. For many bottomhole assemblies, efficiencies approaching 3.0 have been calculated for higher frequencies, although the attenuation effects will limit their usefulness.

These observations point to the importance of retaining drillbit nozzles (or more correctly, the "bit box" or "bit sub") as important conduits for wave transmission modeling insofar as optimizing MWD signals in the drillpipe is concerned. However, they also play an important role in signal production for the annulus. There are two important applications. First, the ability to correlate MWD signals in the annulus with those in the drillpipe (or lack of) is often taken as an indicator of gas influx detection; this is an important safety measurement. Second, the annular signal, which is inherently less noisy than the drillpipe signal, can be used to improve drillpipe signal decoding, since it is not contaminated as much by mudpump or desurger noise. Thus, in some applications, one might aim at optimizing signals in both drillpipe and annulus for the purpose of minimizing bit error, an objective that is not impossible. Not only is the bit sub important – all impedance changes within the drillpipe and drill collar system, as well as different annuli dimensions surrounding drillpipe and drill collar, form important elements of the acoustic channel model. The ratio of nozzle to total area, while dimensionless, is relevant only to hydraulic studies related to viscous pressure loss and not to acoustic transmission directly.

2.2.5 Source modeling subtleties and errors.

Proper source modeling is another area which is incorrectly addressed in all published analyses. It is generally argued that the distance between the MWD source and the drillbit reflector (located at the origin "x = 0") is small compared to a wavelength, so that the source can be modeled as a moving piston located at x = 0. This seems reasonable, and if so, the analysis would apply to all pulsers, whether their disturbance pressure fields are antisymmetric or symmetric with respect to the source. However, the method completely ignores the presence and sign of the downgoing wave discussed above. Thus, the effects of destructive interference (disastrous from a well logging perspective) and constructive interference (significant to signal enhancement without additional power and erosion penalties) cannot be modeled. Whether method-of-characteristics or finite difference approaches are used (to solve the plane wave formulations usually given) is irrelevant. The "x = 0" piston model implicitly assumes very, very low frequencies where downgoing wave effects are extremely simple (this statement, in fact, will be seen from detailed calculations using the more rigorous model valid for all frequencies to be derived later).

The correct theoretical approach requires us to more generally place the source *within* the MWD drill collar away from the bit and not to introduce any piston or solid reflector assumptions that would unrealistically preclude important classes of reflected wave motions. *A priori* assumptions related to drillbit reflections would not be made; instead, general acoustic matching conditions would be invoked at impedance mismatches and "the dice would fall" as they will. However, mathematical complications arise which need to be addressed. When the source is located within the drilling fluid itself, its dynamic properties must be modeled. That is, whether the disturbance pressures (relative to hydrostatic) created by it are symmetric or antisymmetric with respect to source location are critically important – the dipole or monopole nature of signal creation must be accounted for in the mathematical model. For the same telemetry channel and frequency, positive pulsers and mud sirens will obviously create wave patterns – and hence upgoing MWD signal fields – that are completely different from those of negative pulsers.

In summary, the least complicated analytical model of signal generation and wave interference downhole must contain at least the following elements. (1) A six-segment waveguide is required, the basic elements being the drillpipe; the MWD collar (containing a pulser and possibly other logging sensors); a collar beneath it to represent a positive displacement mud-motor (with rubber stators with different acoustic impedance) or, say, resistivity sub with different collar dimensions; a drillbit, bit box or bit sub component; an annulus surrounding the drill collar; and finally, a different annulus surrounding the drillpipe. (2) The "monopole" or "dipole" nature of the source must be specified, that is, whether the disturbance pressure field is symmetric or antisymmetric with respect to the source. And finally, for analysis and modeling purposes, (3) the source must be located *within* the MWD drill collar and allow the propagation of created signals away from it in both directions, as well as the complete transmission of reflected waves through the source itself – this latter requirement is necessary because the long waves created (even at frequencies at high as several hundred Hertz) will effortlessly propagate through valve openings that never completely close, e.g., siren "rotor-stator gaps" (detailed experiments show no evidence of reflections from practical MWD pulsers).

The source strength "Δp," which depends on mechanical design, geometry, drill collar areal cross-section, flow rate, frequency, density and sound speed, is a quantity that is most accurately determined by laboratory differential pressure measurements. However, with care, it can also be inferred from field tests when the only available data is surface data. The value of Δp is not affected by wave motions and the longitudinal geometry of the waveguide (the cross-sectional area will affect local signal strength, however). For telemetry modeling, this strength can be left as "Δp," noting that this will generally depend on frequency. Only the ratios of upgoing drillpipe pressure to Δp, that is "$p_{pipe}/\Delta p$," and upgoing annular pressure to Δp, that is, "$p_{annulus}/\Delta p$," are dynamically significant. Again, we caution that Δp may vary with frequency in a manner that depends on the mechanical design of the valve. The objective of the analysis in this chapter is to identify the conditions under which $p_{pipe}/\Delta p$ is weak, or on a positive note, the conditions under which it can be optimized, so that signal enhancement without the usual anticipated penalties related to power and erosion are realized. We also recognize, of course, that any increase in $p_{pipe}/\Delta p$, if it is obtained at a higher frequency, is subject to higher frequency-dependent attenuation. The hope is that increases in signal amplitude more than offset decreases incurred along the acoustic path. A complete telemetry analysis will consider both the downhole model addressed in this chapter, plus the subject of irreversible thermodynamic losses over large distances as well as the frequency dependence of Δp on mechanical design, both of which are treated later.

2.2.6   Flowloop and field test subtleties.

In deep wells containing attenuative drilling mud, the up and down-going waves created by the source will interact to form *standing waves* within the drill collars, the lower annulus and the bit box, while upgoing *propagating waves* will be found in the drillpipe and in upper annulus (this statement strictly applies to a system excited by a constant frequency source) provided the travel path is long. This observation also applies if the downward waves reflected from mudpumps and desurgers (and from the mud pit) have attenuated enough that they can be ignored – in shallow wells, we emphasize, this outgoing wave condition may be inapplicable and standing wave assumptions in both pipe and annulus may be required. In this chapter, we assume that propagating waves do in fact exist, and use such "radiation" or "outgoing wave" conditions as boundary conditions for the acoustic math model – that is, the drillpipe and upper annulus both extend to infinity and are each in themselves semi-infinite waveguides – or possibly, that they are both finite in length, but that their reflections toward the pulser attenuate significantly enough that they are not dynamically relevant. We implicitly assume deep wells.

The same careful observations apply to flow loop testing and interpretation. In several flow loops operated by petroleum organizations, lengths of 5,000 to 15,000 feet are available and have been used for MWD telemetry testing. Typically, water is used as the test fluid since it is both clean and requires significantly less pump power. Positive displacement pumps act at one end of the loop while open reservoir conditions apply at the other. These boundary conditions are *not* the ones encountered operationally in real wells – they support systems of standing waves as opposed to radiating waves. Often a valve or downstream choke near the reservoir is adjusted to control mean pressure levels in order to prevent cavitation. This valve, while nearly closed, is never completely closed because pumped fluid must be allowed to pass. This opening, per the drillbit-pulser discussion above, permits signals to propagate through it, since long waves will invariably never "see" the valve.

Because flowloops are finite as opposed to infinite despite their mile-long length, pulsers positioned within almost all flowloops that the author is familiar with will create systems of standing waves. These waves – and not propagating waves that attenuate with distance – are the ones measured experimentally and whose data must be reinterpreted in the context of wave motions encountered in the well. As pulser frequency changes within the flowloop, the nodes and anti-nodes of the standing wave move. A pressure transducer located at a fixed position will measure amplitude changes related both to attenuation and node movement that are difficult to distinguish if a single transducer is used – or if multiple transducer data is not properly interpreted. In several publications, measured pressure changes that occurred with frequency increases have been incorrectly identified with attenuation – a term that, in the scientific literature, is usually reserved for irreversible thermodynamic losses. In fact, for a given Δp, the measured pressure at a given location versus frequency increase does not always decrease – it may increase, or otherwise vary non-monotonically with periodic character, but importantly, always in a manner dependent on the sound speed of the fluid, the length of the flow loop and its end boundary conditions.

We emphasize that the measurement of Δp across a positive pulser or mud siren, say using differential pressure transducers, is correct and perfectly legitimate whether the flowloop is finite or infinite because the continuous background wave field (containing even the most complicated reflections) is subtracted out at both sides. However, the measurement of pressure itself needs interpretation because it is subject to the vagaries of reflection – care must be taken to determine if it is in fact a propagating wave result, a standing wave anomaly, or a wave that has been contaminated by reflections.

Many field reports obtained at different service companies have provided confusing and contradictory reports related to high-data-rate signal generation. Invariably, workers failed to distinguish between positive versus negative pulser sources and failed to note the type of drillbit used or the length of the hole. As we have discussed, downhole reflection patterns depend on source type, the manner of reflection at the drillbit, not to mention mud sound speed, pulser frequency and the details of the bottomhole assembly. The parameters discussed thus far explain the sources of confusion encountered by workers in MWD telemetry. Measured standpipe signals have often varied significantly from one rig to another without explanation. Sometimes, extremely weak signals are found with tools that are functioning perfectly mechanically. However, in all cases, differences can be reconciled when bottomhole assembly geometries, the lengths of the boreholes, and other acoustic parameters, are considered within the framework of a comprehensive and rigorous mathematical model.

Again, flowloop testing can be dangerous when pressures (not to be confused with Δp's) are to be measured, and particularly so, when the pulser is placed in a uniform flow loop and not a replica of the bottomhole assembly. Likewise, "real world" field test results are likely to lead to confusing conclusions if not interpreted properly – and similarly, successful high-data-rate MWD telemetry will not be optimal unless preliminary job planning is performed to determine the range of desirable frequencies and also the suite of useful telemetry schemes for any particular rigsite operating scenario.

2.2.7   Wind tunnel testing comments.

Analogous test interpretation problems may arise with wind tunnel testing, that is, the use of hundred-feet versus thousand-feet long wind tunnels for MWD telemetry acoustic testing. For instance, different pressure patterns are obtained accordingly as the fluid source is positive displacement piston driven (that is, a solid reflector) or acts as a centrifugal pump (an open-end reflector). While such wind tunnels are ideal for Δp source strength testing and preliminary signal processing instrumentation and software design, particularly because such tests are convenient and inexpensive, care must be taken to understand the physical phenomenon actually modeled and how it relates to telemetry in a deep well. For example, if outgoing wave conditions are to be modeled at both inlet and outlet, enough damping along all propagation paths (say, using soft acoustic liners) and sufficient length are required to eliminate end reflections. In the next section, we develop a detailed waveguide model for the complete MWD telemetry channel. In later chapters, however, models for finite loops are designed for laboratory application.

2.3   Downhole Wave Propagation Subtleties

We have discussed the complexities related to source signal generation in a continuous wave environment from a downhole perspective. In Figure 2.4, where constant frequencies are assumed, single arrows denote pure propagating wave motions while double-arrows indicate the existence of standing waves.

Diagram (a) illustrates the situation encountered downhole for deep wells, while Diagrams (b), (c) and (d) illustrate other situations that may be appropriate to shallow wells or wind tunnel tests (the darkened cross-sections indicate that alternative boundary conditions might apply, e.g., shortened wind tunnel lengths, shallow wells, open reservoirs or orifice plates). In this chapter, we develop and discuss model results for (a) only, although similar results for the other configurations are easily obtained by modifying one or two end boundary conditions. In addition, our model applies to sirens and poppet dipole valves and not negative pulsers or monopoles.

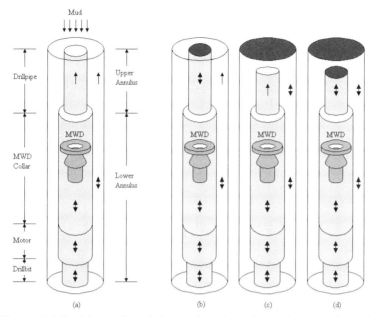

Figure 2.4. Positive and mud siren pulsers in various telemetry channels.

2.3.1   Three distinct physical problems.

So far we have refrained from discussing wave propagation through the entire drillpipe length itself and also the complicated problems associated with reflections at the surface. We now address these issues and explain why these problems can, at least from a modeling perspective, be considered separately. For the downhole problem in Figure 2.4a, the typical length scale in the axial direction is about 200 feet, that is, the distance from the drillbit to the top of the highest collar. Over this distance, the effects of attenuation (meaning irreversible thermodynamic loss) are unimportant and can be neglected. This is not to say that zeros in pressure are not found; they are, but they are related to reversible wave cancellations. With this simplification, the problem for Figure 2.4a can be solved analytically, using algebraic manipulation software or numerical matrix solvers, as we will demonstrate later in this chapter.

What the drillpipe "sees" from afar is not the complicated interactions implicit in Figure 2.4a, but only the net pressure wave $p_{pipe}/\Delta p$ that ultimately leaves the MWD collar – in this sense, the drillpipe problem, primarily one for simple attenuation, is a separate one from that in the foregoing paragraph. In Figure 2.4a, drilling mud is shown moving downward; it interacts with an oscillating positive pulser or rotating mud siren to create signals of strength Δp which depend on flow rate, valve geometry, sound speed "c," rotation and so on.

We assume that Δp is independently known from short wind tunnel measurement. Only Δp and c appear explicitly in the drillpipe acoustic discussion.

The pressure initially entering the drillpipe may be denoted by $T_{eff}\,\Delta p$ where T_{eff} is a dimensionless acoustic "transmission efficiency" factor for source strength production whose properties are the subject of this report. Again, it is 0.5 for a uniform pipe, almost unity for present low-frequency positive pulsers, but it can be much greater than one for an optimized high frequency tool. Even in the ideal case when surface reflections and related complications are not problematic, we note that $T_{eff}\,\Delta p$ is *not* the signal that ultimately arrives at the surface standpipe. This initial signal is corrupted by two mechanisms. First, the wave is attenuated as it travels through the drillpipe by a dissipation factor "exp(-αx)" that varies exponentially with frequency. This irreversible thermodynamic loss arises from the effects of fluid viscosity, heat transfer and local pipe expansion – what is lost cannot be resurrected. In addition, weak periodic reflections are set up which is associated with pipe joints along the transmission path. These two phenomenon define the drillpipe problem.

As explained above, the downhole signal generation problem is distinct from that in the drillpipe. Analogously, the problem of surface reflection modeling – in particular, the extraction of signals from a wavefield corrupted by reflections, desurger distortion and mudpump noise – is likewise distinct from the drillpipe problem. What the surface problem "sees" is simply the reduced MWD input signal $T_{eff}\,\Delta p\,exp(-\alpha x)$ as a "black box" input plus the entire gamut of reflections obtained at the surface. When a single pressure transducer is introduced in the standpipe, it most likely will not measure "$T_{eff}\,\Delta p\,exp(-\alpha x)$," but that, plus a complicated set of reflections and signal distortions. The overall objective is to recover the information encapsulated in $\Delta p(t)$ itself through multiple sets of problems in an environment characterized by high signal-to-noise ratio. The physics described justifies our separate treatment of three key problems, namely, those for downhole signal generation, drillpipe attenuation, and surface reflection deconvolution. We now return to the downhole source problem and consider it in detail, the subject of the present chapter.

2.3.2 Downhole source problem.

It is important to understand the differences between wave propagation when drillpipe and annulus are both long, as in Figure 2.4a, versus situations where either or both are terminated close to the source, as in Figures 2.4b,c,d. This is necessary in order to model field situations for proper telemetry job planning and tool system design, and also, to correctly interpret laboratory data, for instance, using fixtures as suggested in Figures 2.4b,c,d for extrapolation to field situations similar to Figure 2.4a.

The schematic in Figure 2.4a shows a six-segment waveguide consisting of (1) a long drillpipe, (2) an MWD drill collar possibly containing other sensors, (3) the drill collar housing a positive displacement mud motor with possible impedance mismatches due to the presence of the rubber stator – this waveguide section can be used to alternatively model, say, a resistivity-at-bit sub, (4) the drillbit nozzles or the bit sub, (5) a "bottom annulus" surrounding the drillbit, bit sub and both drill collars, and finally, (6) a long "upper annulus" that extends to the surface. If the pulser has operated at a given frequency for some time, standing waves (denoted by double-arrows) are set up at all finite bounded sections while propagating waves (denoted by single-arrows) traveling away from the source appear in the semi-infinite drillpipe and upper annulus.

The assumptions implicit in Figure 2.4a are subtle. Consider first the drillpipe. It is assumed in the model that all waves move upward to the surface and do not return to the bottom. But at the surface, they will reflect at desurgers and mudpumps and travel downward – more than likely, though, any waves that reach the bottomhole assembly will be so weak that our simple upgoing wave assumption applies (if not, a second, smaller standing-wave pattern is created). This will not be so for very strong sources, in which case multiple reflections will be found; when this is so in wind tunnel testing, the "cure" is an appropriate reduction in signal amplitude, which is easily accomplished by decreasing wind speed. Similar considerations apply to the annulus and reflections from surface facilities. Now, both of these upgoing waves will damp as they travel upward, but since their reflections are assumed to be so weak that their influences on the standing wave pattern are unimportant, the simple fact that the only waves that leave the bottomhole assembly are waves that travel upward suffices. This represents the so-called "radiation" or "outgoing wave" condition used in physics. Again, for the purposes of modeling wave interactions in the bottomhole assembly (such as constructive and destructive interference effects in signal enhancement or destruction), the local effects of irreversible thermodynamic attenuation are not important.

Now consider the alternative scenarios shown in Figures 2.4b, 2.4c and 2.4d where dark gray caps indicate terminations where strong reflections are possible. Sections bounded by these terminations will now contain standing wave patterns as opposed to propagating waves and the complete wave pattern found in the bottomhole assemblies will differ from that of Figure 2.4a. In particular, they will depend upon the nature of the reflector. If the termination acts as a solid reflector, the acoustic pressure locally doubles; if it acts as an open-end reflector, the acoustic pressure will locally vanish; finally, if it is an elastic reflector, i.e., a desurger, acoustic reflections will distort in shape and change in size depending on how wave amplitude and frequency interact with effective mass-spring-damper parameters.

Again, the same MWD hardware (that is, bottomhole assembly and tool configuration) operating identically in Figures 2.4a,b,c,d will not produce identical acoustic wave patterns because of boundary condition differences. Also, changing the mud (or, "c") will affect each differently. In some cases, the upgoing wave is strong; in others, it is not. In MWD field tool testing, the effects of the complete acoustic channel must be understood. Simply having pressure transducers in the same location along the standpipe means very little, since the upgoing signals can vary widely depending on borehole length, frequency and mud sound speed, and the locations of surface components like mudpumps and desurgers. With these physical arguments explained clearly, we now turn to a mathematical expression of these ideas and provide exact analytical solutions. We emphasize that computational finite difference and finite element solutions to the general formulation are the worst way to proceed, since numerical dispersion and diffusion errors lead to effective local sound speeds that differ significantly from actual ones, thus implying large phase errors. Great effort was expended to obtain analytical solutions; owing to the complexity of the linear system derived below, recourse to computerized algebraic manipulation methods was necessary, as will be described in detail.

2.4   Six-Segment Downhole Waveguide Model

In order to develop a realistic and accurate acoustic waveguide simulator for the bottomhole assembly, the source and the annulus, it is important to understand every component that affects the combined wave that eventually leaves the MWD drill collar. We have already discussed pressure symmetries and antisymmetries for source-field modeling, appropriate acoustic impedance treatment at areal and material changes, and the proper application of radiation conditions. In this chapter we neglect the effects of repeating drillpipe joints because they are not significant – these have not proven to be detrimental to field operation even in long wells where the initial signals are weak. One final modeling assumption is discussed, related to the events near the drillbit where the directions of the waveguide and the local mud flow both reverse.

We specifically refer to the lowest portion of the borehole occupied by the drill collar, drill bit and the lower annulus. What happens when a downward traveling wave passes through the bit nozzles and turns up into the annulus? One should not confuse hydraulic viscous losses with acoustic losses. For instance, the flow of fluid through a sharp bend will lead to large pressure losses, requiring higher pump power; however, the transmission of sound, particularly sound associated with long waves, through that same fluid and bend will be nearly lossless. How efficient is sound transmission through a bend? Fortunately, this question can be answered because, for frequencies several hundred Hertz or below, our waves are in fact long compared to the cross-sectional dimensions of the pipe or annulus. For this, we draw upon an exact solution of Lippert (1954, 1955) obtained in his well known studies of long wave reflection from ninety-degree bends.

His classic paper provides plots for both measured and calculated reflection and transmission coefficients, versus the dimensionless quantity $2fa/c$, where f is the frequency in Hertz, *a* is the width of the conduit, and c is the sound speed. Lippert shows that long, low frequency sound waves are transmitted through such bends with very minimal reflection. If we take f = 50 Hz (which is high by present MWD standards), and assume $a = 0.5$ ft, with a sound speed c of 5,000 ft/sec, we find that for $2fa/c = 0.01$, more than 99% of the incident power will pass through the bend. In other words, *elbows are effectively straight for acoustical purposes*. It is interesting that an increase of f to 500 Hz increases $2fa/c$ to 0.1; this increases the reflection coefficient to 0.05 and decreases the transmission coefficient to 0.99, which is still extremely close to unity. In conclusion, for frequencies up to several hundred Hertz, a loss-free wave assumption is perfectly reasonable – channels with bends, for the purpose of acoustical analysis, can be assumed as straight without loss of generality, as shown in Figure 2.5, thus significantly simplifying the mathematical formulation and its solution. Further discussion and other essential acoustic ideas are found in the classic book of Morse and Ingard (1968).

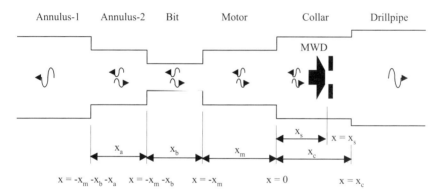

Figure 2.5. "Unwrapped" downhole MWD waveguide structure, to be compared with illustration in Figure 2.4a.

2.4.1   Nomenclature.

A_p ...   Internal cross-sectional area in drillpipe

A_c ...   Internal cross-sectional area in MWD drill collar

A_m ...   Internal cross-sectional area in mud motor (exclude metal rotor)

A_b ...   Internal cross-sectional area in "bit passage"

A_{a2} ...   Cross-sectional area, annulus surrounding drill collar

A_{a1} ...   Cross-sectional area, annulus surrounding drillpipe

B_{mud} ...   Bulk modulus in mud

B_{mm} ...   Bulk modulus in mud motor (weighted average, rubber and mud)

c_{mud} ...   Speed of sound in mud

c_{mm} ...   Sound speed in mud motor (weighted average, rubber and mud)

f ...   Frequency of dipole source (in Hertz)

p ...   Acoustic pressure

Δp ...   Dipole signal strength (for positive pulser or siren, not explicitly used, since it is "p/Δp" that is actually considered here)

ρ_{mud} ...   Mass density of mud

ρ_{mm} ...   Mass density in mud motor (weighted average, rubber and mud)

t ...   Time

u ...   Lagrangian displacement of fluid element from equilibrium

ω ...   Circular frequency of dipole source

x ...   Axial coordinate, "x = 0" at mud-motor/drill-collar interface

x_s ...   Distance to MWD acoustic dipole source, located in drill collar

x_c ...   Drill collar length

x_m ...   Length of mud motor

x_b ...   Length of "bit passage" beneath mud motor

x_a ...   Length of annulus, from floor of hole to pipe-collar interface

The foregoing long-wave property is significant to MWD telemetry modeling. Since the passage of a long acoustic wave *down* the drillpipe and drill collar and then *up* the annulus can be regarded as the kinematic equivalent of two consecutive 90° bends separated by a short "drill bit passage," we can topologically "unwrap" the collar-bit-annulus telemetry channel to approximate Figure 2.4a by the equivalent one-dimensional waveguide in Figure 2.5. In this representation, the origin "x = 0" is taken as the interface between mud motor and MWD drill collars; at the far left, a single left-going wave travels up the annulus, while at the far right, a single right-traveling wave travels up the drillpipe. The lengths and cross-sectional dimensions in Figure 2.5 do not represent actual geometries, but are intended for display purposes only; they are not drawn according to scale. The drillbit, mud motor and MWD collar lengths are, respectively, x_b, x_m and x_c.

High-data-rate mud pulse telemetry is governed by the equations of classical acoustics, e.g., Morse and Ingard (1968). The role of the wave equation is well known, governing fluctuations in quantities such as mass density, pressure, temperature and entropy; not only does the equation apply to the quantities cited, but it applies to more abstract (but often used) mathematical entities such as the "displacement potential" and the "velocity potential." The particular choice of dependent variable to be used is crucial to obtaining useful solutions. Elementary textbooks emphasize, for example, that the selected variable must accommodate the type of boundary conditions used; cases in point are the reflection conditions corresponding to the open or closed ends of an organ pipe. In MWD telemetry, the modeling problem is more complicated: the dependent variable used in our case must also accommodate the "dipole nature" of the source, that is, it must describe in a natural manner the antisymmetric disturbance pressures about the acoustic source generating the downhole signal.

The simplest example of an MWD dipole source is created by placing a low-frequency woofer in a pipe filled with stagnant fluid (e.g., air), with the face of the speaker lying in the cross-sectional plane of the pipe. When the speaker is excited by electric current, its cone oscillates, thereby creating sound waves that propagate in both directions. The nearfield mechanics of the speaker are interesting. When the created pressure on one side is high relative to ambient levels, the pressure on the opposite side is low by the same amount, and vice-versa. Several diameters away from the speaker, the three-dimensional details associated with cone geometry vanish – the resulting so-called "plane wave" does not vary with cross-section.

In other words, the created pressure (or, disturbance pressure relative to hydrostatic levels) is antisymmetric with respect to the source point. Because this pressure field is always antisymmetric, the difference in pressure levels across the source point, at any instant in time, will typically be nonzero. Hence, we say that a *jump* or *discontinuity in pressure* (that is, "delta-p") in pressure exists. We emphasize that this is not to be confused with a statically unchanging pressure differential, e.g., the viscous wake behind a bluff body, which does not propagate as sound; of course, since such a static drop is associated with a change in surface hydrostatic pressure, it can be used to encode information, although at extremely low data rates.

The strength of this jump is determined by events that fall outside the mechanics of acoustic wave propagation. In the foregoing example, the strength of the speaker signal is fixed by the amplitude of the electrical signal fed to the wave generator and the contours of the cone. In MWD, the strength of the created dipole signal will depend the geometry of the poppet valve or mud siren, the frequency of oscillation, the hydraulics of the problem by way of the flow rate and density of the drilling fluid, and so on. The (long) wave, upon reflection and re-reflection from boundaries, will pass freely through the source because rotor-stator gaps are never fully closed; only the jump in pressure can be specified with certainty in any formulation, since exact pressure levels at the source, which arise from interference and re-reflections, will depend on the geometry and boundary conditions associated with the waveguide assumed.

In summary, the mathematical formulation must permit convenient implementation of farfield boundary conditions, but it must allow free passage of pressure waves through the source point, as well as permitting us to represent the MWD source by a "jump in pressure" without specifying the exact local pressure (the exact source pressure will, of course, be available from the solution of the boundary value problem). Before proceeding, we again emphasize that a jump in pressure, or delta-p, is not necessary for signal creation. For example, two oppositely-facing speakers placed in close contact so that their outer rims touch, would model a negative pulser. These will obviously generate pressures that are symmetric with respect to the source point. However, they are associated with a zero delta-p, and instead execute monopole-type "breathing volume oscillations." The dipole sources considered in this book instead generate propagating signals associated with displacement changes and antisymmetric pressures produced without simultaneous volume creation.

2.4.2   Mathematical formulation.

In this section, we will derive the boundary value problem formulation describing the coupled acoustical interactions discussed above. This formulation consists of governing partial differential equations, dipole source model, matching conditions at impedance junctions, and farfield radiation conditions. Again, the objectives of the model are two-fold. We wish to design a model that allows us to study the conditions under which the drillpipe signal is good or bad and the extent to which improvements in MWD telemetry can be made. Second, we desire to obtain predictive values for strong annular MWD signal because their correlation with drillpipe signals can be useful in monitoring gas influx or in providing improved signal detection.

2.4.2.1 Dipole source, drill collar modeling.

A number of physical quantities were cited earlier for potential use as candidate dependent variables; however, none of these fulfill the requirements discussed. It turns out that the "Lagrangian displacement" variable $u(x,t)$ familiar to seismologists provides the flexibility needed to simultaneously model impedance, farfield, and source "delta-p" boundary conditions. The quantity "u" is a *length* representing the lineal displacement of a fluid element from its equilibrium position, which arises as a result of propagating compression and expansion waves that momentarily displace it from its initial position. This length, which varies with space and time, satisfies the one-dimensional wave equation "$\rho_{mud} u^c_{tt} - B_{mud} u^c_{xx} = 0$," where ρ_{mud} and B_{mud} represent, respectively, the mud density and bulk modulus. When desirable, in this book, subscripts will be used to denote partial derivatives, in order to simplify the presentation.

We will add to this classical equation the right-hand-side forcing function "Δp" $\delta(x-x_s)$ where we refrain from conjecturing the meaning of "Δp" for now. We do, however, note that $\delta(x-x_s)$ represents the Dirac delta function – that is, a concentrated force – situated at the source point $x = x_s$; in physical terms, the assumption that the imposed excitation lies at a single point in space requires that the dimensions of the MWD pulser be small compared with a typical wavelength, a requirement presently satisfied by all commercial tools. In doing so, we obtain the inhomogeneous partial differential equation

$$\rho_{mud}\, u^c_{tt} - B_{mud}\, u^c_{xx} = \text{``}\Delta p\text{''}\ \delta(x-x_s) \tag{2.1}$$

The superscript "c" refers to the MWD drill collar, of course, which houses the siren source (we discuss drill collar acoustics first in order to show why the Lagrangian displacement is the proper dependent variable for use in the remainder of the analysis). Note that Equation 2.1 does not include damping. Again, this is not to say that attenuation is unimportant; it is, but attenuation is only significant over large spatial scales, as would be the case when signals travel through the drillpipe. In the present section, which deals with interference dynamics near the source point, interactions act only over short distances on the order of the bottomhole assembly length and the mathematically complicating effects of damping can be justifiably neglected.

Let us now multiply Equation 2.1 throughout by the differential element dx, and integrate the resulting equation from "$x_s - \varepsilon$" to "$x_s + \varepsilon$," where ε is a small positive length on the order of the source dimensions. Since u(x,t) and its time derivatives are continuous through the source point, that is, there is no "tearing" in the fluid, we obtain the "jump condition"

$$- B_{mud}\, u^c_x\, (x_s+\varepsilon,t) + B_{mud}\, u^c_x\, (x_s-\varepsilon,t) = \text{``}\Delta p\text{''} \tag{2.2}$$

since the integral of the delta function is unity. If we note that the acoustic pressure is in general defined by

$$p = - B\, u_x, \tag{2.3}$$

it follows that "Δp" $= p^c_x\, (x_s+\varepsilon,t) - p^c_x\, (x_s-\varepsilon,t)$. In other words, "$\Delta p$" is the *jump* in *acoustic pressure* through the MWD source and it appears naturally *only* in a fluid displacement formulation using u(x,t). This jump is created not only by our aforementioned speakers, but by poppet valves and mud sirens which stop the oncoming flow in one direction, thus creating high pressure, while permitting the flow at the opposite position to pull away, thereby producing low relative pressure. It is also physically clear that the created pressures on either sides of the dipole source must be equal and opposite in strength, measured relative to hydrostatic background levels. We emphasize that this Δp is not to be confused with a statically unchanging pressure differential, e.g., the viscous wake behind a bluff body, which does not propagate as sound and does not possess the required antisymmetries; of course, since such a static drop is associated with a change in surface hydrostatic pressure, it can be used to encode information, although at low data rates. Incidentally, delta functions are used to represent point excitations for jump ropes and violin strings undergoing transverse motions when similar displacement formulations are used.

We note that the source strength Δp for mud sirens and positive pulsers used for continuous wave generators depends on flowrate, the mud sound speed, the mud density, the geometric details of the pulser, and finally, on the manner in which the pulser is driven. It does not depend on the wave propagation itself, and in this sense, can be regarded as a known prescribed quantity for acoustic modeling purposes. The properties of Δp are difficult to assess analytically. Instead, it is simpler to measure them directly in mud flow loops or in wind tunnels, an analysis area we develop more fully later in Chapter 9.

As a digression, we emphasize again that a "Δp" is not necessary for MWD signal generation and transmission, as we have discussed for negative pulsers – in fact, $\Delta p = 0$ identically. To model negative pulsers, the forcing function in Equation 2.1 would take the form "Δp" $[\delta(x-x_s + \varepsilon) - \delta(x-x_s - \varepsilon)]$ where "Δp" would refer instead to the acoustic pressure drop across the drill collar to the formation. The difference in delta functions ensures that no net pressure drop is occurs in the axial direction. More precisely, one models the negative pulser source as a "couple," that is, as the derivative of the delta function in a formulation for $u(x,t)$. To simplify the mathematics for negative pulsers, one alternatively uses "velocity potentials" for $\phi(x,t)$ where the axial acoustic velocity is $v = \partial\phi/\partial x$ and pressure is related to its time derivative. In the MWD drill collar, the statement $\rho_{mud} \, \phi^c_{tt} - B_{mud} \, \phi^c_{xx} = B$ "Δv" $\delta(x-x_s)$ would now indicate that a discontinuity in Δv is the natural math model for negative pulsers.

We also observe that a numerical solution to Equation 2.1, say based on finite differences or finite elements, would lead to very inaccurate results, given the point force excitation used. Even high-order schemes are complicated by large truncation errors which lead to dissipation and inaccuracies in modeling phase effects. The results, for instance, give local sound speeds that may be different than those characteristic of the actual mud. Inaccurate phase modeling would render all conclusions for constructive and destructive wave interference useless. We therefore pursue a completely analytical approach.

2.4.2.2 Harmonic analysis.

We have identified "Δp" as the transient acoustic pressure source, which can take on any functional dependence of time; this dependence, again, is dictated by pulser geometry, oncoming flow rate, fluid density, and the telemetry scheme selected. In this chapter, we physically represent "Δp" in the frequency domain as the product of a harmonic function $e^{i\omega t}$ and a signal strength $\Delta p(\omega)$ that may depend on frequency, but the exact dependence will vary with valve design, flow rate, and so on. Of course, this separation of variables is not as limiting as it initially appears; any transient signal can be reduced to superpositions of harmonic components using Fourier integral methods. Together with the assumption

$$\text{``}\Delta p\text{''} = \Delta p \; e^{i\omega t} \tag{2.4}$$

we consistently assume, still restricting our discussion to the MWD drill collar, a Lagrangian displacement of the form

$$u^c(x,t) = U^c(x)e^{i\omega t} \tag{2.5}$$

Substitution of Equations 2.4 and 2.5 in Equation 2.1 leads to a simple ordinary differential equation, better known as the one-dimensional Helmholtz equation governing the modal function $U^c(x)$,

$$U^c_{xx}(x) + (\omega^2/c_{mud}^2) \; U^c = - (\Delta p/B_{mud}) \; \delta(x-x_s) \tag{2.6}$$

which is satisfied by all fluid elements residing in the drill collar (here, we have introduced the sound speed defined by $c_{mud} = \sqrt{(B_{mud}/\rho_{mud})}$). This real inhomogeneous differential equation can be solved using standard Green's function or Laplace transform techniques (had we allowed attenuation, a more complicated complex equation would have been obtained). Its general solution, to be given later, can be represented in the usual manner as the superposition of a homogeneous solution (with two arbitrary integration constants) satisfying $U^c_{xx}(x) + (\omega^2/c_{mud}^2) \; U^c = 0$, and a particular solution (without free constants) satisfying Equation 2.6 exactly. Now that the fundamental mathematical ideas have been discussed in the context of drill collar analysis, let us consider the remaining elements of the waveguide.

2.4.2.3 Governing partial differential equations.

We have discussed in detail the properties of the differential equation governing the acoustics within the drill collar. The equations governing other sections of the one-dimensional waveguide are similar, although simpler, because they are not associated with MWD sources. Again, using separation of variables like $u(x,t) = U(x)e^{i\omega t}$, we obtain a sequence of Helmholtz equations for our waveguide sections. For clarity and completeness, going from the right to the left of Figure 2.5, these are

$$U^p_{xx}(x) \; + \; (\omega^2/c_{mud}^2) \; U^p \;\; = \;\; 0 \tag{2.7a}$$

$$U^c_{xx}(x) \; + \; (\omega^2/c_{mud}^2) \; U^c \;\; = \;\; - (\Delta p/B_{mud}) \; \delta(x-x_s) \tag{2.7b}$$

$$U^{mm}_{xx}(x) + \; (\omega^2/c_{mm}^2) \;\;\; U^{mm} \; = \;\; 0 \tag{2.7c}$$

$$U^b_{xx}(x) + (\omega^2/c_{mud}^2) \, U^b = 0 \tag{2.7d}$$

$$U^{a2}_{xx}(x) + (\omega^2/c_{mud}^2) \, U^{a2} = 0 \tag{2.7e}$$

$$U^{a1}_{xx}(x) + (\omega^2/c_{mud}^2) \, U^{a1} = 0 \tag{2.7f}$$

We observe that c_{mud} appears in all of the above equations, except Equation 2.7c, which contains c_{mm}. This quantity is just the speed of sound in the mud motor passage, which must be further obtained as a suitable weighted average of rubber, mud, and possibly, steel rotor properties (note that use of our Helmholtz model will preclude the modeling of signal shape distortions possible at downhole rubber interfaces). More than likely, laboratory investigation will be required to determine c_{mm}, the corresponding bulk modulus B_{mm} and the equivalent density ρ_{mm} (as in the case of drilling mud, the relationship of the form $c_{mm} = \sqrt{(B_{mm}/\rho_{mm})}$ applies). We also emphasize that the attenuative nature of the hard rubber that makes up the stators in the mud motor is not important at MWD frequencies less than 100 Hz and typical motor lengths less than 100 ft. Now, the general solutions for Equations 2.7a – 2.7f can be easily written down in closed analytical form, and in particular, are obtained as

$$U^p(x) = C_1 \exp(-i\omega x/c_{mud}) \tag{2.8a}$$

$$U^c(x) = C_2 \cos \omega x/c_{mud} + C_3 \sin \omega x/c_{mud} \tag{2.8b}$$
$$+ 0 \; \textit{if} \; x < x_s, \; or$$
$$- \{c_{mud} \Delta p/(\omega B_{mud})\} \sin \omega(x-x_s)/c_{mud} \; \textit{if} \; x > x_s$$

$$U^m(x) = C_4 \cos \omega x/c_{mm} + C_5 \sin \omega x/c_{mm} \tag{2.8c}$$

$$U^b(x) = C_6 \cos \omega x/c_{mud} + C_7 \sin \omega x/c_{mud} \tag{2.8d}$$

$$U^{a2}(x) = C_8 \cos \omega x/c_{mud} + C_9 \sin \omega x/c_{mud} \tag{2.8e}$$

$$U^{a1}(x) = C_{10} \exp(+i\omega x/c_{mud}) \tag{2.8f}$$

Observe that sines, cosines, and complex exponentials have been used in the solutions given by Equations 2.8a – 2.8f. Let us explain the motivation behind the exact choices made. Note that the typical homogeneous differential equation, e.g., $U_{xx}(x) + (\omega^2/c_{mud}^2) \, U = 0$, has two *real* linearly independent solutions, namely, the usual $\sin \omega x/c_{mud}$ and $\cos \omega x/c_{mud}$, but that equivalent solutions are also given by the *complex* mathematical expressions $\exp(+i\omega x/c_{mud})$ and $\exp(-i\omega x/c_{mud})$. Direct substitution in the governing equation, of course, demonstrates that both of these solution pairs are valid.

The exact representation useful in any particular instance, however, depends on the nature of the wave propagation found in the waveguide section under consideration. Let us consider, for example, the acoustic field in the drillpipe. Recall that we had used the separation of variables $u(x,t) = U(x)e^{i\omega t}$. The assumption taken in Equation 2.8a, namely, $U^p(x) = C_1 \exp(-i\omega x/c_{mud})$, implies that the time-dependent solution is $u(x,t) = C_1 \exp\{i\omega(t-x/c_{mud})\}$ which is the representation for a propagating wave. Similar comments apply to Equation 2.8f for the annulus. In contrast, the "sine" and "cosine" representations naturally describe standing waves inside the waveguide.

2.4.2.4 Matching conditions at impedance junctions.

Continuity of volume velocity requires that the product between waveguide cross-sectional area "A" and longitudinal velocity $\partial u(x,t)/\partial t$ remain constant through an *impedance junction*, that is, any point through which an acoustic impedance mismatch exists. Since we have chosen the Lagrangian displacement as the dependent variable, the area-velocity product takes the form $A\partial u(x,t)/\partial t$. Because $u(x,t) = U(x)e^{i\omega t}$, this quantity is $A\partial\{U(x)e^{i\omega t}\}/\partial t$ or $iAU\omega e^{i\omega t}$. And furthermore, since the coefficient $i\omega e^{i\omega t}$ is the same on both sides of an impedance change despite changes in the modal function U, it follows that the continuity of volume velocity, at least for time harmonic disturbances, requires only that we enforce the continuity of product AU (hence, we have "$A_1U_1 = A_2U_2$"). Continuity of the acoustic pressure $p = - B\,\partial u(x,t)/\partial x$, on the other hand, requires that the derivative quantity BU'(x) remain invariant (consequently, we obtain "$B_1U'_1 = B_2U'_2$"), where B represents the bulk modulus. Although these impedance matching conditions superficially involve real quantities only, the equations for the coefficients C which arise from substitution of Equations 2.8a – 2.8f are complex, because combinations of sinusoidal and complex exponential solutions have be used to fulfill boundary conditions and radiation conditions. Hence, the coefficients C will, in general, be complex in nature. For a more detailed discussion on boundary and matching conditions, the reader is referred to the classical acoustics books of Morse and Ingard (1968) and Kinsler *et al* (2000).

Let us now give the algebraic equations that result from the assumed sinusoidal or exponential forms for U and the required matching conditions (refer to our nomenclature list for all symbol definitions). After some algebra, a detailed set of coupled linear complex equations is obtained, which is solved analytically and exactly in closed form and then evaluated numerically (phase errors of the type found in finite difference and finite element methods are not obtained in the present approach). In the following summary, the particular impedance junction considered is listed and underlined, and followed, respectively, by matching conditions obtained for volume velocity and pressure.

<u>Collar-Pipe Interface:</u>

$$\{1 - i \tan \omega x_c/c_{mud}\} C_1 - \{A_c/A_p\}C_2 - \{(A_c/A_p) \tan \omega x_c/c_{mud}\}C_3 =$$
$$- \{A_c c_{mud} \Delta p \sin \omega(x_c-x_s)/c_{mud}\}/\{A_p \omega B_{mud} \cos \omega x_c/c_{mud}\} \qquad (2.9a)$$

$$\{-\tan \omega x_c/c_{mud} - i\} C_1 + \{\tan \omega x_c/c_{mud}\}C_2 - C_3 =$$
$$- \{c_{mud} \Delta p \cos \omega(x_c-x_s)/c_{mud}\}/\{\omega B_{mud} \cos \omega x_c/c_{mud}\} \qquad (2.9b)$$

<u>Mud Motor-Collar Interface:</u>

$$C_2 - \{A_m/A_c\}C_4 = 0 \qquad (2.10a)$$

$$C_3 - \{(C_{mud}B_{mm})/(C_{mm}B_{mud})\}C_5 = 0 \qquad (2.10b)$$

<u>Bit Passage-Mud Motor Interface:</u>

$$\{A_m \cos \omega x_m/c_{mm}\}C_4 - \{A_m \sin \omega x_m/c_{mm}\}C_5$$
$$- \{A_b \cos \omega x_m/c_{mud}\}C_6 + \{A_b \sin \omega x_m/c_{mud}\}C_7 = 0 \qquad (2.11a)$$

$$\{(B_{mm}/C_{mm}) \sin \omega x_m/c_{mm}\}C_4 + \{(B_{mm}/C_{mm}) \cos \omega x_m/c_{mm}\}C_5 \qquad (2.11b)$$
$$- \{(B_{mud}/C_{mud}) \sin \omega x_m/c_{mud}\}C_6 - \{(B_{mud}/C_{mud}) \cos \omega x_m/c_{mud}\}C_7 = 0$$

<u>Annulus (2) - Bit Passage Interface:</u>

$$C_6 - \{\tan \omega(x_m+x_b)/c_{mud}\}C_7 - \{A_{a2}/A_b\}C_8$$
$$+ \{(A_{a2}/A_b) \tan \omega(x_m+x_b)/c_{mud}\}C_9 = 0 \qquad (2.12a)$$

$$\{\tan \omega(x_m+x_b)/c_{mud}\}C_6 + C_7 - \{\tan \omega(x_m+x_b)/c_{mud}\}C_8 - C_9 = 0 \qquad (2.12b)$$

<u>Annulus (1) - Annulus (2) Interface:</u>

$$C_8 - \{\tan \omega(x_m+x_b+x_a)/c_{mud}\}C_9$$
$$+ (A_{a1}/A_{a2})\{- 1 + i \tan \omega(x_m+x_b+x_a)/c_{mud}\}C_{10} = 0 \qquad (2.13a)$$

$$\{\tan \omega(x_m+x_b+x_a)/c_{mud}\}C_8 + C_9$$
$$+ \{- \tan \omega(x_m+x_b+x_a)/c_{mud} - i\}C_{10} = 0 \qquad (2.13b)$$

2.4.2.5 Matrix formulation.

The foregoing equations define a 10×10 system of linear algebraic equations in the *complex* unknowns C_1, C_2, ... , C_{10}. An analytical solution is possible with the use of algebraic manipulation software – it can, of course, be derived more laboriously by hand. The equation system can be represented efficiently if we rewrite our equations in a more transparent matrix form that highlights the structure of the coefficient terms. When we do this, we straightforwardly obtain an equation of the form [S] [C] = [R] which can be easily interpreted by computer algebra algorithms, that is,

$$
\begin{vmatrix}
S_{1,1} & S_{1,2} & S_{1,3} \\
S_{2,1} & S_{2,2} & S_{2,3} \\
& S_{3,2} & & S_{3,4} \\
& & S_{4,3} & & S_{4,5} \\
& & & S_{5,4} & S_{5,5} & S_{5,6} & S_{5,7} \\
& & & S_{6,4} & S_{6,5} & S_{6,6} & S_{6,7} \\
& & & & & S_{7,6} & S_{7,7} & S_{7,8} & S_{7,9} \\
& & & & & S_{8,6} & S_{8,7} & S_{8,8} & S_{8,9} \\
& & & & & & & S_{9,8} & S_{9,9} & S_{9,10} \\
& & & & & & & S_{10,8} & S_{10,9} & S_{10,10}
\end{vmatrix}
\begin{vmatrix}
C_1 \\ C_2 \\ C_3 \\ C_4 \\ C_5 \\ C_6 \\ C_7 \\ C_8 \\ C_9 \\ C_{10}
\end{vmatrix}
=
\begin{vmatrix}
R_1 \\ R_2 \\ 0 \\ 0 \\ 0 \\ 0 \\ 0 \\ 0 \\ 0 \\ 0
\end{vmatrix}
$$

$$\text{(2.14)}$$

where the elements of **S** and **R** are defined by the real and complex quantities given below. For the elements of the coefficient matrix **S**, we have

$$S_{1,1} = 1 - \mathbf{i}\,\tan \omega x_c/c_{mud} \qquad\qquad (2.15\text{a})$$

$$S_{1,2} = -\, A_c/A_p \qquad\qquad (2.15\text{b})$$

$$S_{1,3} = -\,(A_c/A_p)\,\tan \omega x_c/c_{mud} \qquad\qquad (2.15\text{c})$$

$$S_{2,1} = -\,\tan \omega x_c/c_{mud} \;\; -\, \mathbf{i} \qquad\qquad (2.15\text{d})$$

$$S_{2,2} = \tan \omega x_c/c_{mud} \qquad\qquad (2.15\text{e})$$

$$S_{2,3} = -1 \qquad\qquad (2.15\text{f})$$

$$S_{3,2} = 1 \qquad\qquad (2.15\text{g})$$

$$S_{3,4} = -\, A_m/A_c \qquad\qquad (2.15\text{h})$$

$$S_{4,3} = 1 \qquad\qquad (2.15\text{i})$$

$$S_{4,5} = -\,(C_{mud}B_{mm})/(C_{mm}B_{mud}) \qquad\qquad (2.15\text{j})$$

$$S_{5,4} = A_m \cos \omega x_m/c_{mm} \qquad\qquad (2.15\text{k})$$

$$S_{5,5} = -\, A_m \sin \omega x_m/c_{mm} \qquad\qquad (2.15\text{l})$$

$$S_{5,6} = -\, A_b \cos \omega x_m/c_{mud} \qquad\qquad (2.15\text{m})$$

$$S_{5,7} = A_b \sin \omega x_m/c_{mud} \qquad\qquad (2.15\text{n})$$

$$S_{6,4} = (B_{mm}/C_{mm})\,\sin \omega x_m/c_{mm} \qquad\qquad (2.15\text{o})$$

$$S_{6,5} = (B_{mm}/C_{mm})\,\cos \omega x_m/c_{mm} \qquad\qquad (2.15\text{p})$$

$$S_{6,6} = - (B_{mud}/C_{mud}) \sin \omega x_m/c_{mud} \qquad (2.15q)$$

$$S_{6,7} = - (B_{mud}/C_{mud}) \cos \omega x_m/c_{mud} \qquad (2.15r)$$

$$S_{7,6} = 1 \qquad (2.15s)$$

$$S_{7,7} = - \tan \omega(x_m+x_b)/c_{mud} \qquad (2.15t)$$

$$S_{7,8} = - A_{a2}/A_b \qquad (2.15u)$$

$$S_{7,9} = (A_{a2}/A_b) \tan \omega(x_m+x_b)/c_{mud} \qquad (2.15v)$$

$$S_{8,6} = \tan \omega(x_m+x_b)/c_{mud} \qquad (2.15w)$$

$$S_{8,7} = 1 \qquad (2.15x)$$

$$S_{8,8} = - \tan \omega(x_m+x_b)/c_{mud} \qquad (2.15y)$$

$$S_{8,9} = -1 \qquad (2.15z)$$

$$S_{9,8} = 1 \qquad (2.15a')$$

$$S_{9,9} = - \tan \omega(x_m+x_b+x_a)/c_{mud} \qquad (2.15b')$$

$$S_{9,10} = (A_{a1}/A_{a2})\{- 1 + i \tan \omega(x_m+x_b+x_a)/c_{mud}\} \qquad (2.15c')$$

$$S_{10,8} = \tan \omega(x_m+x_b+x_a)/c_{mud} \qquad (2.15d')$$

$$S_{10,9} = 1 \qquad (2.15e')$$

$$S_{10,10} = - \tan \omega(x_m+x_b+x_a)/c_{mud} - i \qquad (2.15f')$$

whereas, for the elements of the forcing function **R**, we have

$$R_1 = - \{A_c c_{mud} \Delta p \sin \omega(x_c-x_s)/c_{mud}\}/\{A_p \omega B_{mud} \cos \omega x_c/c_{mud}\} \qquad (2.16a)$$

$$R_2 = - \{c_{mud} \Delta p \cos \omega(x_c-x_s)/c_{mud}\}/\{\omega B_{mud} \cos \omega x_c/c_{mud}\} \qquad (2.16b)$$

2.4.2.6 Matrix inversion.

It is important to observe from Equation 2.14 that the coefficient matrix **S** is both *sparse* (that is, most of its terms are identically zero, and therefore need not be stored by a custom designed algorithm) and *banded* (in other words, each equation contains a limited, closely clustered number of unknowns, with the overall nonzero elements of **S** located within a narrow diagonal band). To mathematicians anyway, these properties imply significant computational advantages, especially in view of the complex nature of the coefficients underlying the overall system. Although the solution to the foregoing system is trivial on workstations and mainframe computers, we emphasize that practical solutions even on modern personal computers are not always possible, since stack overflow is typically encountered in the execution of the required complex transcendental arithmetic.   A computer program was therefore written and structured to accommodate memory limitations expected of desktop machines; the program, written in standard Fortran for portability, delivers almost instantaneous acoustic solutions on all desktop computers, and, as noted earlier, can be used to optimize MWD source placement and telemetering frequency. Obvious checks were used to validate the production code against programming error. For example, in the uniform waveguide limit where all cross-sectional areas are identical and materials do not change, calculated pressures immediately to the left and right of the source point are found, as required, to be equal and opposite; their difference, of course is found to be the inputted Δp. Other checks included agreement with transmission and reflection coefficients obtained for simple waveguide geometries for which simple analytical solutions were available.

2.4.2.7 Final data analysis.

Once the numerical values for the elements of [C] are available from matrix inversion, the primary quantities of physical interest can be obtained by straightforward post-processing.   The Lagrangian displacement and acoustic pressure of fluid particles in the drillpipe are obtained by taking real parts as indicated in Equations 2.17a – 2.17c,

$$u^p(x,t) = Re\ U^p(x)\ e^{i\omega t} \tag{2.17a}$$

$$p^p(x,t) = -B_{mud}\ Re\ dU^p(x)/dx\ e^{i\omega t} \tag{2.17b}$$

$$= Re\ \{(i\omega B_{mud}/c_{mud})\ C_1\ exp\ i\omega(t-x/c_{mud})\} \tag{2.17c}$$

noting that the argument "t - x/c_{mud}" signifies an upward propagating wave. For the borehole annulus, the corresponding "t + x/c_{mud}" solutions traveling away from the drillbit nozzles are

$$u^{al}(x,t) = Re\ U^{al}(x)\ e^{i\omega t} \tag{2.18a}$$

$$p^{al}(x,t) = -B_{mud}\ Re\ dU^{al}(x)/dx\ e^{i\omega t} \tag{2.18b}$$

$$= Re\ \{(-i\omega B_{mud}/c_{mud})\ C_{10}\ exp\ i\omega(t+x/c_{mud})\} \tag{2.18c}$$

While pressure levels themselves are important to signal-to-noise determination in signal processing, as well as to surface transducer selection and placement, very often some measure of signal generation efficiency is required, for different reasons. For example, one might reasonably ask, "How optimal is the created signal that is propagating up the drillpipe for a given Δp?" This question is important in applying downhole constructive interference to maximizing carrier wave strength, a technique that is all the more significant because there is no valve erosion or turbine power penalty associated with this type of signal enhancement. On the other hand, suppose that MWD signals are desired, which are to travel up the annulus (such signals have been used for gas influx detection, e.g., the inability to cross-correlate with drillpipe waveforms suggests the possible existence of un-dissolved gas bubbles). For such applications, we might ask the complementary question, "How optimal is the created signal that is propagating up the annulus for a given pulser Δp?"

A convenient dimensionless measure, in either case, is obtained by normalizing the acoustic pressure by the pulser Δp; the absolute value of this ratio, we have termed the "transmission efficiency." In designing an optimized pulser, it is this quantity that we seek to maximize. We emphasize, however, that high values of $p/\Delta p$ alone may not suffice. Since greater (thermodynamic) attenuation is found at higher frequencies, the increase in source strength may not always be enough to enable transmission to the surface; these increases must offset increases in attenuation found at higher frequencies. Whereas Δp depends largely on flow rate and pulser geometry, the transmission efficiency is completely independent of Δp and depends on waveguide geometry, sound speed, and pulser location and frequency only.

In an idealized situation where the dipole source resides in an infinite pipe without areal or material discontinuities, so that reflections and impedance mismatches are entirely ruled out, it is clear that half of this Δp signal propagates uphole while the remaining half travels downhole. This physical fact can be easily deduced from D'Alembert's formula in mathematics but it is also apparent from symmetry considerations. The theoretical value for transmission efficiency is identically 0.5; indeed, this simple value for both pipe and annular wave solutions serves as a critical software and programming check in the said limit. It turns out, as detailed calculations show, that at low frequencies, e.g., those typical of existing positive pulsers, a value of unity is approached as a result of constructive interference reflections at the bit – we had previously explained why this was so using purely physical arguments for dipole sources. However, this unit value by no means represents perfection – at higher frequencies, depending on pulser location and BHA details, transmission efficiencies exceeding 1.0 and approaching 3.0 are possible. In Chapter 10, a prototype design for a 10 bits/sec system is offered, in which the transmission efficiency is 1.7 assuming typical bottomhole assemblies and muds.

The transmission efficiencies corresponding to Equations 2.17c and 2.18c are easily determined from division by Δp and taking absolute values. Since the absolute value of $\exp i\omega(t \pm x/c_{mud})$ is exactly unity, it follows that

$$\textit{Transmission efficiency}_{1,10} = |p/\Delta p| = \{\omega B_{mud}/(|\Delta p|c_{mud})\} |C_{1,10}|$$

$$= \{\omega B_{mud}/(|\Delta p|c_{mud})\} \sqrt{(C_{1,10}C_{1,10}^*)} \qquad (2.19)$$

where the asterisk denotes complex conjugates.

Transmission efficiency plays a strong role in later calculations, i.e., for a given BHA and mud sound speed, the dependence of $p/\Delta p$ on position in the drill collar and on frequency is of interest. In practice, mud properties and BHA are not within the control of the MWD operator. However, the complete range of frequencies can be "swept," with signals evaluated at the surface, to determine optimal values for phase or frequency shift keying.

2.5   An Example: Optimizing Pulser Signal Strength

2.5.1   Problem definition and results.

We now consider a practical example and examine signal strength optimization by constructive wave interference, which is relevant to carrier wave enhancement for phase-shift-keying (PSK) schemes or frequency selection for frequency-shift-keying (FSK) methods. The reader new to acoustic methods should bear in mind the two examples from classical physics illustrated in Figure 2.6. At the top, an incident P_0 signal doubles at the reflector, whereas for the telescoping waveguide, an incident signal quadruples (the area ratio is two and the length of the extension is a quarter wavelength). The first amplification method was in fact successfully tested on the standpipe by the author using one hundred feet of hydraulic hose and subsequented patented (see Chapter 6). Our point is this: signal amplification is as natural as signal loss.

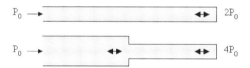

Figure 2.6. Classical wave amplification solutions.

What happens with the bottomhole assembly in Figure 2.7? For our input variables, we assumed water as the drilling fluid, and neoprene rubber as the mud motor stator medium. In calculating the cross-sectional area available to wave propagation for the mud motor, we subtracted out that corresponding to the metal rotor, whose acoustic impedance greatly exceeds that of water or rubber. Also, the presence of the MWD tool body within the MWD collar was ignored for simplicity; its effect is simply to reduce the collar area. In our calculations, the position of the MWD pulser x_s was incremented every foot, starting with $x_s = 1$ ft immediately above the motor, to $x_s = 22$ ft just below the drillpipe; also, transmission frequencies ranging from 1 Hz to 50 Hz were taken in single Hertz increments. The 50×22 matrix of runs required approximately five seconds of computing time on a typical personal computer, making the algorithm practical for MWD telemetry job planning, say, in applications where ideal frequencies are to be identified. Representative numerical results are given in Figure 2.8a below, while the surface plots in Figures 2.8b and 2.8c are based on the entire set of 1,100 data points.

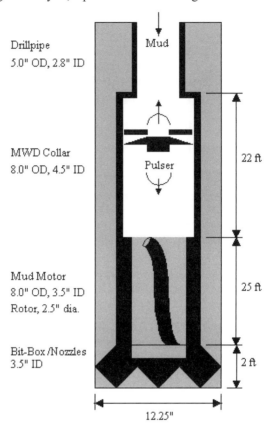

Figure 2.7. "Typical" bottomhole assembly with mud motor.

Insofar as the signal propagating up the drillpipe is concerned, the positive effects of constructive interference always obtain at 1 Hz or less; enough of the downward wave propagating toward the bit apparently reflects upward to positively superpose in phase with new upgoing waves. As noted earlier, the transmission efficiency associated with low data rate transmissions is typically 90% or more and source position is unimportant. But at 12 Hz, the signal decreases as the pulser moves away from the mud motor; however, this spatial trend reverses itself at high frequencies. The magnitudes of the pressure wave in the borehole annulus show consistently low values, typically 1-6 % of the pulser Δp, with higher values achieved at the lower frequencies. These results are consistent with numbers quoted by workers in MWD gas influx detection. We caution that the above results apply *only* to the bottomhole assembly studied in Figure 2.7, and then, only to our use of water as the drilling fluid; the reader should *not* generalize these results to other situations.

P/ΔP in Drillpipe

Hz	Source Position Xs(ft)				
	1	5	10	15	22
1	0.9422	0.9411	0.9397	0.9383	0.9363
12	0.7485	0.6084	0.4309	0.2536	0.0744
20	0.2628	0.0810	0.1687	0.3998	0.7142
30	0.0308	0.1750	0.4183	0.6466	0.9248
40	0.0894	0.3047	0.5551	0.7694	0.9828
50	0.1961	0.4372	0.6970	0.8861	0.9974

P/ΔP in Upper Annulus

Hz	Source Position Xs(ft)				
	1	5	10	15	22
1	0.05932	0.05928	0.05924	0.05921	0.05920
12	0.08162	0.07647	0.07108	0.06725	0.06516
20	0.06470	0.05813	0.05056	0.04458	0.04104
30	0.04583	0.04053	0.03357	0.02727	0.02305
40	0.03245	0.02904	0.02374	0.01822	0.01398
50	0.03154	0.02956	0.02447	0.01819	0.01266

Figure 2.8a. Transmission efficiency p/Δp in drillpipe and annulus.

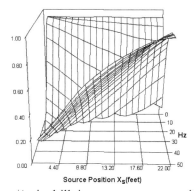

Figure 2.8b. $P_{drillpipe}/\Delta p$ in drillpipe versus source position and frequency.

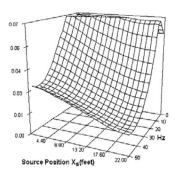

Figure 2.8c. $P_{annulus}/\Delta p$ in annulus versus source position and frequency.

2.5.2   User interface.

In order to streamline routine analyses, a convenient Windows graphical user interface is employed as shown in Figure 2.9a. The menu provides simple text input boxes. Clicking "Simulate" launches the solver and automatically produces the color graphical results for drillpipe and annulus shown in Figures 2.9b,c. Figure 2.9b clearly demonstrates that for low data rate pulsers, the transmission efficiency is close to unity and not sensitive to pulser position within the MWD drill collar. At 12 Hz, however, destructive interference can reduce the wave energy available to support a viable carrier frequency. Although changes in the bottomhole assembly, in pulser position within the collar and in mud sound speed are not within the control of the operator, the choice of frequency is, at least for operators of a next generation tool being considered. The software can be used, as in Chapter 10, to evaluate optimized higher frequencies which support higher data rate.

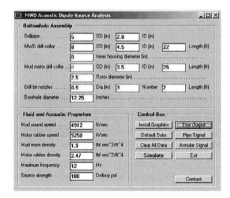

Figure 2.9a. Graphical user interface.

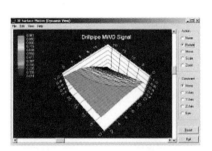
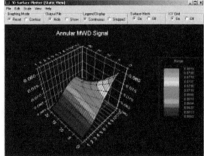

Figures 2.9b,c. MWD signal entering drillpipe (left) and annulus (right).

2.5.3 Constructive interference at high frequencies.

We emphasize that a high drillpipe p/Δp by itself will not guarantee higher data rate. In the final analysis, the increase in starting pressure must offset any frequency-dependent increase in attenuation over the drillpipe acoustic path. But what are the limits of pressure optimization? The exact results for Figure 2.6 suggest that amplifications exceeding unity may be possible, but calculations have never (until now) been undertaken for waveguides with the geometric complexity of Figure 2.7. Are the increases 10%, 20% or more? With expectations of only modest increases, this author ran the acoustic simulator, each time only increasing the range of analysis frequencies incrementally. But each time, optimistic results encouraged additional increases in the frequency range. Again, the results obtained are exact and do not contain numerical dissipation or phase errors typical of discretization schemes. Results for drillpipe p/Δp, plotted against pulser position and frequency, are offered next.

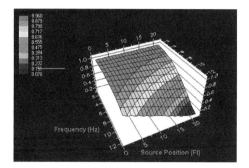

Figure 2.10a. Drillpipe p/Δp to 12 Hz.

Figure 2.10a again shows that, say at 1 Hz and below, source position is not important and all signals are equally strong. The (red) transmission efficiency is almost unity. In fact, a wave description of the problem is not necessary. But at 12 Hz, destructive interference will reduce the carrier frequency range available for MWD transmission if the pulser is not positioned properly. The range of frequencies considered is increased to 24 Hz in Figure 2.10b. One clearly observes an intermediate band of frequencies for which the transmission efficiency is low (highlighted in blue). However, near 24 Hz, a transmission efficiency of unity is almost achieved for a narrow range of source positions. Similar results are found in Figure 2.10c. In Figure 2.10d, the range of frequencies studied increases to 75 Hz. At low frequencies near 1 Hz, the transmission efficiency at all pulser locations is approximately unity as before, although the color mapping is now green. Interestingly, high (red) transmission efficiencies near *two* are now found for a limited band of frequencies greater than 60 Hz for a range of pulser locations. Similar results are found in Figure 2.10e in which the frequency analysis range is increased to 100 Hz.

82 MWD Signal Analysis, Optimization and Design

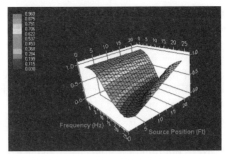

Figure 2.10b. Drillpipe p/Δp to 24 Hz.

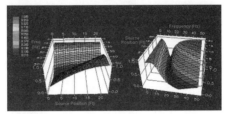

Figure 2.10c. Drillpipe p/Δp to 50 Hz.

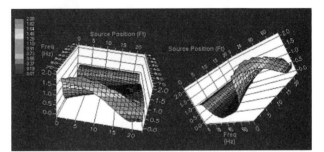

Figure 2.10d. Drillpipe p/Δp to 75 Hz.

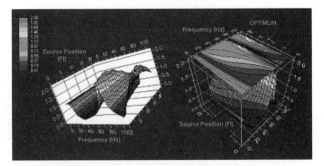

Figure 2.10e. Drillpipe p/Δp to 100 Hz.

2.6   Additional Engineering Conclusions

In Figures 2.10f,g,h, the frequency range is increased to 1,000 Hz. Two results in this unfolding drama are surprising. First, the maximum transmission efficiency increases to about 2.6, which is desirable. But even more important is the number of good frequencies (strong signals) and their close proximity to bad frequencies (weak signals). In a frequency-shift-keying (FSK) scheme, one need not alternate between good frequencies and zero Hertz (that is, complete siren stoppage, which imposes unreasonable system power demands associated with siren rotor inertia). Alternating, say, between 50 and 60 Hz would increase data rate (since more time is available for frequency cycling) while simplifying mechanical design. From Figure 2.10h, there is no shortage of red peaks.

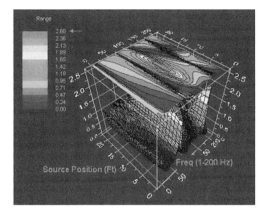

Figure 2.10f. Drillpipe p/Δp to 200 Hz.

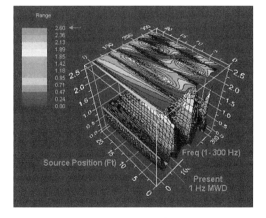

Figure 2.10g. Drillpipe p/Δp to 300 Hz.

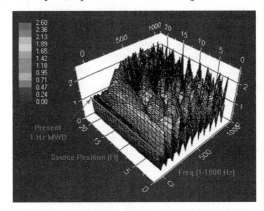

Figure 2.10h. Drillpipe p/Δp to 1,000 Hz.

The implications for MWD telemetry are clear. Phase-shift-keying (PSK) not only confuses surface detectors with "ghost reflections" created at the drillbit: they typically reduce the wave amplitudes needed for transmission over large distances. On the other hand, frequency-shift-keying (FSK) schemes in which 0's and 1's are transmitted by alternating between frequencies with high and low transmission efficiency offer significant advantages. First, constructive wave interference is used to enhance signal strength without incurring power or erosion penalties as would be the case in mechanical methods, e.g., mud sirens now increase signal only by reducing rotor-stator gap.

Second, the multiplicity of available frequencies allows significantly higher transmission data rates, say, by switching rapidly between 100 Hz, 110 Hz and 120 Hz. Third, the closeness of these frequencies implies that the mechanical power requirements needed to change from one rotation rate to nearby close ones are small, and this is particularly so with low-torque, rotor-downstream sirens such as those described in Chin and Trevino (1988) and Chin (2004). Finally, the large number of available optimal frequencies suggests that next-generation tools with multiple sirens should be investigated. Numerous "conversations" can be multiplexed along the channel using methods standard in communications theory.

Over the years, petroleum industry practitioners have accepted a maximum threshold of about 25 Hz for MWD transmissions based on flawed assumptions. The results of flow loop tests, for one, have not been interpreted properly, in the sense that commingled destructive wave interference and true attenuation effects have never been separately analyzed. Adding to the confusion, long flow loops, as will be explained in Chapter 9, offer boundary conditions that differ from those in real wells – they tend to support irrelevant standing wave systems rather than provide test platforms for evaluating novel telemetry concepts.

Software reference. The numerical engine and software interface for the foregoing model reside in C:\MWD-00. The Fortran source code is TENEQ22.FOR while the Visual Basic interface code in SOURCE.VBP calls the Fortran engine.

2.7   References

Oppenheim, A.V. and Schafer, R.W., *Digital Signal Processing*, Prentice-Hall, New Jersey, 1975.

Oppenheim, A.V. and Schafer, R.W., *Discrete-Time Signal Processing*, Prentice-Hall, New Jersey, 1989.

3
Harmonic Analysis:
Elementary Pipe and Collar Models

In Chapter 2, we developed a comprehensive six-segment downhole acoustic waveguide model in order to understand interactions between the pulser, bottomhole assembly and the geometry of the borehole. The harmonic analysis is useful for detailed analysis and in designing hardware and telemetry schemes for signal optimization. In other applications, simpler models suffice, e.g., when it is not possible to characterize the environment accurately. Also, simple models descriptive of idealized geometries may be useful for interpreting tests made in specially designed laboratory facilities. In this chapter, we continue our study of wave motions under harmonic excitation, that is, those underlying "frequency shift keying" (FSK) schemes. The "simple" geometric models considered here consist of, at most, a finite length drill collar and a semi-infinite drillpipe. For clarity, we present our results in two categories: (A) those in which the waveguide is uniform in area (i.e., drillpipe only) and (B) solutions for which an MWD signal originating in a drill collar travels up the drillpipe. In the latter case, the source can be located anywhere within the collar, the frequency (and, hence, wavelength) is general, and the collar cross-sectional area (that is, the collar area minus that of the central hub) may be less than, equal to, or greater than that of the drillpipe.

3.1 Constant area drillpipe wave models

Here we develop four mathematical models for which the MWD source resides in a "drillpipe only" environment without a drill collar. The source is situated away from the drillbit end of the pipe. The drillbit may be a solid or open reflector. As in Chapter 2, we caution against visual judgement; whether the bit is solid or open depends on acoustic frequency and geometric characteristics like bit box lengths, and not, for instance, on "how small" the nozzles appear visually.

We discuss several solutions to the acoustic wave equation applicable to wave propagation where the signals are created by dipole sources within the acoustic channel. These solutions correspond to the waveguides shown in Figure 3.A.1. In order to illustrate key ideas, uniform cross-sectional areas are assumed for simplicity (for more detailed analysis, use the six-segment waveguide model of Chapter 2). Again we take the Lagrangian displacement u(x,t) as the dependent variable, where t is time and x is the propagation coordinate. Then, away from the source, in the absence of attenuation, we have

$$\partial^2 u/\partial t^2 - c^2\, \partial^2 u/\partial x^2 = 0 \qquad\qquad (3.A.1)$$

$$c^2 = B/\rho \qquad\qquad (3.A.2)$$

$$p = - B\, \partial u/\partial x \qquad\qquad (3.A.3)$$

where c is the sound speed, B is the bulk modulus, ρ is fluid density and p is the acoustic pressure.

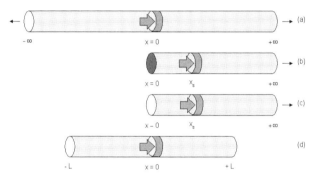

Figure 3.A.1. Different propagation modes.

3.1.1   Case (a), infinite system, both directions.

The "infinite-infinite" system in Figure 3.A.1a is considered. Both ends are far away and reflections do not return to the dipole source located at x = 0. As our system is linear, we consider a Fourier component of the excitation pressure having frequency ω. We seek separable solutions of the form

$$u(x,t) = X(x)\, e^{\,i\omega t} \qquad\qquad (3.A.4)$$

in which case

$$d^2 X(x)/dx^2 + \omega^2/c^2\, X = 0 \qquad\qquad (3.A.5)$$

This has linearly independent solutions of the form sin $\omega x/c$, cos $\omega x/c$, $e^{i\omega x/c}$ and $e^{-i\omega x/c}$, different combinations of which may be used to satisfy boundary conditions. For this example, we include all the mathematical details in order to illustrate the method. For the example in Figure 3.A.1a, complex exponentials are appropriate and we assume

$$u_1 = C_1 \, e^{\,i\omega(x/c + t)} + C_2 \, e^{\,i\omega(-x/c + t)}, \, x < 0 \qquad (3.A.6)$$

$$u_2 = C_3 \, e^{\,i\omega(x/c + t)} + C_4 \, e^{\,i\omega(-x/c + t)}, \, x > 0 \qquad (3.A.7)$$

That u_1 and u_2 represent left and right-going waves, respectively, requires $C_2 = 0$ and $C_3 = 0$. This illustrates the implementation of so-called "radiation" or "outgoing wave" conditions in classical physics. Continuity of fluid displacement requires

$$u_1(x,t) = u_2(x,t) \text{ at } x = 0 \qquad (3.A.8)$$

If the dipole source produces a Δp in the form $p_2 - p_1 = p_s \, e^{i\omega t}$, then

$$B \, \partial u_1/\partial x - B \, \partial u_2/\partial x = p_s \, e^{i\omega t} \text{ at } x = 0 \qquad (3.A.9)$$

These conditions lead to the pressure solutions

$$p_1(x,t) = - B \, \partial u_1/\partial x = - \tfrac{1}{2} \, p_s \, e^{\,i\omega(x/c + t)}, \, x < 0 \qquad (3.A.10)$$

$$p_2(x,t) = - B \, \partial u_2/\partial x = + \tfrac{1}{2} \, p_s \, e^{\,i\omega(-x/c + t)}, \, x > 0 \qquad (3.A.11)$$

These equations state the obvious fact that a Δp pulse of strength p_s splits into two waves that travel in opposite directions having equal and opposite strengths $- \tfrac{1}{2} \, p_s$ and $+ \tfrac{1}{2} \, p_s$. For this "infinite-infinite" system, nothing else happens that is of physical interest. It is interesting to note that $|p_1| \, /\Delta p = |p_2| \, /\Delta p = 0.5$ identically for this ideal infinite-infinite system, a property explained previously.

3.1.2   Case (b), drillbit as a solid reflector.

Here we consider the wave motion in Figure 3.A.1b. This applies when the drillbit nozzles are very small and when the drillbit is firmly pressing against hard rock. The problem is defined on $x > 0$ and the dipole source is now located at $x = x_s$. For $0 < x < x_s$, we take u_1 with linear combinations of sin $\omega x/c$ and cos $\omega x/c$ to model standing waves, but for $x > x_s$, our u_2 is proportional to $e^{\,i\omega(-x/c + t)}$ to represent a non-reflecting propagating wave traveling to the right. Since a solid reflector satisfies $u = 0$ at $x = 0$, and Equations 3.A.8 and 3.A.9 now apply at $x = x_s$, we find the solution

$$p_2(x,t) = + i \, \{2 \sin (\omega x_s/c)\} \, \tfrac{1}{2} \, p_s \, e^{\,i\omega(-x/c + t)}, \, x > x_s \qquad (3.A.12)$$

3.1.3   Case (c), drillbit as open-ended reflector.

Here we consider the wave motion in Figure 3.A.1c. Although we often think of a drillbit as a solid reflector because the nozzle area is small compared to the cross-sectional area, this is not true most of the time. In Case (a), we found that $|p_1| \, /\Delta p = |p_2| \, /\Delta p = 0.5$ in the absence of reflections. In Chapter 2 for our six-segment waveguide, we found that at very low frequencies, source position was unimportant in the drill collar and $p_{pipe} \, /\Delta p \approx 0.95$, approximately, or almost 1.0. This result, as explained in that write-up, is consistent with modeling the drillbit as an open-ended reflector. The problem here is defined on

$x > 0$ and the dipole source is again located at $x = x_s$. For the domain $0 < x < x_s$, we take u_1 with linear combinations of sin $\omega x/c$ and cos $\omega x/c$ to model standing waves, but for $x > x_s$, our u_2 is proportional to e $^{i\omega(-x/c\ +\ t)}$ to represent a non-reflecting propagating wave traveling to the right. Since an open-ended reflector satisfies $\partial u/\partial x = 0$ at $x = 0$, and Equations 3.A.8 and 3.A.9 apply at $x = x_s$, we find the solution

$$p_2(x,t) = \{2 \cos (\omega x_s/c)\} \; \tfrac{1}{2} \; p_s \; e^{\; i\omega(-x/c\ +\ t)}, \; x > x_s \qquad (3.A.13)$$

3.1.4   Case (d), "finite-finite" waveguide of length 2L.

The illustration in Figure 3.A.1d shows a dipole source centered at $x = 0$ in a waveguide of length 2L. We will assume open-ended reflectors satisfying $\partial u/\partial x = 0$ at $x = \pm L$ and discuss its physical significance later. Since standing waves are found at both sides of the source, linear combinations of sin $\omega x/c$ and cos $\omega x/c$ are chosen on each side to represent the displacement $u(x,t)$. Use of our Equations 3.A.8 and 3.A.9 at the source $x = 0$ leads to the solutions

$$p_1(x,t) = - \{p_s/(2 \tan \omega L/c)\} \; [\sin \omega x/c + (\tan \omega L/c) \cos \omega x/c] \; e^{i\omega t}$$
on $- L < x < 0$ \qquad (3.A.14)

$$p_2(x,t) = - \{p_s/(2 \tan \omega L/c)\} \; [\sin \omega x/c - (\tan \omega L/c) \cos \omega x/c] \; e^{i\omega t}$$
on $0 < x < + L$ \qquad (3.A.15)

3.1.5   Physical Interpretation.

Here we address the physical meanings and implications of the solutions obtained in Cases (a) – (d). These solutions may not represent the detailed telemetry channel discussed in Chapter 2, but they facilitate physical understanding and enable us to build acoustic test fixtures whose data may be interpreted unambiguously and accurately. More on testing and evaluation methods appears later in Chapter 9.

Case (a). Again, our exact solution states that a Δp pulse of strength p_s splits into two waves that travel in opposite directions having equal and opposite strengths $- \tfrac{1}{2} \; p_s$ and $+ \tfrac{1}{2} \; p_s$. That is, we have $p_1(x,t) = - \tfrac{1}{2} \; p_s \; e^{\; i\omega(x/c\ +\ t)}, \; x < 0$ and $p_2(x,t) = + \tfrac{1}{2} \; p_s \; e^{\; i\omega(-x/c\ +\ t)}, \; x > 0$. These solutions are the result of not having reflections – the waves on both sides travel away from the source and do not return to the point of origin.

Case (b). When the drillbit is modeled as a solid reflector, we have $p_2(x,t) = + i \; \{2 \sin (\omega x_s/c)\} \; \tfrac{1}{2} \; p_s \; e^{\; i\omega(-x/c\ +\ t)}, \; x > x_s$. This important solution indicates that the pressure is the product of an interference factor "$2 \sin (\omega x_s/c)$" and the no-reflection "$\tfrac{1}{2} \; p_s \; e^{\; i\omega(-x/c\ +\ t)}$" solution of Case (a). This factor is the result of constructive or destructive interference. It states that we can expect constructive interference if $\omega x_s/c = \pi/2, 3\pi/2, 5\pi/2$ and so on, and destructive interference if $\omega x_s/c = \pi, 2\pi, 3\pi$ and so on.

It is interesting to note that the maximum amplitude increase is a factor of 2.0 for this simple model – our six-segment waveguide results indicate that factors exceeding 2.0 are also possible for more complicated bottomhole assembly geometries. It is even more important that multiple frequencies exist which give this maximum constructive interference. These frequencies are uniformly separated by increments of $\omega x_s/c = \pi$. If frequency shift keying is used to telemeter results, one or more of these optimal frequencies (each associated with different amplitudes) can be used to achieve not just 0's and 1's, but 0's, 1's, 2's, 3's and so on. A simple binary scheme is also possible. Adjacent to each "good" frequency f_{good} (strong signal) is a "bad" frequency f_{bad} (weak signal), as our equations show. *Importantly*, to convey 0's and 1's, it is *not* necessary to frequency shift between f_{good} and 0 Hz, that is, bring the siren to a complete stop, since mechanical inertia demands on the drive motor may be significant. Instead, one might alternate between f_{good} and f_{bad}.

Case (c). When the drillbit is an open-ended reflector, we have instead $p_2(x,t) = \{2 \cos (\omega x_s/c)\}$ $\frac{1}{2}$ p_s $e^{i\omega(-x/c + t)}$, $x > x_s$. In this case, we have the same maximum constructive interference factor of 2.0 as in Case (b). Maximum constructive interference is now obtained at $\omega x_s/c = 0$, π, 2π, 3π and so on, and destructive interference is achieved at $\omega x_s/c = \pi/2$, $3\pi/2$, $5\pi/2$ and so on. Now, we may justifiably say that all of this is confusing – in practice, how do we know if we have Case (b) or Case (c)? The answer is interesting – *it does not matter*. The important conclusion is that, whatever the model, whether we have a solid reflector, open reflector or the more complete six-segment waveguide, multiple frequencies exist that yield good constructive interference and that these frequencies are close. We do not need a math model at the rigsite to compute these. Periodically, drilling operations can stop and the MWD pulser can "sweep" a range of frequencies (that is, slowly change from 1 Hz to 200 Hz, say). We can "listen" at the standpipe (being careful to subtract out the effects of surface reflections) to look for good and bad frequencies. Measured signals on the standpipe will contain the effects of surface reflections, which may also be good or bad. These FSK frequencies can be used as explained above.

Case (d). It is easily verified that $p_1 = 0$ at $x = -L$ and $p_2 = 0$ at $x = +L$. Also, the acoustic pressure solution is antisymmetric with respect to the source position $x = 0$, where Δp is maintained. This is an important solution for wind tunnel determination of Δp. It allows placement of the piezoelectric transducer anywhere, at any position "x" for measurement of the "p," which includes the effects of all reflections needed to set up the standing wave. Then, depending on whether Equation 3.A.14 or 3.A.15 is used, we can solve for the p_s representing "delta-p" in "$p_2 - p_1 = p_s e^{i\omega t}$" directly.

It is important that we understand how different sources of experimental error may arise in our use of Case (d) results. In particular:

(1) We have mathematically assumed that "x = 0" was our dipole source. In practice, the siren stator-rotor and electric drive system may be as much as 1 foot long and is not "x = 0." To minimize measurement error, the use of a longer wind tunnel is preferred, say 100-200 feet. It is not necessary to use the very long, 2,000 ft wind tunnel described later for evaluating many telemetry concepts – a simple 100-200 ft pipe with a blower and a piezoelectric pressure transducer suffices.

(2) If the dipole source were an electric speaker, the pressures p_1 and p_2 will be perfectly antisymmetric because there are no other sources of noise. However, when a mud siren is used, there is turbulent flow noise upstream of the stator, and downstream of the rotor, there exists turbulence noise plus strong pressure oscillations due to a swirling vortex motion imparted by the turning rotor. Measured pressures will be acoustic ones calculated in this book, plus these additional sources of noise. In order to measure acoustic Δp's properly, the noise on each side of the siren should ideally be filtered. The noise upstream of the stator is probably simpler to filter out – one might attempt a white noise or a Gaussian noise filter. The noise downstream of the rotor is more challenging. There is a periodic component due to the acoustics, but also a periodic component of the same frequency due to vortex motions (to visualize these motions, one might use the "ball in cage" method described in Chapter 9). This vortex component can be removed by placing flow straighteners just downstream of the rotor. These straighteners might occupy an axial distance of, say, 3-4 inches. A differential pressure transducer can be used to obtain Δp directly, with one end of the transducer placed just upstream of the stator and the other downstream of the flow straighteners. Of course, the method of Case (d) assumes that a differential transducer is not used, and that p_s is to be calculated from "p" obtained at a location "x."

(3) Again, the recommendation is to use piezoelectric pressure data upstream of the stator as it is less noisy and does not contain the swirling effects of the flow downstream of the rotor. However, it should not be located at the ends $x = \pm L$ because acoustic pressure vanishes at this locations; it is preferable, for example, to select an upstream location midway to the blower, say, $x = -\frac{1}{2} L$, in order to avoid downstream noise associated with the swirling rotor vortex flow having the same frequency. Turbulence noise may need to be filtered before calculating p_s from Equation 3.A.14.

3.2   Variable area collar-pipe wave models

In MWD wave propagation, two characteristic cross-sectional areas are important, namely, the area of the drillpipe and the drill collar area minus that of the central hub. When these areas differ, reverberant fields are created within the MWD collar that produce noisy upgoing signals. Ideally, area mismatches are eliminated in the mechanical design; when this is so, the uniform pipe models discussed above apply.

3.2.1   Mathematical formulation.

When area mismatches cannot be avoided, our models must be able to predict the reverberant fields excited by the dipole source. Here we consider the waveguide geometry of Figure 3.B.1 where an MWD drill collar is present. The cross-sectional area of the collar may be less than, equal to, or greater than that of the drillpipe because the collar center usually contains a central housing with mechanical components. The waves generated by the siren dipole source will reverberate within the collar, leading to constructive or destructive interference, depending on geometric and fluid acoustic properties. The wave shown at the right of Figure 3.B.1 is the MWD signal that ultimately leaves the simplified bottomhole assembly assumed. Both solid and opened "drillbit" left-end reflectors are considered here.

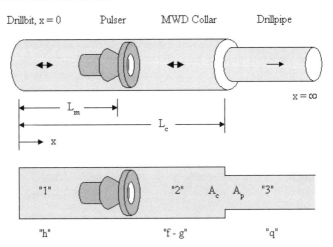

Figure 3.B.1. Collar-pipe acoustic two-section waveguide.

The functions h, f, g and q above refer to a general transient formulation considered later and may be ignored here. In this chapter, we consider harmonic oscillations only. In the above, A_c and A_p are cross-sectional areas for the collar and pipe, with A_c referring to the collar area minus the area of the central hub.

$$\partial^2 u/\partial t^2 - c^2\, \partial^2 u/\partial x^2 = 0 \tag{3.B.1}$$

$$c^2 = B/\rho \tag{3.B.2}$$

$$p = -\,B\,\partial u/\partial x \tag{3.B.3}$$

$$u(x,t) = X(x)\,e^{\,i\omega t} \tag{3.B.4}$$

$$d^2 X(x)/dx^2 + \omega^2/c^2\,X = 0 \tag{3.B.5}$$

$$X_1(x) = \mathbf{A}\,\sin \omega x/c + \mathbf{B}\,\cos \omega x/c \tag{3.B.6}$$

$$X_2(x) = \mathbf{C}\,\sin \omega x/c + \mathbf{D}\,\cos \omega x/c \tag{3.B.7}$$

$$X_3(x) = \mathbf{E}\,\exp(-i\omega x/c) \tag{3.B.8}$$

Equations 3.B.6 and 3.B.7 allow standing waves to form in Sections 1 and 2, while Equation 3.B.8 for Section 3 states that $u_3 = \mathbf{E}\,e^{\,i\omega(t-x/c)}$ which implies that the wave is traveling to the right without reflection. This wave will attenuate, but only at a distance far from the dimensions shown in Figure 3.B.1. At $x = L_c$, continuity of volume velocity $(A_c u_2 = A_p u_3)$ and of acoustic pressure $(\partial u_2/\partial x = \partial u_3/\partial x)$ require that

$$\mathbf{C}\,\sin \omega L_c/c + \mathbf{D}\,\cos \omega L_c/c = (A_p/A_c)\,\mathbf{E}\,\exp(-i\omega L_c/c) \tag{3.B.9}$$

$$\mathbf{C}\,\cos \omega L_c/c - \mathbf{D}\,\sin \omega L_c/c = -\,i\,\mathbf{E}\,\exp(-i\omega L_c/c) \tag{3.B.10}$$

At $x = L_m$, continuity of displacement yields

$$\mathbf{A}\,\sin \omega L_m/c + \mathbf{B}\,\cos \omega L_m/c = \mathbf{C}\,\sin \omega l_{-m}/c + \mathbf{D}\,\cos \omega L_m/c \tag{3.B.11}$$

The pulser develops a pressure discontinuity at $x = L_m$ of the form

$$p_2 - p_1 = \Delta p = P_s \exp(i\omega t) \tag{3.B.12}$$

Since $p = -\,B\,\partial u/\partial x$, we have

$$\mathbf{A}\,\cos \omega L_m/c - \mathbf{B}\,\sin \omega L_m/c - \mathbf{C}\,\cos \omega L_m/c + \mathbf{D}\,\sin \omega L_m/c = cP_s/(\omega B)$$
$$\tag{3.B.13}$$

At the drillbit $x = 0$, the assumption of an open reflector, i.e., **Case (e)**, requires $\partial u_1/\partial x = 0$ or $\mathbf{A} = 0$, while the assumption of a solid reflector, that is, **Case (f)**, requires $u_1 = 0$ or $\mathbf{B} = 0$, In either event, we have five equations for the five complex unknowns $\mathbf{A}$, $\mathbf{B}$, $\mathbf{C}$, $\mathbf{D}$ and $\mathbf{E}$ which can be solved exactly in closed analytical form. Some algebra shows that

$$\mathbf{E}_{solid} = \{(cP_s)/(\omega B)\}\,\sin \omega L_m/c\,\exp(+i\omega L_c/c)\,/\,\{(A_p/A_c)\,\cos \omega L_c/c + i\,\sin \omega L_c/c\} \tag{3.B.14}$$

$$\mathbf{E}_{open} = -\,\{(cP_s)/(\omega B)\}\,\cos \omega L_m/c\,\exp(+i\omega L_c/c)/\,\{(A_p/A_c)\,\sin \omega L_c/c - i\,\cos \omega L_c/c\} \tag{3.B.15}$$

Since $p_3(x,t) = -\,B\,\partial u_3/\partial x = i(B\omega/c)\,\mathbf{E}\,e^{\,i\omega(t-x/c)}$, then

$p_{3,solid}(x,t)/P_s = i \sin \omega L_m/c \exp (+i\omega L_c/c) / \{(A_p/A_c) \cos \omega L_c/c + i \sin \omega L_c/c\} \ e^{\ i\omega(t-x/c)}$

$$(3.B.16)$$

$p_{3,open}(x,t)/P_s = - i \cos \omega L_m/c \exp (+i\omega L_c/c)/ \{(A_p/A_c) \sin \omega L_c/c - i \cos \omega L_c/c\} \ e^{\ i\omega(t-x/c)}$

$$(3.B.17)$$

The quantity $| P_3(L_c,t)/P_s |$ provides a dimensionless measure of signal optimization due to constructive wave interference. Recall that in an "infinite-infinite" system without area change, a Δp pulse of strength p_s splits into two waves that travel in opposite directions having equal and opposite signal strengths $- \frac{1}{2} p_s$ and $+ \frac{1}{2} p_s$. The complicated factors shown above represent wave interference factors accounting for reflections at the drillbit and the collar-pipe junction. The factors involve both amplitude and phase changes.

3.2.2   Example calculations.

Here we describe typical results and software capabilities. Our simulations assume a drillpipe ID of 4 inches; an MWD drill collar having an ID of 6 inches, an inner hub with a 3 inch diameter, and an axial length of 30 feet; a mud sound speed of 4,000 ft/sec; and, finally, a maximum frequency of 300 Hz. Figures 3.B.2a to 3.B.2f display the MWD signal entering the drillpipe as a function of source position in the collar and excitation frequency.

These displays are automatically generated – the entire process requires seconds on personal computers. Results for "open" and "solid reflector" are both computed. Dynamic views allowing rotation of the figures about various axes and static views supporting contour plots are both supported. Again, the following work assumes sinusoidal excitations whereas Chapters 4 and 5 allow fully transient waveforms.

Case (e), two-part waveguide, open-ended reflector . . .

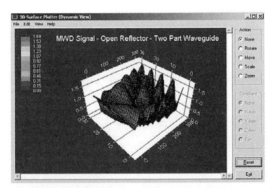

Figure 3.B.2a. MWD signal, open reflector, dynamic rotatable view.

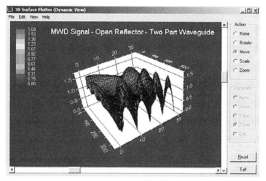

Figure 3.B.2b. MWD signal, open reflector, dynamic rotatable view.

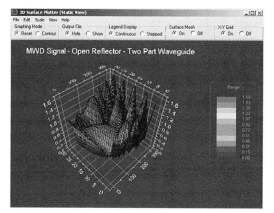

Figure 3.B.2c. MWD signal, open reflector, static view.

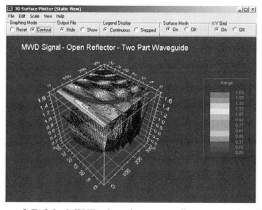

Figure 3.B.2d. MWD signal, open reflector, static view.

Case (f), two-part waveguide, solid-end reflector . . .

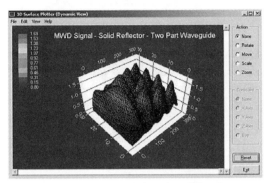

Figure 3.B.2e. MWD signal, solid reflector.

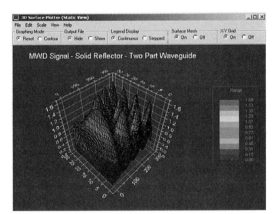

Figure 3.B.2f. MWD signal, solid reflector.

Software reference. Computer programs are not given for the constant area models in Cases (a), (b), (c) and (d) since the solutions already appear analytically in closed form. For the variable area problem, both solid and open drillbit reflector models are hosted in C:\MWD-00\TENEQ22-SUBSET-3.FOR. Plots similar to those shown above are automatically produced by the Fortran code. An interface does not exist. Parameters are changed by editing Fortran source code directly.

3.3 References

Oppenheim, A.V. and Schafer, R.W., *Digital Signal Processing*, Prentice-Hall, New Jersey, 1975.

Oppenheim, A.V. and Schafer, R.W., *Discrete-Time Signal Processing*, Prentice-Hall, New Jersey, 1989.

4
Transient Constant Area
Surface and Downhole Wave Models

Overview

In Chapters 2 and 3, we assumed harmonic or sinusoidal $\Delta p(t)$ excitations, for the purposes of analysis; these two chapters considered, respectively, comprehensive six-segment waveguides and simpler pipe-alone and collar-pipe models for *downhole* signal analysis and optimization. The results are directly applicable to "frequency shift keying" (FSK) telemetry. They are also useful in determining the extent to which constructive wave interference methods might be employed in signal enhancement, since destructive signal cancellations associated with random phase-shifting are not present. We focused on a particular downhole "forward" problem in Chapters 2 and 3, that is, the mathematical properties of the total created pressure field when a harmonic Δp is prescribed at a specific source location at a given frequency. The strength of Δp, that is, its dependence on flow rate, valve geometry, rotation rate, mud properties, and so on, is measured separately in flow loops or wind tunnels.

In this chapter, we introduce the study of *uphole* models. These methods apply at the surface and are principally developed to model reflections of the net upgoing MWD signal at the desurger and mudpump, and to evaluate single and multi-transducer echo cancellation and pump noise removal signal processing methods. Several uphole methods are presented and different models are evaluated and compared computationally against each other. The recovery of the upcoming Δp signal from noisy signals contaminated by surface reflections and mudpump action defines the "surface inverse problem." The transient models here and in Chapter 5 are "true transient" models derived in the time domain and do not rely on FFT and similar constructions.

We will also continue our investigation of downhole signal generation, studying the interaction between the wave traveling directly uphole from the source and that propagating down toward the drillbit and then reflecting upward. The transmitted signal, i.e., the sequence of 0's and 1's that is a consequence of the $\Delta p(t)$ pressure differential created by the *position-encoded* MWD pulser that travels up the drillpipe, however, is not $\Delta p(t)$. Instead, it is the cumulative "confused" signal consisting of up and downgoing waves noted. For downhole problems, our objective is the recovery of the $\Delta p(t)$ signal from the "confused" signal waveform entering the drillpipe – this signal is *really* complicated because a completely transient signal not restricted to sinusoidal motions is now permitted in our analysis. We term this the "downhole inverse problem."

For the downhole inverse problem in this chapter, the telemetry channel is assumed to be constant in area and is terminated at the "left" drillbit end – differences in MWD collar and drillpipe area are ignored. This assumption is not unrealistic: the MWD drill collar in practice contains a central hub (say, three inches in diameter) to which mechanical siren, turbine and alternator components are attached, and the resulting mismatch in collar and drillpipe cross-sectional area is often small (refer to Figure 5.0 in Chapter 5 and its discussion). If it isn't, it *should* be – area mismatches result in inefficient acoustic reverberations that additionally "scramble" the $\Delta p(t)$ wave into unrecognizable upgoing pressure waveforms that are difficult to deconvolve. This constant area model is introduced for simplicity only: the downhole inverse problem allowing large general changes in cross-sectional area is studied in detail in Chapter 5. The methods in this chapter are listed below. The particular source code used appears within the individual write-ups.

.
Contents

- **Method 4-1**. Upgoing wave reflection at solid boundary, single transducer deconvolution using delay equation, no mud pump noise.

- **Method 4-2**. Upgoing wave reflection at solid boundary, single transducer deconvolution using delay equation, with mud pump noise.

- **Method 4-3**. Directional filtering – *difference* equation method (two transducers required).

- **Method 4-4**. Directional filtering – *differential* equation method (two or more transducers required).

- **Method 4-5**. Downhole reflection and deconvolution at the bit, waves created by MWD dipole source, bit assumed as perfect solid reflector (very, very small drillbit nozzles).

- **Method 4-6**. Downhole reflection and deconvolution at the bit, waves created by MWD dipole source, bit assumed as perfect open-end, that is, zero acoustic pressure reflector (typical nozzle sizes).

4.1 Method 4-1. Upgoing wave reflection at solid boundary, single transducer deconvolution using delay equation, no mud pump noise (software reference, XDUCER*.FOR).

4.1.1 Physical problem.

Consider an upgoing pressure wave originating from downhole, containing encoded mud pulse information. It travels up the standpipe and is assumed to reflect at a solid reflector, the mudpump piston (this model does not apply to centrifugal pumps), and then propagates downward. Both incident and reflected waves are found in the surface standpipe where pressure transducers are installed. We wish to extract the upgoing wave from the total signal. We will not consider mudpump noise, random noise, or other noise sources here, as these subjects are deferred to Chapter 6. Multiple transducer methods are available to remove downward reflection and other down-going signals as given in Methods 4-3 and 4-4. Here, we describe a single transducer method that assumes a solid reflector – convenient for standpipe use when two transducers cannot be installed. The solid reflector in Figure 4.1a represents the pump piston.

Schematic

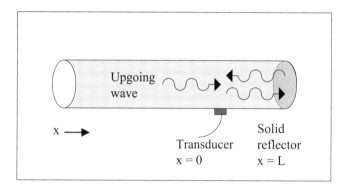

Figure 4.1a. Wave reflection at solid reflector.

The surface reflector is assumed to be solid for the single transducer approaches of Methods 4-1 and 4-2. This applies to positive displacement pumps only – an excellent assumption for almost all mud pumps that is validated by numerous experiments. For centrifugal pumps, an open end, zero acoustic pressure boundary condition would apply; centrifugal pumps and free-end conditions are not treated in this book because they are not very common, although an exact analysis would only require minor theory and software changes. However, our multiple-transducer deconvolution models in Methods 4-3 and 4-4 apply to both solid ($u = 0$) and open-end ($u_x = 0$) reflectors.

Also note that the upcoming wave, which is closely correlated to Δp(t) *only* at low frequencies, does not typically represent the train of desired logging 0's and 1's since it contains "ghost reflections" generated downhole. Again, downhole near the MWD signal source, waves are created that travel uphole; in addition, waves are created which travel downhole and then reflect upward. Thus, the net signal traveling uphole from the MWD tool is not the siren position-encoded Δp(t) but the superposition of an intended signal and its unwanted ghost reflections. Surface signal processing only removes undesired surface effects. From the signal obtained from surface processing, the models derived in Methods 4-5 and 4-6 must be used next to extract the intended signal, that is, Δp(t), which contains the encoded 0's and 1's containing logging information. All of these filters must be supplemented by additional ones, introduced in Chapter 6, for other noise mechanisms. Taken together, the complete system of filters provides the basic foundation of a signal processor applicable to very high data rates.

4.1.2 Theory.

Let $u(x,t)$ denote the Lagrangian fluid displacement variable for the acoustic field. Then, the function

$$u(x,t) = F(t - x/c) - F(t + x/c - 2L/c) \qquad (4.1a)$$

represents the superposition of an upgoing wave $F(t - x/c)$ and a downgoing wave $- F(t + x/c - 2L/c)$, with

$$u(L,t) = F(t - L/c) - F(t + L/c - 2L/c) = 0 \text{ at } x = L \qquad (4.1b)$$

Note that we have assumed $u = 0$ at the pump piston face $x = L$ (the pressure is measured at the standpipe location $x = 0$). The piston speed is very small compared to the sound speed in the fluid and can be ignored. If we had assumed a centrifugal pump, the boundary condition at $x = L$ would have been $\partial u/\partial x = 0$. If p denotes the acoustic pressure and B is the bulk modulus, then

$$p(x,t) = -B \, \partial u/\partial x = + (B/c) \, [F'(t - x/c) + F'(t + x/c - 2L/c)] \qquad (4.1c)$$

At the transducer $x = 0$, $p(0,t) = (B/c) [F'(t) + F'(t - 2L/c)]$ states that the total measured pressure $p(0,t)$ is the sum of an upward contribution $+ (B/c)F'(t)$ and the delayed function $+ (B/c)F'(t - h)$ having the same sign because a solid reflector has been assumed, where $h = 2L/c$ is the known roundtrip time delay from the transducer to the solid reflector. In more physically meaningful nomenclature, we can rewrite Equation 4.1c as

$$P_{upgoing}(t) + P_{upgoing}(t-h) = P_{measured}(t) \qquad (4.1d)$$

The deconvolution problem is stated as follows. When the total pressure $P_{measured}(t)$ is available at a single transducer as a discrete array of values at different instances in time, and $P_{upgoing}(t)$ vanishes at $t < 0$, find the function

$P_{upgoing}(t)$. This simple "delay equation" can be solved in closed analytical form and its solution is easily implemented digitally. Before proceeding with example solutions, we emphasize the physical assumptions.

Equation 4.1d applies to solid reflectors only and assumes that the functional form and amplitude of the reflection is unchanged. This type of reflection does not necessarily apply to desurgers, for which there may be shape distortion and damping, particularly at low frequencies and high amplitudes (detailed discussions are offered in Chapter 6). Thus, it is only applicable to positive displacement mudpumps with piston reflectors, and only when the roundtrip signal attenuation between the transducer and the pump pistons is negligible, a condition satisfied in practice. It is a trivial matter to handle attenuation by inclusion of a fractional scale factor in Equation 4.1d if required, for instance, if significant attenuation due to the rotary hose is present.

In summary, if the pump is a centrifugal pump and attenuation is negligible, the above equation is replaced by

$$P_{upgoing}(t) - P_{upgoing}(t-h) = P_{measured}(t) \qquad (4.1e)$$

because an acoustically-opened (zero pressure, "$\partial u/\partial x = 0$") boundary condition applies instead of the solid reflector "$u = 0$." If attenuation is not negligible, the above equations are replaced by $P_{upgoing}(t) + \delta P_{upgoing}(t-h) = P_{measured}(t)$ and $P_{upgoing}(t) - \delta P_{upgoing}(t-h) = P_{measured}(t)$, respectively, where $0 < \delta < 1$ is a positive loss factor, with $\delta P_{upgoing}(t-h)$ describing the reduced pressure at the transducer after the roundtrip travel time $t = h$ (the constant δ is measured separately).

Note that the distance L from the standpipe to the mudpump is typically 50 ft or less, since space on offshore rigs is limited (the total length of the standpipe is about 30 ft). If water or brine is used as the drilling fluid, then approximately, $c = 5,000$ ft/sec and $h = 2L/c = 2(50)/5,000$ ft or 0.02 sec. For typical drilling muds, $c = 3,000$ ft/sec and the delay time h may increase to 0.03 or 0.04 sec. Results for typical field numbers are presented next. Again, no other noise excepting reflection at the solid piston is assumed in these examples.

4.1.3   Run 1. Wide signal – low data rate.

The upgoing signal assumed is a rectangular pulse with rounded edges, with higher R values increasing the sharpness of the corners. This is constructed as shown below using two hyperbolic tangent functions. In this book, the Fortran code used is shown in Courier font, as seen below.

```
C      CASE 1. BROAD PULSE WIDTH
C      Clearly see upgoing and reflected pulses enhance the signal
       A = 1.0
       R = 20.0
       G = A*(TANH(R*(T-0.100))-TANH(R*(T-0.600)))/2.
```

- Roundtrip delay time = 0.02 sec
- Sampling time = 0.01 sec = 10 ms
- Pulse width is 0.600 – 0.100 or 0.500 sec.

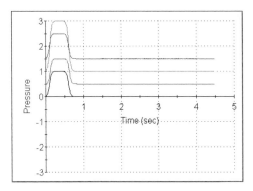

Figure 4.1b. Wide signal – low data rate.

Fortran source code XDUCER*.FOR is used to produce the results in Method 4-1, Run 1. Figure 4.1b shows pressure measurements at a single standpipe transducer. The incident upgoing assumed signal (black) is a broad pulse with a width of about 0.5 sec. The reflection (red) is the reflection obtained at a solid reflector with no attenuation assumed; there is very little shifting of the red curve relative to the black curve, since the total travel distance to the piston is very short. The transducer will measure the superposition of incident and reflected signals which broadly overlap. This superposition appears in the green curve – about twice the incident signal due to constructive wave interference, it does not cause any problems and actually enhances signal detection. The blue curve is the signal extracted from data using only the green curve and the algorithm described above. This blue curve clearly recovers the black incident wave very successfully. This is a low data rate run, typical of existing MWD systems. (For readers of the black and white version of this book, colors above are, respectively, black, red, green and blue, starting from the bottom curve.) The previous run is next repeated with finer sampling time, in particular,

- Roundtrip delay time = 0.02 sec
- Sampling time = 0.005 sec = 5 ms

Again, the superposition of incident and reflected waves enhances signal detection, noting that the time pulse width (0.5 sec, for 1 or 2 bits/sec) is large compared to the roundtrip delay time (0.02 sec). Reflections do not cause confusion, and both coarse and fine sampling times are acceptable. The computed results are shown in Figure 4.1c.

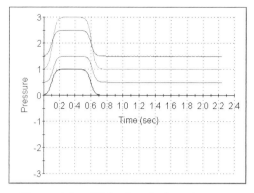

Figure 4.1c. Wide signal – low data rate.

4.1.4   Run 2. Narrow pulse width – high data rate.

```
C     CASE 2.  NARROW PULSE WIDTH
C     Clearly see interference between upgoing and reflected pulses
      A = 20.0
      R = 100.0
      G = A*(TANH(R*(T-0.100))-TANH(R*(T-0.101)))/2.
C
```

Due to the properties of the hyperbolic tangent function used above to model a real-world pulse, emphasizing that it is not the Heaviside step function, the "0.101 – 0.100" time difference of 0.001 sec is not directly related to pulse width (a step function would yield unrealistic infinite derivatives at the front and back ends of the pulse). For pulse width, refer directly to the black upgoing wave. The pulse width, from Figure 4.1d, is small and indicates high data rate.

- Roundtrip delay = 0.02 sec
- Sampling time = 0.001 sec = 1 ms

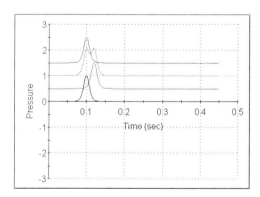

Figure 4.1d. Narrow pulse width – high data rate (with widened green trace).

Source code XDUCER*.FOR is used for Method 4-1, Run 2. Here, the incident upgoing black signal is about 0.05 sec wide, nominally a 20 bit/sec run. The delayed red reflected signal is also shown. The green is the superposition of the upgoing and reflected signals. It is clearly much wider than the black signal, and shows the interference of the upgoing and downgoing signals. In fact, the green line falsely conveys the presence of two signals, and definitely, would confuse the transducer into "thinking" that there really are two signals. This widening is not good for high-data-rate telemetry and is also known as "intersymbol interference." The top blue curve uses data from the green curve only, and recovers the black curve successfully. The above run was repeated for finer sampling, in particular,

- Roundtrip delay = 0.02 sec
- Sampling time = 0.0005 sec = 0.5 ms

Again, when the pulse width is narrow compared to the delay time, the standpipe transducer will "see" two pulses instead of one, or if the two are overlapping, one large signal with two "camel humps." Results appear in Figure 4.1e. The top blue curve recovers the black signal very nicely. Both sampling rates show same phenomena.

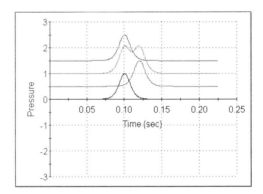

Figure 4.1e. Narrow pulse width – high data rate.

4.1.5   Run 3. Phase-shift keying or PSK.

We consider first a 12 Hz low carrier frequency. This is the frequency used to transmit siren information at 3 bits/sec or less. The input source code shown below is used to create the black curve in Figure 4.1f.

```
C      CASE 3.  PHASE-SHIFTING (F = FREQUENCY IN HERTZ)
       PI = 3.14159
       A = 0.25
       F = 12.
C      G = A*SIN(2.*PI*F*T)
```

```
C      Means 2*PI*F cycles in 2*PI secs, or F cycles per sec, F is
Hz.
C      One cycle requires time PERIOD = 1./F
       PERIOD = 1./F
       IF(T.GE.0.0.   AND.T.LE.    PERIOD) G = +A*SIN(2.*PI*F*T)
       IF(T.GE.PERIOD.AND.T.LE.2.*PERIOD) G = -A*SIN(2.*PI*F*T)
       IF(T.                 GE.2.*PERIOD) G = 0.
```

- Roundtrip delay = 0.02 sec
- Sampling time = 0.001 sec

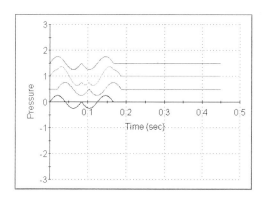

Figure 4.1f. 12 Hz low frequency carrier wave with phase shift keying.

Source code XDUCER*.FOR is used to produce the results of Method 4-1, Run 3. A 12 Hz carrier wave is assumed. We phase-shift after one cycle; we continue wave generation and turn off the pulser after two wave cycles. Note the black upgoing clean signal in Figure 4.1f and the delayed reflected red signal. The green signal combines both incident black and reflected red signals, and appears somewhat like the black signal, except for two indentations on the bottom (one for incident phase-shift and the second for the reflected phase-shift). The "apparent phase shift" seen in the green signal is a somewhat stretched valley. The green signal is a little wider than the original black signal with taller humps and can be very confusing. The blue one is the deconvolved signal using only input from the green data, and recovers the black one very nicely, to include both the phase shift and the silent tail end of the signal.

Next, we consider a 24 Hz carrier wave, the higher carrier frequency typically used in PSK schemes. This may transmit information at up to 6 bits/sec if attenuation is not problematic. In practice, the Δp siren signal strength is weak and the transmitted signal attenuates over large distances. This target data rate is almost never used in deep wells.

- Roundtrip delay = 0.02 sec
- Sampling time = 0.001 sec

A sampling time of 0.001 sec is used in Figure 4.1g. The green curve does not look like the black. The blue curve is recovered from the green data, and is almost identical to the original black curve. The recovery is very successful.

```
C       CASE 3.   PHASE-SHIFTING (F = FREQUENCY IN HERTZ)
        PI = 3.14159
        A = 0.25
        F = 24.
C       G = A*SIN(2.*PI*F*T)
C       Means 2*PI*F cycles in 2*PI secs, or F cycles per sec, F is
Hz.
C       One cycle requires time PERIOD = 1./F
        PERIOD = 1./F
        IF(T.GE.0.0.   AND.T.LE.   PERIOD) G = +A*SIN(2.*PI*F*T)
        IF(T.GE.PERIOD.AND.T.LE.2.*PERIOD) G = -A*SIN(2.*PI*F*T)
        IF(T.              GE.2.*PERIOD) G = 0.
```

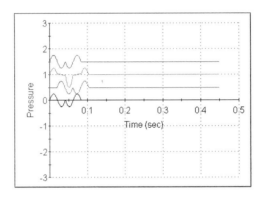

Figure 4.1g. 24 Hz low frequency carrier wave with phase shift keying.

Next we continue with 24 Hz, but instead use finer time sampling, in particular, selecting

- Roundtrip delay = 0.02 sec
- Sampling time = 0.0005 sec

In Figure 4.1h, the sampling time of 0.0005 sec is used. Both Figures 4.1g and 4.1h show that the green superposed signal is distorted relative to the original black one. There are two positive crests in the signal, as there are in the original black signal, and a single bump on the bottom. Also, the green signal is wider than the black. Intersymbol interference is strong. Also observe how well the algorithm using the blue signal recovers the black. Next, in Runs 4 and 5, let us examine very high data rates.

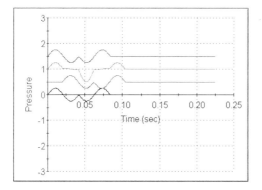

Figure 4.1h. 24 Hz low frequency carrier wave with phase shift keying.

4.1.6   Runs 4 and 5. Phase-shift keying or PSK, very high data rate.

In this example, we consider a high data rate 48 Hz carrier wave not presently used commercially by any service company.

```
C       CASE 3.  PHASE-SHIFTING (F = FREQUENCY IN HERTZ)
        PI = 3.14159
        A = 0.25
        F = 48.
C       G = A*SIN(2.*PI*F*T)
C        Means 2*PI*F cycles in 2*PI secs, or F cycles per sec, F is
Hz.
C        One cycle requires time PERIOD = 1./F
        PERIOD = 1./F
        IF(T.GE.0.0.   AND.T.LE.   PERIOD) G = +A*SIN(2.*PI*F*T)
        IF(T.GE.PERIOD.AND.T.LE.2.*PERIOD) G = -A*SIN(2.*PI*F*T)
        IF(T.               GE.2.*PERIOD) G = 0.
```

- Roundtrip delay = 0.02 sec
- Sampling time = 0.00025 sec

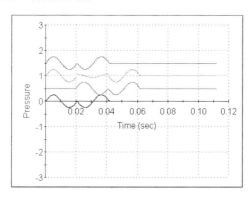

Figure 4.1i. New high data rate 48 Hz carrier with PSK.

In Figure 4.li, the green signal is 50% wider than the black signal and is showing severe distortion compared to the original black signal. However, our algorithm using the blue data recovers the black curve successfully.

Finally, we ambitiously increase our carrier wave frequency to 96 Hz in order to test the robustness of our scheme. The following parameters are used.

- Roundtrip delay = 0.02 sec
- Sampling time = 0.0002 sec

This very, very high data rate run with narrow black pulse shows how the incident and reflected red wave can superpose in the green signal to appear as a longer duration signal with additional fictitious phase shifts. This effect is detrimental to high-data-rate continuous wave telemetry. However, using the green data only, the blue curve recovers the black curve very nicely. Results are shown in Figure 4.1j.

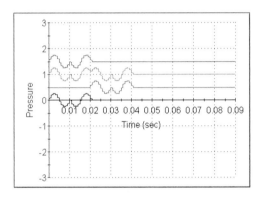

Figure 4.1j. Very, very high 96 Hz carrier with PSK.

4.2   Method 4-2. Upgoing wave reflection at solid boundary, single transducer deconvolution using delay equation, with mud pump noise (software reference, HYBRID*.FOR).

4.2.1 Physical problem.

This single transducer approach is similar to Method 4-1, except that the mudpump noise measured *at the pump* is delayed and subtracted from the standpipe transducer signal (which now records both upgoing and reflected signals, plus mudpump noise). Once the pump noise is subtracted, the algorithm is identical to Method 4-1. This capability is listed as a separate method only because of subtle differences in software structure. As in Method 4-1, this method does not handle wave shape distortion induced by the desurger.

4.2.2 Software note.

In Method 4-1, we did not allow pump noise; the purpose was to evaluate the echo cancellation or deconvolution method under ideal circumstances. Method 1 utilizes the XDUCER*.FOR software series. Method 4-2 here *with* mudpump noise utilizes our HYBRID*.FOR software series, a modification of the single transducer XDUCER*.FOR model to include downward traveling pump noise. By pump noise, we mean traveling pressure waves excited by the pump pistons and not turbulence or other noise associated with surface facilities.

Schematic

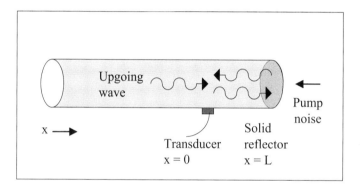

Figure 4.2a. Wave reflection at solid reflector, with pump noise.

4.2.3 Theory.

In this section, consider the reflection of an upgoing wave at a solid pump piston, to include the effect of pump noise propagated in the downgoing direction from the piston location.   Again a single transducer method is assumed. Let $p_{measured}(t)$ now denote all pressures recorded by the standpipe transducer, upgoing and downgoing, which would include mud pump noise. We also assume that another pressure transducer independently records the mud pump noise function $N(t)$ *at the pump*, where "t" is the same time instant at which the transducer records its signals. This is, in a sense, a two transducer method. However, from an operational viewpoint, it only requires a single transducer located on the standpipe, which is more convenient (and less dangerous) for rigsite operations. The second transducer can be placed near the mudpump and will not inconvenience any drillers on the rig floor. Then, if p(t) is the upgoing signal, and p(t-h) is the downgoing pressure reflected at the solid piston, where "h" is the roundtrip delay time, it follows that

$$p(t) + p(t-h) = p_{measured}(t) - N(t - h/2) \tag{4.2}$$

applies. Again, the measured pressure $p_{measured}(t)$ contains pump noise, the latter of which is then subtracted from the right side, leaving the noise-free equation considered in Method 4-1. The interesting question is, "How large can N be until the signal recovery is bad?" In other words, we want to explore the role of the "signal to noise" (or, "S/N") ratio in signal recovery. Usually, "noise" may be, for instance, white or Gaussian, which generally degrades signal enhancement procedures; however, as we will show, mudpump noise is not entirely harmful. Since its direction and acoustic properties can be well characterized, its effects may be subtracted out as demonstrated below.

4.2.4   Run 1. 12 Hz PSK, plus pump noise with S/N = 0.25.

The source code below generates an MWD signal with an amplitude of 0.25, while the pump noise function possesses an amplitude of 1.0 – thus, the signal-to-noise ratio is 0.25, which is very small. The pump noise function is defined in the Fortran block FUNCTION PUMP(T) while the MWD signal appears in FUNCTION G(T).

```
      FUNCTION G(T)
C
C     Test MWD upward waveforms stored here
C     CASE 3.  PHASE-SHIFTING (F = FREQUENCY IN HERTZ)
      PI = 3.14159
      A = 0.25
      F = 12.
C     G = A*SIN(2.*PI*F*T)
C      Means 2*PI*F cycles in 2*PI secs, or F cycles per sec, F is
Hz.
C     One cycle requires time PERIOD = 1./F
      PERIOD = 1./F
      IF(T.GE.0.0.   AND.T.LE.   PERIOD) G = +A*SIN(2.*PI*F*T)
      IF(T.GE.PERIOD.AND.T.LE.2.*PERIOD) G = -A*SIN(2.*PI*F*T)
      IF(T.                 GE.2.*PERIOD) G = 0.
      RETURN
      END

      FUNCTION PUMP(T)
      PUMP = 1. * SIN(2. * 3.14159 * 15. * T )
      RETURN
      END
C
```

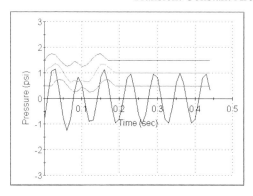

Figure 4.2b. 12 Hz carrier, phase shift in presence of large pump noise.

- Roundtrip delay time = 0.02 sec
- Sampling time = 0.01 sec

For readers of the black and white version of this book, in Figure 4.2b, curve colors are black, red, green and blue, respectively, from the bottom to top.

- Black is the combination of the upgoing MWD signal and the downward pump noise. Because the pump noise amplitude is four times larger, the black line does not even appear as if it contains an MWD signal.

- Red is the reflection of the upward MWD signal at the surface solid reflector.

- Green is the sum of the upgoing signal and reflected downward signal, with the pump noise subtracted out (the pump noise is measured at the pump, delayed, and then the subtraction is performed).

- Blue is the recovered upgoing signal as assumed (12 Hz with a phase shift), which appears just as the red signal does (the blue signal is recovered from green data which is a longer stretched pulse). Note the small signal-to-noise of 0.25 assumed.

4.2.5   Run 2. 24 Hz PSK, plus pump noise with S/N = 0.25.

We next repeat the calculation of Run 1, but double the carrier frequency to 24 Hz. The signal and pump noise functions are identical to those of Run 1, except that for the signal, a value of F = 24 Hz is used instead of F = 12 Hz (we are therefore running a 24 Hz carrier with a phase shift). Similar comments as those for Run 1 apply to our very successful recovery, as shown in the blue signal of Figure 4.2c.

- Roundtrip delay time = 0.02 sec
- Sampling time = 0.01 sec

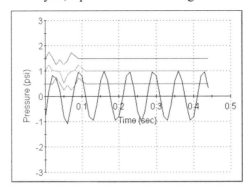

Figure 4.2c. 24 Hz carrier, phase shift with large pump noise.

4.3   Method 4-3.   Directional filtering – difference equation method requiring two transducers   (software reference, 2XDCR*.FOR).

4.3.1 Physical problem.

We next consider two-transducer surface signal processing in Methods 4-3 and 4-4. In this section, we have an upgoing pressure signal containing encoded downhole MWD mud pulse information (plus downhole "ghost reflections" produced at the pulser) which travels through the standpipe. The signal continues to the mudpump, where it will reflect at a solid reflector at the pistons for a positive displacement pump, and at an acoustic free-end for a centrifugal pump. *All types of pumps are permissible for Methods 4-3 and 4-4. Waves can also reflect at the desurger where, depending on the amplitude and frequency and the mass-spring-damper properties of the desurger, reflected waves can distort in both amplitude and shape.* In other words, we consider the very general and difficult problem in which no information about the mudpump and the desurger will be required – a very powerful signal processing method.

The two-transducer algorithms of Method 4-3 and 4-4, unlike some schemes, do not require any knowledge of the nature of the mudpump noise signature or the properties of the desurger distortion and are powerful in these regards. Surface sound speed is required, which can be determined by simple timing tests near the rig. Unlike published two-transducer methods, they need not assume periodic waves and apply generally to transient motions, e.g., there are no limiting restrictions related to the usual quarter-wavelength properties. Note that Method 4-3 is based on a *difference* equation, while Method 4-4 is based on a *differential* equation. In the difference equation model, there are no restrictions on transducer separation, although practical "not too close" and "not too far away" considerations apply. In the differential equation model, the usual notions on derivatives hold, which in the acoustic context, of course, suggest that transducer separations should be a small fraction of the wavelength.

Schematic

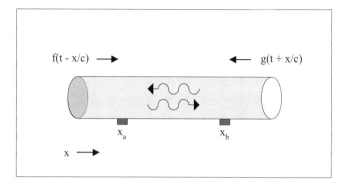

Figure 4.3a. General bi-directional waves.

4.3.2 Theory.

The pressure wave taken here is the sum of two waves traveling in opposite directions. Let "f" denote the incident wave traveling from downhole (again, this upgoing wave is not the "clean" Δp signal, but the intended signal plus ghost reflections associated with reflections at the drillbit and collar-pipe junction).   Then, "g" will denote reflections of any type at the mudpump (positive displacement pistons and centrifugal pumps both allowed), plus reflections at the desurger with any type of shape distortion permitted, plus the mudpump noise itself. In general, we can write

$$p(x,t) = f(t - x/c) + g(t + x/c) \qquad (4.3a)$$

where c is the measured speed of sound at the surface. Two transducers are assumed to be placed along the standpipe. Note that the impedance mismatch between standpipe and rubber rotary hose does introduce some noise, however, since this effect propagates downward, it is part of the "g" which will be filtered out in its entirety. The rotary hose does not introduce any problem with our approach.   Now let x_a and x_b denote any two transducer locations on the standpipe. At location "b" we have

$$p(x_b,t) = f(t - x_b/c) + g(t + x_b/c) \qquad (4.3b)$$

If we define $\tau = (x_b - x_a)/c > 0$, it follows that

$$p(x_b,t - \tau) = f\{t + (x_a - 2x_b)/c\} + g(t + x_a/c) \qquad (4.3c)$$

But at location "a" we have

$$p(x_a,t) = f(t - x_a/c) + g(t + x_a/c) \qquad (4.3d)$$

Subtraction yields

$$p(x_a,t) - p(x_b, t - \tau) = f(t - x_a/c) - f\{t + (x_a - 2x_b)/c\} \qquad (4.3e)$$

Without loss of generality, we set $x_a = 0$ and take x_b as the positive transducer separation distance

$$f(t) - f(t - 2x_b/c) = p(0,t) - p(x_b, t - x_b/c) \qquad (4.3f)$$

or

$$f(t) - f(t - 2\tau) = p(0,t) - p(x_b, t - \tau) \qquad (4.3g)$$

The right side involves subtraction of two measured transducer pressure values, with one value delayed by the transducer time delay τ, while the left side involves a subtraction of two unknown (to-be-determined upgoing) pressures, one with twice the time delay or 2τ. The deconvolution problem solves for $f(t)$ given the pressure values on the right and is solved by our 2XDCR*.FOR code.

Method 4-3 is extremely powerful because it eliminates any and all functions $g(t + x/c)$, that is, all waves traveling in a direction opposite to the upgoing wave. Thus, $g(t + x/c)$ may apply to mudpump noise, reflections of the upgoing signal at the mudpump, and reflections of the upgoing signal at a desurger, regardless of distortion or phase delay, reflections from the rotary hose connections, and so on. The functional form of the downgoing waves need not be known and can be arbitrary. This is not to say that all downward moving noise sources are removed. For example, fluid turbulence noise traveling downward with the drilling fluid is not acoustic noise will not be removed and may degrade the performance of Method 4-3. Additional noise sources and filters are considered in Chapter 6. The order in which filters are applied will affect the outcome of any signal processing, and it is this uncertainty that provides the greatest challenge in signal processor design. The model in Equation 4.3g is solved exactly, that is, analytically in closed form, and is implemented in our 2XDCR*FOR software series. In Figure 4.3b, black, red, green and blue curves appear, respectively, from the bottom to top. The color coding conventions for our graphical output results are as given as follows.

- Black curve ... upward MWD signal
- Red curve ... XNOISE function
- Green curve ... sum of MWD and XNOISE
- Blue curve ... deconvolved signal

4.3.3   Run 1. Single narrow pulse, S/N = 1, approximately.

Carefully note that the amplitude "A" in the MWD signal function SIGNAL refers to the difference of two hyperbolic tangent functions, whereas the amplitude "AMP" in the XNOISE function refers to a sinusoidal function. They are not the same.

```
      FUNCTION SIGNAL(T)
C     MWD upward wave signal function
C     Train of pulses, 0.5 sec width, 0.5 sec separation
C     CASE 2.   NARROW  PULSE  WIDTH  (Considered in  single  xducer
method)
C     Clearly see interference between upgoing and reflected pulses
      A = 10.0
      R = 100.0
      SIGNAL = A*(TANH(R*(T-0.100))-TANH(R*(T-0.101)))/2.
      RETURN
      END
C
      FUNCTION XNOISE(T)
C     Mud pump noise function may also include reflected MWD
C     signal, but it is not necessary to add the wave reflection to
C     the total noise to demonstrate directional filtering.
C     FRQPMP = Hertz freq of pump noise, propagates downward
      PI = 3.14159
      FRQPMP = 15.
      AMP = 0.25
      XNOISE = AMP*SIN(2.*PI*FRQPMP*T) + 0.0*SIGNAL(T)
      RETURN
      END
```

- Time delay between transducers = 0.010 sec = 10 ms
- 30 ft standpipe length/3,000 ft/sec mud sound speed = 0.01 sec, taken as sampling time in code.

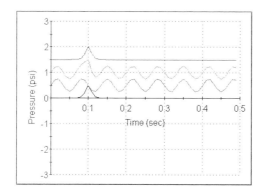

Figure 4.3b. MWD signal with pump noise (S/N = 1, approximately).

Note that the single black upgoing MWD pulse shown in Figure 4.3b is about 0.05 sec wide, representing a nominal 20 bits/sec, and the red continuous mud pump signal is roughly the same frequency. This is a difficult test case. Because both MWD and pump frequencies are about the same, the use of conventional frequency-based filtering will fail. The green superposition of the upgoing MWD signal and the pump noise shows a continuous wave signal with a small bump at the left, with the single signal buried in the longer trace. The blue curve shows how the black single pulse is successfully recovered from the green data using Equation 4.3g. The silence obtained everywhere else also demonstrates the success of our filtering. Note the signal and noise have roughly the same amplitude, so that S/N = 1, approximately.

4.3.4   Run 2. Very noisy environment.

In this run, we increase the pump noise significantly, so that the MWD signal is not visible to the naked eye in the green signal. The time delay between transducers is 0.01 sec or 10 ms, as before. Here, the signal to noise ratio is about 1:4 and the same recovery success is achieved.

```
        FUNCTION SIGNAL(T)
C       MWD upward wave signal function
C       Train of pulses, 0.5 sec width, 0.5 sec separation
C       CASE 2.    NARROW  PULSE  WIDTH  (Considered  in  single  xducer
method)
C       Clearly see interference between upgoing and reflected pulses
        A = 10.0
        R = 100.0
        SIGNAL = A*(TANH(R*(T-0.100))-TANH(R*(T-0.101)))/2.
        RETURN
        END
C
        FUNCTION XNOISE(T)
C       Mud pump noise function may also include reflected MWD
C       signal, but it is not necessary to add the wave reflection to
C       the total noise to demonstrate directional filtering.
C       FRQPMP = Hertz freq of pump noise, propagates downward
        PI = 3.14159
        FRQPMP = 15.
C       AMP = 0.25
        AMP = 1.
        XNOISE = AMP*SIN(2.*PI*FRQPMP*T) + 0.0*SIGNAL(T)
        RETURN
        END
```

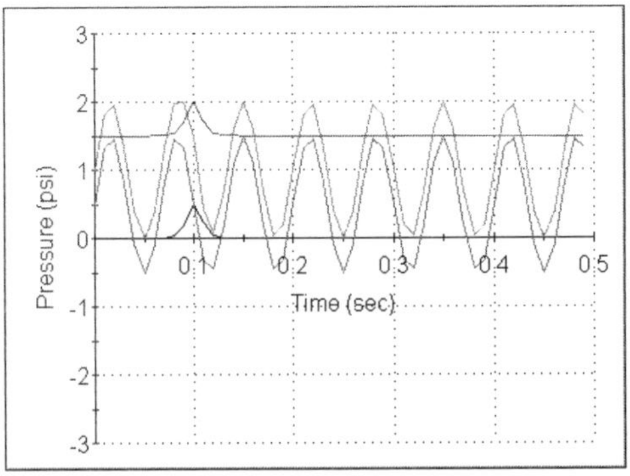

Figure 4.3c. Very noise environment, S/N about 0.25.

4.3.5   Run 3. Very, very noisy environment.

Here we increase the noise level significantly to determine how well the method continues to work. The time delay between transducers is 0.01 sec or 10 ms, as before. Here, a S/N of approximately 1/8 is assumed, and from our Figure 4.3f, the blue curve again replicates the black pulse very successfully.

```
      FUNCTION SIGNAL(T)
C     MWD upward wave signal function
C     Train of pulses, 0.5 sec width, 0.5 sec separation
C     CASE  2.   NARROW  PULSE  WIDTH  (Considered  in  single  xducer
method)
C     Clearly see interference between upgoing and reflected pulses
      A = 10.0
      R = 100.0
      SIGNAL = A*(TANH(R*(T-0.100))-TANH(R*(T-0.101)))/2.
      RETURN
      END

      FUNCTION XNOISE(T)
C     Mud pump noise function may also include reflected MWD signal,
C     but it is not necessary to add the wave reflection to the
C     total noise to demonstrate directional filtering.
C     FRQPMP = Hertz freq of pump noise, propagates downward
      PI = 3.14159
      FRQPMP = 15.
C     AMP = 0.25
      AMP = 2.
      XNOISE = AMP*SIN(2.*PI*FRQPMP*T) + 0.0*SIGNAL(T)
      RETURN
      END
```

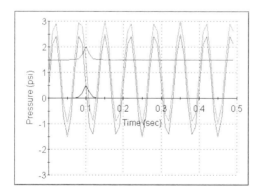

Figure 4.3d. Very, very noisy environment.

4.3.6 Run 4. Very, very, very noisy environment.

Here, the noise level is increased further! The time delay between transducers is 0.01 sec or 10 ms, as before. The S/N ratio is assumed to be 1/16, which is very, very small. Excellent signal recovery is apparent in Figure 4.3e, with blue and black signals being identical.

```
      FUNCTION SIGNAL(T)
C     MWD upward wave signal function
C     Train of pulses, 0.5 sec width, 0.5 sec separation
C     CASE 2.  NARROW PULSE WIDTH (Considered in single xducer
method)
C     Clearly see interference between upgoing and reflected pulses
      A = 10.0
      R = 100.0
      SIGNAL = A*(TANH(R*(T-0.100))-TANH(R*(T-0.101)))/2.
      RETURN
      END
C
      FUNCTION XNOISE(T)
C     Mud pump noise function may also include reflected MWD
C     signal, but it is not necessary to add the wave reflection to
C     the total noise to demonstrate directional filtering.
C     FRQPMP = Hertz freq of pump noise, propagates downward
      PI = 3.14159
      FRQPMP = 15.
C     AMP = 0.25
      AMP = 4. [Twice the above example, nothing else changed]
      XNOISE = AMP*SIN(2.*PI*FRQPMP*T) + 0.0*SIGNAL(T)
      RETURN
      END
```

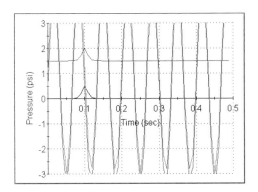

Figure 4.3e. Very, very, very noisy environment.

4.3.7   Run 5. Non-periodic background noise.

In Runs 1-4, we assumed a periodic wave to model (a Fourier component of) the mudpump noise. However, we can easily model non-sinusoidal distortions caused by the desurger (Chapter 6 provides more realistic models). Numerous non-sinusoidal functions have been tested successfully. To demonstrate the general nature allowed for the XNOISE function, we assume a straight line time function to locally represent a large scale slowly varying noise function. This can be an idealization of a signal distorted by the desurger.

```
      FUNCTION SIGNAL(T)
C     MWD upward wave signal function
C     Train of pulses, 0.5 sec width, 0.5 sec separation
C     CASE  2.   NARROW  PULSE  WIDTH  (Considered  in  single  xducer
method)
C     Clearly see interference between upgoing and reflected pulses
      A = 10.0
      R = 100.0
      SIGNAL = A*(TANH(R*(T-0.100))-TANH(R*(T-0.101)))/2.
      RETURN
      END
C
      FUNCTION XNOISE(T)
C     Mud pump noise function may also include reflected MWD signal,
C     but it is not necessary to add the wave reflection to the
C     total noise to demonstrate directional filtering.
C     FRQPMP = Hertz freq of pump noise, propagates downward
      PI = 3.14159
      FRQPMP = 15.
C     AMP = 0.25
      AMP = 5.
C     XNOISE = AMP*SIN(2.*PI*FRQPMP*T) + 0.0*SIGNAL(T)
      XNOISE = AMP*T
      RETURN
      END
```

The time delay between transducers is 0.01 sec or 10 ms. In Figure 4.3f, the black curve is the upgoing pulse, the red is a linearly growing non-periodic noise function in time that may be representative of other more general forms of noise, and the green is the superposition of the upgoing MWD signal and the noise function. The blue curve shows how the black signal is successfully recovered from the green input data using Method 4-3.

The reader may again ask, "Why are we so successful at recovering signals even with signal-to-noise ratios less than 10%, when the signal processing literature is much less optimistic?" The reason: conventional signal processing deals typically with random noise. Our noise is not random, but a propagating wave with wave-like properties – noise that a "smart algorithm" such as ours can remove. If our noise also contains random or other types of noise, then that noise must be separately removed before or after application of our filters.

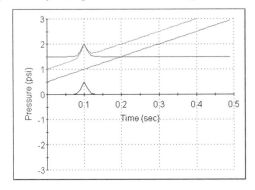

Figure 4.3f. MWD pulse in non-periodic noise (desurger distortion field).

4.4 Method 4-4. Directional filtering – differential equation method requiring two transducers (software reference, SAS14D*.FOR, Option 3 only).

4.4.1 Physical problem.

The problem considered here is identical to that of Method 4-3. However, the signal processing method is based on a *differential* as opposed to a *difference delay* equation. "Differential" refers to "differential equation based" while "difference" refers to algebraic equations with non-infinitesimal separations. We emphasize that, in practice, the drillpipe pressure field consists of slowly varying hydrostatic and rapidly changing wave components. The effects of the former on Methods 4-3 and 4-4 are small, thus simplifying signal processing.

Schematic

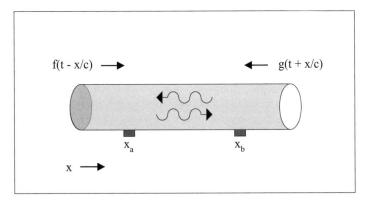

Figure 4.4a. General bi-directional waves.

4.4.2   Theory.

Our derivation at first follows the author's United States Patent No. 5,969,638 entitled "Multiple Transducer MWD Surface Signal Processing." But the present method, which includes a robust integrator to handle sharp pressure pulses, substantially changes the earlier work, which is incomplete; the full formulation is developed here. In Method 4-3, we used time delayed signals for which there was no restriction on time delay size. For Method 4-4, we invoke time and space derivatives; thus sampling times should be small compared to a period and transducer separations should be small compared to a wavelength.

As in Method 4-3, the representation of pressure as an upgoing and downgoing wave is still very general, and all waves in the downward direction are removed with no information required at all about the mudpump, the desurger or the rotary hose. Note that "c" is the mud sound speed at the surface and should be measured separately. In our derivation, expressions for time and space derivatives of $p(x,t)$ are formed, from which the downgoing wave "g" is explicitly eliminated, leaving the desired upgoing "f." The steps shown are straightforward and need not be explained.

$$p(x,t) = f(t - x/c) + g(t + x/c) \tag{4.4a}$$

$$p_t = f' + g' \tag{4.4b}$$

$$p_x = -c^{-1} f' + c^{-1} g' \tag{4.4c}$$

$$cp_x = -f' + g' \tag{4.4d}$$

$$p_t - cp_x = 2f' \tag{4.4e}$$

$$f' = \tfrac{1}{2}(p_t - cp_x) \tag{4.4f}$$

Equation 4.4f for "f," which is completely independent of the downgoing wave "g," however, applies to the time derivative of the upgoing signal $f(t - x/c)$ and not "f" itself. Thus, if a square wave were traveling uphole, the derivative of the signal would consist of two noisy spikes having opposite signs. This function must be integrated in order to recover the original square wave, and at the time the patent was awarded, a robust integration method was not available. The required integration is not discussed in the original patent, where it was simply noted that both original and derivative signals in principle contain the same information. In our recent SAS14D software, a special integration algorithm is given to augment the numerical representation in Equation 4.4f. The success of the new method is demonstrated below in Figure 4.4b.

At first glance, the two-transducer delay approach in Method 4-3 seems to be more powerful because it does not require time integration, and since it does not involve derivatives, there are no formal requirements for sampling times to be small and transducer separations to be close. However, in any practical high-data-rate application, the latter will be the case anyway, e.g., slow sampling rates will not capture detailed data. Thus, Method 4-4 is no more restrictive than Method 4-3. However, the present method is powerful in its own right because the presence of the $\partial p/\partial x$ derivative implies that one can approximate it by more than two (transducer) values of pressure at different positions using higher-order finite difference formulas – in operational terms, one can employ multiple transducers and transducer arrays to achieve higher accuracy. Similarly, the presence of $\partial p/\partial t$ means that one can utilize more than two time levels of pressure in processing in order to achieve high time accuracy. The required processing in space and time is inferred from the use of finite difference formulas in approximating the derivatives shown and numerous such computational molecules are available in the numerical analysis literature. The illustrative calculation used below, however, assumes only two transducers and pressures stored at two levels in time.

4.4.3   Run 1. Validation analysis.

SAS14D presently runs with only Option 3 fully tested and other options are under development. We explain below some hard-coded assumptions and interpret computed results. The software model assumes the following upgoing MWD signal, as noted in output duplicated below.

```
Internal MWD upgoing (psi) signal available as
P(x,t) = +     5.000 {H(x-  150.000-ct) - H(x-  400.000-ct)}
         +    10.000 {H(x-  600.000-ct) - H(x- 1000.000-ct)}
         +    15.000 {H(x- 1400.000-ct) - H(x- 1700.000-ct)}
```

This upgoing MWD wave signal is presently hard-coded. H is the Heaviside step function. At time $t = 0$, the pressure $P(x,0)$ contains three rectangular pulses with amplitudes (1) 5 for $150 < x < 400$, (2) 10 for $600 < x < 1000$, and (3) 15 for $1400 < x < 1700$. Thus, the pulse widths and separations, going from left to right, are

- 400 – 150   = 250 ft
- 600 – 400   = 200 ft
- 1000 – 600   = 400 ft
- 1400 – 1000 = 400 ft
- 1700 – 1400 = 300 ft

The average spatial width is about 300 ft. If the sound speed is 5,000 ft/sec (as assumed below) then the time required for this pulse to displace is 300/5,000 or 0.06 sec. Since sixteen of these are found in one second, this represents 16 bits/sec, approximately. Below we define the noise function, which propagates in a direction opposite to the upgoing signal. For our upgoing signal we have 16 bits/sec. In our noise model below, we assume a sinusoidal wave (for convenience, though not a requirement) of 15 Hz, amplitude 20 (which exceeds the 5, 10, 15 above). These approximately equal frequencies provide a good test of effective filtering based on directions only. Note that conventional methods to filter based on frequency will not work since both signal and noise have about the same frequency. Again, the sound speed (assuming water) is 5,000 ft/sec. For the MWD pulse, the far right position is 1,700 ft. We want to be able to "watch" all the pulses move in our graphics, so we enter "1710" (>1700 below). We also assume a transducer separation of 30 ft. This is about 10% of the typical pulse width above, and importantly, is the length of the standpipe; thus, we can place two transducers at the top and bottom of the standpipe. Recall that Method 4-4 is based on derivatives. The meaning of a derivative from calculus is "a small distance." Just how small is small? The results seem to suggest that 10% of a wavelength is small enough.

```
Units:   ft, sec, f/s, psi ...
Assume canned MWD signal? Y/N:  y
Downward propagating noise (psi) assumed as
N(x,t) = Amplitude * cos {2π f (t + x/c)} ...
o  Enter noise freq "f"  (hz):  15
o  Type noise amplitude (psi):  20
o  Enter sound speed c (ft/s):  5000
o  Mean transducer x-val (ft):  1710
o  Transducer separation (ft):  30
```

Note, the noise amplitude is not small, but chosen to be comparable to the MWD amplitudes, although only large enough so that all the line drawings fit on the same graphical display. The method actually works for much larger amplitudes.

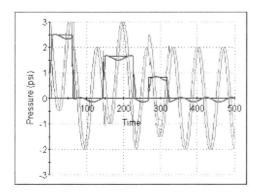

Figure 4.4b. Recovery of three step pulses from noisy environment.

After SAS14D executes, it creates two output files, SAS14.DAT and MYFILE.DAT. The first is a text file with a "plain English" summary. The second is a data file used for plotting. To plot results, run the program FLOAT32, which will give the results in Figure 4.4b where an index related to time is shown on the horizontal axis (the software will be fully integrated at a future date).

In Figure 4.4b, black represents the clean upgoing MWD three-pulse original signal. Red is the recovered pulse – this result is so good that it partially hides the black signal. The green and blue lines are pressure signals measured at the two pressure transducers, here separated by thirty feet. From these green and blue traces, one would not surmise that the red line can be recovered (only minor "bumps" indicate that the green and blue lines differ).

4.4.4 Run 2. A very, very noisy example.

The results above are very optimistic, but the algorithm is even more powerful than they suggest. In the run below, a noise amplitude of 200 is assumed, so that the signal-to-noise ratio ranges from 0.025 to 0.075 depending on the pulse interval. Simulation inputs are shown below and calculated results are displayed in Figure 4.4c. The recovery of the three-pulse signal is remarkable despite the high noise level. In practice, this directional filter will be used in concert with conventional frequency and random noise filters.

```
Units:   ft, sec, f/s, psi ...
Assume canned MWD signal? Y/N:   y
Downward propagating noise (psi) assumed as
N(x,t) = Amplitude * cos {2π f (t + x/c)} ...
o   Enter noise freq "f"  (hz):   5
o   Type noise amplitude (psi):   200
o   Enter sound speed c (ft/s):   5000
o   Mean transducer x-val (ft):   1700
o   Transducer separation (ft):   30
```

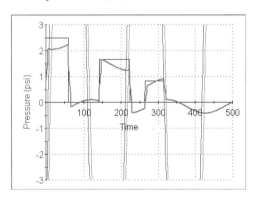

Figure 4.4c. Recovery of three step pulses from very noisy environment.

The typical frequency spectrum in Figure 6-8b shows mudpump noise in the 0–25 Hz range, with lower frequencies associated with higher amplitudes. The results in Figure 4.4b assume large amplitude 15 Hz pump noise, while Figure 4.4c assumes very large amplitude 5 Hz pump noise. In both cases, signal recovery is remarkable. We emphasize that time domain integrations, and not discrete Fourier transforms, are used in both of our approaches.

4.4.5   Note on multiple-transducer methods.

The excellent abilities afforded by multiple transducer methods in canceling pump noise and MWD reflections from the pistons and desurger have been demonstrated in numerous simulations. Minimal information is required – only an accurate determination of surface mud sound speed is required. The previous discussions focused on two-transducer methods for simplicity, but their extension to three or more is straightforward. For example, from Equation 4.4f or $f' = \frac{1}{2}(p_t - cp_x)$, improvements to p_x and p_t imply better filtering. The spatial derivative was calculated from $p_x = (p_2 - p_1)/\Delta + O(\Delta)$ where Δ is transducer separation and $O(\Delta)$ is the error. Use of a high-order three-point formula for three-transducer application, for instance, leads to better spatial accuracy. Similar considerations apply to p_t, that is, more levels of data storage can produce higher time accuracy.

A problem with multiple transducers is their effect on the standpipe. Drillers are reluctant to tap additional holes since this decreases its structural integrity. One possible solution is to drill only two holes but to connect these two by a longer hydraulic hose or metal pipe – and then tap, say, three or more holes into it. The pressures on this "bypass loop" are used for signal processing. This idea is described in U.S. Patent No. 5,515,336 awarded to the lead author in 1996. From the Abstract, "An acoustic detector in a mud pulse telemetry system includes a bypass loop in parallel with a section of the main mud line that supplies drilling mud to a drill string. The detector includes a pair of pressure sensing ports in the bypass line, and one or more pressure transducers for detecting the pressure at different locations in the bypass loop so that the differential pressure can be measured. The bypass loop has a small internal passageway relative to the main mud supply line and may include a constriction so as to create two regions in the passageway that differ in cross sectional areas. Forming the pressure sensing ports in the regions of differing cross sectional areas allows the pressure transducers to more precisely detect the mud pulse signals. Because of its relatively small cross sectional area, only a small fraction of the drilling mud flows through the bypass loop. The bypass loop may thus be constructed of hydraulic hose and a relatively small rigid body having a central through bore." Some diagrams are reproduced in Figure 4.4d.

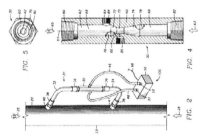

Figure 4.4d. Bypass loop for multiple transducer implementation.

4.5 Method 4-5. Downhole reflection and deconvolution at the bit, waves created by MWD dipole source, bit assumed as perfect solid reflector (software reference, DELTAP*.FOR).

In Methods 4-1 to 4-4, our objective was to recover *all* signals (good and bad) originating from downhole. This was accomplished successfully by removing surface reflections and pump generated noise. However, we now re-emphasize that the upgoing MWD signal itself is not "clean," but contains "ghost reflections." By ghost reflections, we refer to the downhole signal generation process. When a signal is created at the pulser that travels uphole, a pressure signal of opposite sign travels downhole, reflects at the drillbit, and then propagates upward to interfere with new upgoing signals. Thus, at the surface, the signal arriving first is the intended one, which is followed by a ghost signal or shadow.

For the purposes of evaluating Methods 4-1 to 4-4, it was acceptable to focus on recovering *everything* originating from downhole. But once surface reflections at the pump and desurger, and mudpump generated noise, are removed, we must not forget that the recovered signal itself contains ghost echoes from downhole which must be eliminated through further processing. At high data rates, the signals arriving from downhole are never ideal and will always contain the intended signal plus ghost reflections. At low frequencies of 0-1 bit/sec, the effects of ghost reflections can be very constructive or very destructive depending on the way the drillbit reflects and on the degree of collar-pipe area mismatch. Methods 4-1 to 4-4 will always yield the combined "intended plus ghost" signal. Once this is extracted from standpipe processing, the "intended" and "ghost" components must be separated. In Methods 4-5 and 4-6, we accomplish this objective, assuming a straight telemetry channel without area changes; the former assumes that the drillbit is a solid reflector while the latter assumes an acoustic open-end. Area discontinuity restrictions are completely removed in Chapter 5 to provide very powerful deconvolution tools.

4.5.1 Software note.

The net upgoing pressure signal (designated below by "$p_2(L,t)$") originating from downhole is assumed to be known from Methods 4-1 to 4-4. But this signal is generally not the $\Delta p(t)$ across the MWD pulser that is directly related to position-encoding by the pulser. Method 4-5 solves for $\Delta p(t)$ if $p_2(L,t)$ is known from Methods 4-1 to 4-4 using the DELTAP*.FOR software series.

In this series, the drillbit is treated as a solid reflector. We caution the reader against visual judgments. The "solid reflector" assumption does not always mean "small drillbit nozzles." Reflection characteristics can only be inferred from the results of a comprehensive model such as that in Chapter 2 where the axial and cross dimensions of the bit box and other geometric elements are considered. Of course, if in a field experiment, a test signal reflects with the same sign, that would confirm solid reflector characteristics.

4.5.2 Physical problem.

The key physical ideas are simply stated. Downhole information, meaning the sequence of 0's and 1's describing logging data, is directly encoded in valve positions and thus appears directly in Δp. Recovery of Δp implies recovery of the train of 0's and 1's. However, Δp is never directly measured. It is only indirectly observed in measured pressures as the latter contain the effects of reflections. Removal of surface reflections and mudpump noise solves only part of the problem. Downhole reflections are still prevalent even after Methods 4-1 to 4-4 are applied. Again, these downhole reflections are problematic and only clever surface processing will remove their influence.

When a positive pulse valve opens and shuts, or when a mud siren changes rotation speed based on azimuthal location of the rotor relative to the to indicate 0's and 1's, over-pressure and under-pressure acoustic signals are created at opposite sides of the source which have equal magnitude (this is confirmed in numerous experiments). The downgoing signals reflect at the drillbit and interfere with newly created upgoing signals at the source. These ghost reflections contaminate the intended Δp signal by superposing with tail end of the later upgoing signal. In summary, a single valve action (that is, an "open" or a "close") creates two upgoing pressure pulses that enter the drillpipe to travel to the surface. At the surface, these may reflect at the desurger and at the mudpump, thus forming even more signals that are sensed in the standpipe pressure transducer array. Methods 4-1 to 4-4 handle surface reflections, while Methods 4-5 and 4-6 handle Δp recovery from surface-filtered results.

In terms of the math symbols below, the $p_2(L,t)\}$ remaining after Methods 4-1 to 4-4 are applied must be further processed to extract $\Delta p(t)$, which is the transient function containing the 0's and 1's that are implicit in the position-encoding used. For Methods 4-5 and 4-6, we assume a "dipole" source. By a "dipole source" we mean a pressure source whose created pressures are antisymmetric with respect to source point – this section applies to dipole sources, e.g., positive pulsers and mud sirens, but not negative pulse systems.

4.5.3   On solid and open reflectors.

It is important to clarify at this point possible confusion on the use of "zero acoustic pressure" boundary conditions in modeling open ends. We have stated, and proved mathematically, why solid reflectors like pump pistons double incident wave pressures locally. This is straightforward. What happens at open ends, however, is less clear. The results of detailed and rigorous three-dimensional analyses show that long waves in tubes reflect at open ends with opposite pressure polarity and almost identical magnitude. Hence, "zero acoustic pressure" is found at open ends. Of course, to the lab technician working outside the wind tunnel, the acoustic signature heard is far from zero compared to ambient room conditions. We emphasize, however, that the pressure is "zero" relative to sound pressures *within* the pipe.

Schematic

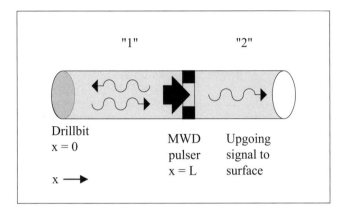

Figure 4.5a. Solid boundary reflection, waves from dipole source.

4.5.4 Theory.

The formulation for the difference delay equation model used is derived here referring to Figure 4.5a. In Section "1," the Lagrangian displacement function is assumed as $u(x,t) = h(t - x/c) - h(t + x/c)$ where "h" is unknown, with $u(0,t) = h(t - 0) - h(t + 0) = 0$ assuming that the bit at $x = 0$ is a solid reflector. Then $p(x,t) = -B \, \partial u/\partial x = + (B/c) [h'(t - x/c) + h'(t + x/c)]$ represents the corresponding acoustic pressure function. In Section "2," we assume radiation conditions, that is, $u(x,t) = f(t - x/c)$, a wave traveling to the right without reflection. The function $p(x,t) = -B \, \partial u/\partial x = + (B/c) f'(t - x/c)$ represents the acoustic pressure in Section "2."

Boundary matching conditions apply at the MWD source point. At the pulser $x = L$, we have $p_2 - p_1 = \Delta p(t)$. The exact time dependencies in the given transient function are determined by the telemetry encoding method used, while the peak-to-peak strength is determined by valve geometry, rotation rate, flow rate, fluid density, and so on. Substitution of the foregoing wave assumptions leads to

$$f'(t - L/c) - h'(t - L/c) - h'(t + L/c) = (c/B) \Delta p(t) \qquad (4.5a)$$

The requirement $u_1 = u_2$ at $x = L$ implies continuity of displacement, for which we obtain $h(t - L/c) - h(t + L/c) = f(t - L/c)$. Thus, taking partial time derivatives, we obtain

$$h'(t - L/c) - h'(t + L/c) = f'(t - L/c) \qquad (4.5b)$$

If we eliminate f' between the Equations 4.5a and 4.5b, the "$t - L/c$" terms cancel, and we obtain $h'(t + L/c) = - \{c/(2B)\} \Delta p(t)$. In terms of the dummy variable $\tau = t + L/c$, we have

$$h'(\tau) = - \{c/(2B)\} \, \Delta p(\tau - L/c) \tag{4.5c}$$

Next we subtract Equation 4.5b from Equation 4.5a, but this time introduce the dummy variable $\tau = t - L/c$, to obtain

$$f'(\tau) = h'(\tau) + \{c/(2B)\} \, \Delta p(\tau + L/c) \tag{4.5d}$$

If we eliminate $h'(\tau)$ between Equations 4.5c and 4.5d, we obtain $\Delta p(\tau + L/c) - \Delta p(\tau - L/c) = (2B/c) \, f'(\tau)$, which we can express in the more meaningful physical form

$$\tfrac{1}{2} \, [\Delta p(\tau + L/c) - \Delta p(\tau - L/c)] = (B/c) \, f'(\tau) \tag{4.5e}$$

We now recast the above equation in a more useful form, taking the dummy variable $\tau = t - L/c$. Since it is clear that $(B/c) \, f'(t - L/c)$ is just the upgoing pressure $p_2(L,t)$, we can write

$$\tfrac{1}{2} \, [\Delta p(t) - \Delta p(t - 2L/c)] = (B/c) \, f'(t - L/c) = p_2(L,t) \tag{4.5f}$$

where $p_2(L,t)$ is the result of using Methods 4-1, 4-2, 4-3 or 4-4 and Δp is to be recovered. Equation 4.5f can be easily interpreted physically. Suppose that a pulser creates a Δp signal. A signal $\Delta p/2$ goes uphole. It simultaneously sends a signal $- \Delta p/2$ going downhole, so that the net pressure differential is Δp. The signal $- \Delta p/2$ travels downward to the bit, which is assumed as a solid reflector, and the signal reflects with an unchanged pressure sign, that is, the contribution $- \tfrac{1}{2} \, \Delta p(t - 2L/c)$, now including the roundtrip time delay $2L/c$. This derivation applies to dipole pulsers where the created MWD pressure is antisymmetric with respect to the source position. Obvious changes will apply to monopole (negative pressure) pulsers which, again, are not considered in this book.

Note that the two terms on the left side of the Equation 4.5f refer to incident and ghost signals, while $p_2(L,t)$ is obtained from measured surface data. This is the "p_{pipe}" of Chapter 2. It is obtained from surface signal processing results of using Methods 4-1, 4-2, 4-3 or 4-4 applied at the standpipe. It is not necessary to know the amount of energy loss incurred from downhole to surface to apply Methods 4-5 and 4-6 since the same attenuation applies to all parts of the signal. Any convenient electronic gain suffices. If we denote $H = 2L/c$ as the roundtrip delay time between the pulser and the solid bit reflector, we can write Equation 4.5f as

$$\Delta p(t) - \Delta p(t - H) = 2 \, p_2(L,t) \tag{4.5g}$$

Detection problems are associated with Equation 4.5g. Consider the use of phase-shift-keying (PSK). When H is "not small," a phase shift propagates uphole while one simultaneously travels downward and reflects upward. Two phase shifts travel up the drillpipe, one real and the other apparent. Solution of Equation 4.5g, which is possible exactly and analytically, is required to determine $\Delta p(t)$ which alone contains the input sequence of 0's and 1's.

We give estimates for typical H values. Consider a typical thirty-feet MWD collar and assume a pulser is positioned at the very top. If the drillbit lies at its bottom (this is almost never the case, since resistivity subs or drilling motors are found adjacent to the bit), and the sound speed is 3,000 ft/sec, then the collar travel time is 30/3,000 or 0.01 sec. In general, thus, H >> 0.02 sec.

4.5.5   Run 1. Long, low data rate pulse.

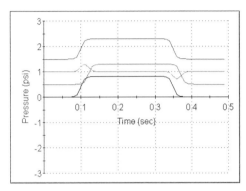

Figure 4.5b. Long, low data rate pulse.

```
C      CASE 1
       A = 0.8
       R = 100.0
       SIGNAL = A*(TANH(R*(T-0.1))-TANH(R*(T-0.35)))/2.
```

For our Method 4-5 figures, black, red, green and blue curves are plotted, respectively, from bottom to top. In this run, the time delay between pulser and drillbit is 0.01 sec, the same as the sampling time. The pulse width is about 0.25 sec. The black curve is the positive over-pressure created ahead of the valve when the poppet valve or the siren closes. At the same time, a negative under-pressure at the valve propagates down toward the bit and reflects upward without a change in sign. This negative pressure is shown flipped over in the red curve so it can be compared with the black one more easily. Green is the actual pressure ahead of the pulser transmitted to the surface and shows positive and negative bumps. Most of the signal has been lost due to destructive wave interference, and what is worse, the single black pulse is replaced by two smaller green pulses. It is this green double-pulse that is detected uphole. If the green data is available, the blue curve shows how the black curve is recovered successfully using the solution to Equation 4.5g.

4.5.6   Run 2. Higher data rate, faster valve action.

The time delay between pulser and bit is 0.01 sec, same as the sampling time. We consider a much higher data rate represented by a narrower pulse as shown in Figure 4.5c. As before, we have very good signal recovery. The pulse width (black curve) is about 0.05 sec, for a data rate of about 20 bits/sec.

```
C      CASE 2
       A = 20.0
       R = 100.0
       SIGNAL = A*(TANH(R*(T-0.100))-TANH(R*(T-0.101)))/2.
```

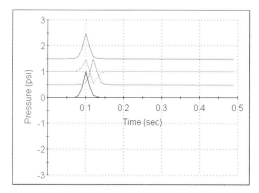

Figure 4.5c. Higher data rate, faster valve action.

4.5.7   Run 3. PSK example, 12 Hz frequency.

Here we consider an example in phase-shift-keying, assuming a carrier frequency of 12 Hz. The time delay between pulser and bit is 0.01 sec, same as the sampling time. The pulse description appears immediately below.

```
C      CASE 3.  PHASE-SHIFTING (F = FREQUENCY IN HERTZ)
       PI = 3.14159
       A = 0.25
       F = 12.
C      G = A*SIN(2.*PI*F*T)
C      Means 2*PI*F cycles in 2*PI secs, or F cycles per sec, F is Hz.
C      One cycle requires time PERIOD = 1./F
       PERIOD = 1./F
       IF(T.GE.0.0.   AND.T.LE.   PERIOD) SIGNAL = +A*SIN(2.*PI*F*T)
       IF(T.GE.PERIOD.AND.T.LE.2.*PERIOD) SIGNAL = -A*SIN(2.*PI*F*T)
       IF(T.              GE.2.*PERIOD) SIGNAL = 0.
```

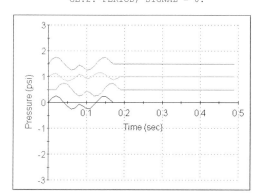

Figure 4.5d. PSK example, 12 Hz frequency.

The black curve is the pressure trace created ahead of the valve. At the same time, a negative trace at the valve propagates down toward the drillbit and reflects upward without a change in sign. This is shown flipped over in the red curve so that it can be compared with the black more easily to show the time delay. Green is the actual pressure ahead of the pulser that is transmitted to the surface and shows the superposition of the two. It is this green pressure trace that is detected uphole. At 12 Hz, the green curve looks somewhat like the black curve, i.e., they both have two large positive crests, except the green curve is stretched out somewhat. A robust receiver algorithm might be able to read this and interpret it correctly. If the green data is available, the blue curve shows how the shorter black curve is recovered successfully.

4.5.8   Run 4. 24 Hz, Coarse sampling time.

Here we increase our carrier frequency to 24 Hz and continue with phase-shift-keying. The time delay between pulser and bit is 0.01 sec, same as the sampling time.

```
C       CASE 3.  PHASE-SHIFTING (F = FREQUENCY IN HERTZ)
        PI = 3.14159
        A = 0.25
        F = 24.
C       G = A*SIN(2.*PI*F*T)
C       Means 2*PI*F cycles in 2*PI secs, or F cycles per sec, F is Hz.
C       One cycle requires time PERIOD = 1./F
        PERIOD = 1./F
        IF(T.GE.0.0.   AND.T.LE.   PERIOD) SIGNAL = +A*SIN(2.*PI*F*T)
        IF(T.GE.PERIOD.AND.T.LE.2.*PERIOD) SIGNAL = -A*SIN(2.*PI*F*T)
        IF(T.               GE.2.*PERIOD) SIGNAL = 0.
```

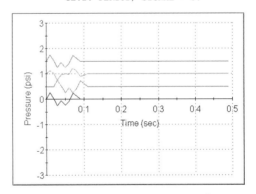

Figure 4.5e. 24 Hz, coarse sampling.

Here, the phase shift is clearly seen in the black signal created ahead of the valve and which propagates uphole. The reversed-sign red signal travels downhole (shown as a positive for easy comparison with the black trace) and reflects at the bit with sign unchanged and travels uphole, appending itself to the wave created ahead of the valve, to produce the green signal. The green signal is the one that is actually sent uphole. Using data from the green line only, we recover the blue signal, which is seen to be identical to the black signal, the intended signal. The green signal does not look like the black signal, but has the appearance of two sine waves separated by an interval of silence.

4.6 Method 4-6. Downhole reflection and deconvolution at the bit, waves created by MWD dipole source, bit assumed as perfect open end or zero acoustic pressure reflector (software reference, DPOPEN*.FOR).

4.6.1 Software note.

In Method 4-5, the drillbit is assumed as a solid reflector, meaning "small drillbit nozzles" subject to cautions previously noted; it can also mean "not so small" nozzles, but drilling into very hard rock. In Method 4-6, larger drillbit nozzles are considered, that is, the drillbit is taken as an acoustically opened zero pressure boundary condition. This is probably the typical situation, as explained in detail in Chapter 2. When "$p_2(L,t)$" is known, $\Delta p(t)$ is solved by our DPOPEN*.FOR model, which is derived from DELTAP*.FOR. The present method also enables excellent signal recovery.

4.6.2   Physical problem.

The physical problem considered here is same as that in Method 4-5 (where the bit is a solid reflector) except that the drillbit is now assumed as an open-ended acoustic reflector.

Schematic

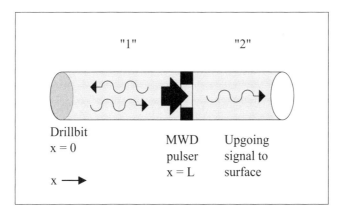

Figure 4.6a. Open end bottom reflection, waves from dipole source.

4.6.3 Theory.

In Section "1," we now take the Lagrangian displacement u(x,t) in the form u(x,t) = h(t − x/c) + h(t + x/c), which no longer vanishes at bit, since it is not solid reflector, but which allows the acoustic pressure to vanish at the bit, that is, p(x,t) = -B ∂u/∂x = + (B/c) [h'(t − x/c) − h'(t + x/c)] = 0 at x = 0. In Section "2," we retain our upgoing wave assumption u(x,t) = f(t − x/c), which implies that p(x,t) = -B ∂u/∂x = + (B/c) f '(t − x/c) or p(L,t) = + (B/c) f '(t − L/c). At the pulser x = L, the enforced pressure drop is $p_2 − p_1 = \Delta p(t)$, a given function dictated by the pulser position-encoding scheme. These assumptions lead to

$$f '(t − L/c) − h'(t − L/c) + h'(t + L/c) = (c/B) \Delta p(t) \qquad (4.6a)$$

Continuity of displacement through the MWD source, that is, $u_1 = u_2$ at x = L, requires that h(t − L/c) + h(t + L/c) = f(t − L/c). Thus, taking partial time derivatives, we have

$$h'(t − L/c) + h'(t + L/c) = f '(t − L/c) \qquad (4.6b)$$

If we eliminate f ' between the two Equations 4.6a and 4.6b, the "t − L/c" terms cancel, and we obtain h'(t + L/c) = + {c/(2B)} Δp(t). In terms of the dummy variable τ = t + L/c,

$$h'(\tau) = + \{c/(2B)\} \Delta p(\tau − L/c) \qquad (4.6c)$$

Next we subtract Equation 4.6b from 4.6a, but this time introduce the dummy variable τ = t − L/c, to obtain

$$f '(\tau) = h'(\tau) + \{c/(2B)\} \Delta p(\tau + L/c) \qquad (4.6d)$$

If we eliminate h'(τ) between Equations 4.6c and 4.6d, we obtain Δp(τ + L/c) + Δp(τ − L/c) = (2B/c) f '(τ) which we can express in the more meaningful physical form

$$½ [\Delta p(\tau + L/c) + \Delta p(\tau − L/c)] = (B/c) f '(\tau) \qquad (4.6e)$$

We now recast the Equation 4.6e in a more useful form, taking the dummy variable τ = t − L/c. Since it is clear that (B/c) f '(t − L/c) is just the upgoing pressure $p_2(L,t)$, we can write, noting $p_2(L,t)$ = + (B/c) f '(t − L/c),

$$½ [\Delta p(t) + \Delta p(t − 2L/c)] = (B/c) f '(t − L/c) = p_2(L,t) \qquad (4.6f)$$

or

$$p_2(L,t) = ½ [\Delta p(t) + \Delta p(t − 2L/c)] \qquad (4.6g)$$

This can be easily interpreted physically. Suppose that a pulser creates a Δp signal. A signal Δp/2 travels uphole. It simultaneously sends a signal − Δp/2 downhole, so that the net pressure differential is Δp. This −Δp/2 travels downward to the bit, which is assumed as an open ended reflector, and the signal reflects with an changed pressure sign, that is, + ½ Δp(t − 2L/c), now including the roundtrip time delay 2L/c. This derivation applies only to dipole pulsers where the created MWD pulser is antisymmetric with respect to the source position. This $p_2(L,t)$ is identical to the "p_{pipe}" in Chapter 2. The two terms on the right side of Equation 4.6g are the incident and ghost waves. We need to extract Δp in order to obtain the 0's and 1's information stream encoded in $\Delta p(t)$. Once Methods 4-1, 4-2, 4-3 or 4-4 are applied at the surface for standpipe signal processing, Methods 4-5 or 4-6 are carried out as needed.

4.6.4   Run 1. Low data rate run.

Here a single near-rectangular pulse is considered. The time delay between pulser and bit is 0.01 sec. In Figure 4.6b, the black curve is the upgoing pressure wave created ahead of the pulser. The red curve is the negative of this (shown reversed so that comparison of the time delay is enhanced), which propagates down to the drillbit open end and reflects with a sign change – it now has same sign as the wave ahead of the valve. Thus, the green signal shows enhanced superposition and the green waveform is a little wider than the black one. Using the green data, the blue curve is recovered, and duplicates the black one successfully. In this run, the pulse width is approximately 0.25 sec, representing a high data rate by conventional standards. From our calculations, it is very obvious that nozzle size is very important, and determines whether the bit is a solid reflector or an acoustic free-end. This in turn dictates the shape of the wave – that is, the all-important green curve – that travels to the surface with all the downhole information.

```
C       CASE 1
        A = 0.8
        R = 100.0
        SIGNAL = A*(TANH(R*(T-0.1))-TANH(R*(T-0.35)))/2.
```

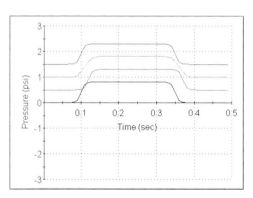

Figure 4.6b. Rectangular pulse run.

Display and source code note: If one examines the black and red curves above, the green superposition result should be twice the amplitude shown, but it is not. This is not an error but a difference in plotting conventions. In Methods 4-1 to 4-4, we dealt with *pressure levels* at the surface. In Methods 4-5 and 4-6, we instead examine Δp's. In the analytical derivations, and referring to DELTAP04.FOR and DPOPEN01.FOR, functions UPSIG and DNSIG are used for the upgoing and downgoing signals, but which are actually Δp functions and not pressures. The green line is defined as the total (UPSIG+DNSIG)/2 or (UPSIG-DNSIG)/2. The TOT function is not Δp, but the actual pressure level that enters the drillpipe. This factor of 2 explains why the black, red and green all have the same amplitudes. Again, because black and red represent plots of Δp's, while green represents pressure itself. "Equation 21" referred to in both Fortran source codes is really "left side PRESSURE" = "half the difference of two DELTA-P functions." This is the result programmed and plotted.

4.6.5   Run 2. Higher data rate.

Here we consider a higher data rate, and study the dynamics of a pulse with a width of approximately 0.05 sec, as shown in Figure 4.6c. The time delay between pulser and drill bit is 0.01 sec. The green signal is the signal that travels up the drillpipe and does not resemble the intended black signal. The top blue signal recovers the bottom black signal very well.

```
C       CASE 2
        A = 20.0
        R = 100.0
        SIGNAL=A*(TANH(R*(T-0.100))-TANH(R*(T-0.101)))/2.
```

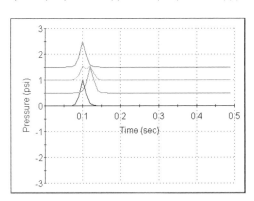

Figure 4.6c. Higher data rate, narrower pulse (note distorted green trace).

4.6.6   Run 3. Phase-shift-keying, 12 Hz carrier wave.

Here we consider a phase-shift keying example. The time delay between pulser and drillbit is 0.01 sec. In Figure 4.6d, the black bottom curve is the intended signal, while the green curve combines the downward wave reflected at the open-end bit.   The black and green traces appear similar, at least qualitatively, but the green is wider. The sharp phase shift in the black appears as a stretched-out wave in the green. Recovery is very good.

```
C    CASE 3.  PHASE-SHIFTING (F = FREQUENCY IN HERTZ)
     PI = 3.14159
     A = 0.25
     F = 12.
C    G = A*SIN(2.*PI*F*T)
C    Means 2*PI*F cycles in 2*PI secs, or F cycles per sec, F is Hz.
C    One cycle requires time PERIOD = 1./F
     PERIOD = 1./F
     IF(T.GE.0.0.   AND.T.LE.   PERIOD) SIGNAL = +A*SIN(2.*PI*F*T)
     IF(T.GE.PERIOD.AND.T.LE.2.*PERIOD) SIGNAL = -A*SIN(2.*PI*F*T)
     IF(T.               GE.2.*PERIOD) SIGNAL = 0.
```

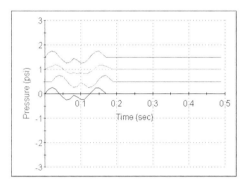

Figure 4.6d. Phase-shift-keying, 12 Hz carrier (blue and black are identical).

4.6.7   Run 4. Phase-shift-keying, 24 Hz carrier wave.

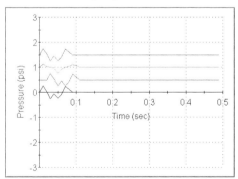

Figure 4.6e. Phase-shift-keying, 24 Hz carrier wave.

The previous simulation is repeated assuming a 24 Hz carrier wave. The time delay between pulser and drillbit is 0.01 sec. As seen from Figure 4.6e, the (blue) recovery of the black signal from the green data is very good.

```
C      CASE 4.  PHASE-SHIFTING (F = FREQUENCY IN HERTZ)
       PI = 3.14159
       A = 0.25
       F = 24.
       PERIOD = 1./F
       IF(T.GE.0.0.    AND.T.LE.    PERIOD) SIGNAL = +A*SIN(2.*PI*F*T)
       IF(T.GE.PERIOD.AND.T.LE.2.*PERIOD) SIGNAL = -A*SIN(2.*PI*F*T)
       IF(T.                 GE.2.*PERIOD) SIGNAL = 0.
```

4.6.8   Run 5. Phase-shift-keying, 48 Hz carrier.

Finally, we repeat the above run assuming a 48 Hz carrier wave. The time delay between pulser and bit is 0.01 sec. As shown in Figure 4.6f, the green pressure traveling up the drillpipe does not at all resemble the intended black signal. However, the blue curve, obtained using only green data alone, successfully replicates the black curve. Recovery is excellent.

```
C      CASE 5.  PHASE-SHIFTING (F = FREQUENCY IN HERTZ)
       PI = 3.14159
       A = 0.7
       F = 48.
       PERIOD = 1./F
       IF(T.GE.0.0.    AND.T.LE.    PERIOD) SIGNAL = +A*SIN(2.*PI*F*T)
       IF(T.GE.PERIOD.AND.T.LE.2.*PERIOD) SIGNAL = -A*SIN(2.*PI*F*T)
       IF(T.                 GE.2.*PERIOD) SIGNAL = 0.
C
```

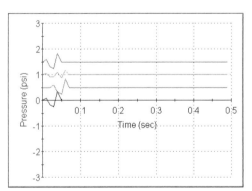

Figure 4.6f. Phase-shift-keying, 48 Hz carrier wave.

This completes our analysis of downhole ghost signals which, for simplicity, assumed that area mismatches between MWD collar and drillpipe do not exist. In Chapter 5, the "real world" internal annular collar area may be greater than, equal to or less than the drill pipe area, thus adding modeling complications. The results are exact and extend the capabilities reported in this chapter. In closing, we emphasize certain "original black" versus "observed green" signal characteristics that we have computed, e.g., see Figures 4.1e,f,g. The green traces appear elongated and distorted relative to the intended black signals. Very often, these effects, referred to as "phase distortion," "leading and trailing edge error," "pulse lengthening," and so on, are attributed to nonlinearities and frequency dispersion in the telemetry channel. In the lead author's opinion, these causes are not very likely; in fact, the green traces shown in this chapter are the sole consequence of reflections occurring at the drillbit – in a very linear assumed system. For this reason, few of the "fixes" offered by more complicated signal processing schemes – relying on incorrect physical explanations – have failed in the field. Readers of the corresponding patents should exercise caution and carefully question underlying assumptions.

4.7   References

Oppenheim, A.V. and Schafer, R.W., *Digital Signal Processing*, Prentice-Hall, New Jersey, 1975.

Oppenheim, A.V. and Schafer, R.W., *Discrete-Time Signal Processing*, Prentice-Hall, New Jersey, 1989.

5

Transient Variable Area
Downhole Inverse Models

In Chapters 2 and 3, we examined the forward problem associated with harmonic $\Delta p(t)$ excitations in general drilling telemetry channels with multiple area changes. These area changes, or acoustic impedance mismatches, are physically important because propagating waves do not transmit unimpeded without reflection. Depending on geometry, sound speed and frequency, constructive and destructive wave interference are possible. Constant frequency excitations are interesting in these respects, but also, they are useful in enhancing frequency-shift-keying telemetry methods. For example, the use of constructive interference can provide signal optimization without the penalties associated with metal erosion and high mechanical power in pulser design.

Chapter 4 dealt with surface reflection signal processing and pump noise removal assuming constant area telemetry channels, an assumption that is acceptable for conventional standpipe applications, focusing on general transient $\Delta p(t)$ functions. It also introduced downhole inverse problems, that is, the recovery of transient $\Delta p(t)$'s from pressures measured at the drillpipe inlet which are contaminated by MWD drill collar reverberations. For these problems, never before discussed in the literature, the mathematics was simplified by assuming constant area. However, in many unavoidable practical field situations, the cross-sectional area in the MWD drill collar and the drillpipe will not be the same, and there could be significant acoustic impedance mismatches. Such effects lead to unwanted reverberations within the drill collar that result in a distorted or "blurry" signals transmitted into the drillpipe.

Again, for Methods 4-1 to 4-4 in Chapter 4, our objective was to recover *all* signals (good *plus* bad) originating from downhole. This was accomplished successfully by removing surface reflections and pump generated noise. However, we now re-emphasize that the upgoing MWD signal itself is not "clean," but contains additional "ghost reflections." By ghost reflections, we refer to the downhole signal generation process. When a signal is created at the pulser that travels uphole, a pressure signal of opposite sign travels downhole, reflects at the drillbit, and then propagates upward to interfere with new upgoing signals. Thus, at the surface, the signal arriving first is the intended one, which is followed by a ghost signal or shadow often with unknown phase and amplitude. We emphasize that this ghost noise *always* exists in real MWD tools, unless, of course, the signal source is actually a piston face coincident with the drillbit – which it, of course, *never* is.

Tool design considerations. We stress the above point with the illustration in Figure 5.0 below. In this book, by "MWD collar area" we mean the cross-sectional area inside the collar, minus the area of the hub upon which the siren is installed. Depending on the tool, this net area may be greater than, equal to or less than the area of the drillpipe. Unless the areas are identical, so that no mismatches exist, reverberations will be created within the collar that *further* distort the ghost-bearing signal that travels up the drillpipe. The associated reverberations introduce "fuzziness" to the upgoing signal.

Area matching

$$A_{MWD\,collar} - A_{hub} = A_{drillpipe}$$

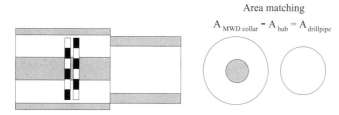

Figure 5.0. MWD collar – drillpipe area mismatch.

Objectives. In this chapter, we develop methods to recover fully transient $\Delta p(t)$ functions from pressure signals measured in the drillpipe (that is, obtained as the result of surface signal processing using Methods 4-1, 4-2, 4-3 or 4-4). Here, arbitrary differences in collar and pipe cross-sectional area are permitted and no restrictions are made as to their relative magnitudes. Method 5-1 assumes that the drillbit is an open-end reflector, while Method 5-2 assumes that it is a solid reflector. Difference delay equation approaches are used and exact solutions provided in software demonstrate perfect consistency between the complicated created signal field and the even more subtle process developed for signal inversion.

5.1 Method 5-1. Problems with acoustic impedance mismatch due to collar-drillpipe area discontinuity, with drillbit assumed as open-end reflector (software reference, collar-pipe-open-16.for).

5.1.1 Physical problem.

The engineering problem considered is shown in Figure 5.1a, where the source represents both positive pulsers and mud sirens. The telemetry channel consists of an MWD drill collar of finite length and a semi-infinite drillpipe. The pulser is located generally within the collar away from the bit, in order to allow for both left and right-going waves – it may execute generally transient $\Delta p(t)$ signals so long as wavelengths are large compared to typical cross-sectional dimensions. We consider two problems. The first determines the transient pressure that enters the drillpipe as a result of reverberations within a drill collar excited by a general $\Delta p(t)$, while the second deals with $\Delta p(t)$ signal recovery when a fully transient drillpipe signal (that is, the result of using Methods 4-1, 4-2, 4-3 or 4-4) is available. Note that Methods 4-1, 4-2, 4-3 and 4-4, Methods 5-1 and 5-2, and an attenuation model for signal propagation along long drillpipes, can be combined to develop a first-generation high-data-rate signal processor capable of handling all reflections.

Schematic

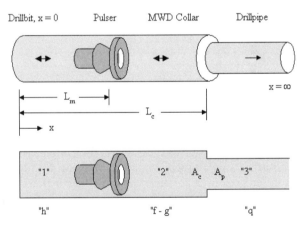

Figure 5.1a. MWD dipole pulser in drill collar and drillpipe system, with drillbit modeled as open-end reflector.

5.1.2 Theory.

In Figure 5.1a, four Lagrangian displacement functions "f," "g," "h" and "q" are introduced. The first three describe reverberant fields in the locations shown, while the last represents a propagating wave in the drillpipe. In Section "1," $0 < x < L_m$, the superposition of up and downgoing waves takes the form $u_1(x,t) = h(t - x/c) + h(t + x/c)$. The corresponding pressure $p_1 = - B \, \partial u_1/\partial x = B/c \, [h'(t - x/c) - h'(t + x/c)]$ vanishes at the assumed open end $x = 0$. In Section "3," with $x > L_c$, we assume a pure upgoing wave $u_3(x,t) = q(t - x/c)$ which satisfies standard radiation conditions. This ignores small reflections at pipe joints whose effects, from field experience, are known to be minimal; also, if reflections (and other noise) at the surface travel downhole, this model assumes that attenuation renders them insignificant by the time they reach the MWD drill collar (thus, modifications may be needed for very shallow wells). The corresponding acoustic pressure is $p_3 = - B \, \partial u_3/\partial x = B/c \, q'(t - x/c)$. In Section "2," $L_m < x < L_c$, we more generally assume a linear superposition of the form $u_2(x,t) = f(t - x/c) + g(t + x/c)$, which supports up and downgoing waves, with acoustic pressure $p_2 = - B \, \partial u_2/\partial x = B/c \, [f'(t - x/c) - g'(t + x/c)]$. So far, we have used the far-left and far-right boundary conditions. Note Sections "1" and "2" support left and right-going waves, while Section "3" only supports a right-going propagating wave.

At the "2-3" collar-pipe junction $x = L_c$, continuity of volume velocity and of pressure require $A_c \, \partial u_2/\partial t = A_p \, \partial u_3/\partial t$ and $\partial u_2/\partial x = \partial u_3/\partial x$, respectively. The A's denote collar and pipe cross-sectional areas. The dummy time independent variables in the resulting algebraic equations can be adjusted so that explicit solutions for "f" and "g" can be obtained in terms of "q." These matching conditions imply that $f'(t - L_m/c) = \frac{1}{2} (A_p/A_c +1) \, q'(t - L_m/c)$ and $g'(t - L_m/c) = \frac{1}{2} (A_p/A_c - 1) \, q'(t - L_m/c - 2L_c/c)$. The first equation states that when $A_p = A_c$, "f" and "q" are identical, since there is no impedance mismatch due to area change; the second result confirms this, stating that the left-going "g" wave vanishes identically.

At the "1-2" junction $x = L_m$, the MWD pulser supports a general transient acoustic pressure difference $\Delta p(t)$ satisfying $p_2(t) - p_1(t) = \Delta p(t)$. In addition, continuity of volume velocity requires $\partial u_1/\partial t = \partial u_2/\partial t$ since cross-sectional area does not change. These conditions together lead to $g'(t - L_m/c) - h'(t - L_m/c) = - \frac{1}{2} (c/B) \Delta p(t - 2L_m/c)$ and also $f'(t - L_m/c) - h'(t - L_m/c) = + \frac{1}{2} (c/B) \Delta p(t)$ (in the foregoing matching conditions, appropriate changes in dummy variables have been used to transform f, g and h arguments to "$t - L_m/c$" form so that independent displacement solutions can be obtained). Subtraction of one equation from the other leads to a relationship connecting "f" to "g" that is independent of the function "h." Then, use of the equations in the above paragraph and replacing "q'" by its equivalent in terms of p_3 yields a fundamental time difference delay equation that is customized in different ways depending on the application. These applications are discussed next.

If reflection signal processing is used at the surface standpipe to remove reflections at the desurger, the rotary hose and the pump, and also to remove pump noise, what remains is a signal proportional to $p_3(t)$, which again, contains the MWD signal and all the downhole reflections at the drillbit and collar-pipe junction. (Note that we have not addressed thermodynamic attenuation along the drillpipe – it is not significant computationally since it is assumed to affect all parts of the signal uniformly, e.g., any electronic gain may be used but it is not numerically important.) From this $p_3(t)$, we wish to extract a fully transient $\Delta p(t)$ which contains the true downhole well logging information. This is our first application. To do this, the equation referred to in the above paragraph is written as follows –

$$\Delta p(t) + \Delta p(t - 2L_m/c) =$$
$$= (A_p/A_c + 1)\, p_3(t - L_m/c) - (A_p/A_c - 1)\, p_3(t - L_m/c - 2L_c/c) \qquad (5.1.1)$$

Equation 5.1.1 is solved using the exact solution and algorithm for difference delay equations discussed in Chapter 4. Once $\Delta p(t)$ is obtained, we have the "0" and "1" information formed by the position-encoding of the pulser.

In the second application, we assume that $\Delta p(t)$ across the pulser is given, and that the upgoing $p_3(t)$ signal is required, say, to estimate the form of the signal that might be obtained at the surface (a model for attenuation must be used to account for non-Newtonian losses along the drillpipe, which may extend miles in directional drilling applications, and this is developed in Chapter 6). Then, Equation 5.1.1 can be formally written in the form

$$p_3(t - L_m/c) - \{(A_p - A_c)/(A_p + A_c)\}\, p_3(t - L_m/c - 2L_c/c) =$$
$$= \{\Delta p(t - 2L_m/c) + \Delta p(t)\}/(A_p/A_c + 1) \qquad (5.1.2)$$

We emphasize that, in this form, Equation 5.1.2 is not completely meaningful physically. To see why, we return to the mathematical formulation and note that the pressure function $p_3(x,t)$ is defined for $x \geq L_c$ only. The argument in the first term on the left side above refers to $x = L_m$ which is less than L_c. To render the equation useful, we need to change dummy variables and shift all arguments to obtain

$$p_3(t - L_c/c) - \{(A_p - A_c)/(A_p + A_c)\}\, p_3(t - 3L_c/c) =$$
$$= \{\Delta p(t - L_m/c - L_c/c) + \Delta p(t + L_m/c - L_c/c)\}/(A_p/A_c + 1) \qquad (5.1.3)$$

Now, we recognize that $p_3(t - L_c/c)$ is the acoustic pressure at the very bottom of the drillpipe ($x = L_c$) just above the MWD drill collar. It is this "P_{pipe}" at $x = L_c$ that travels up the drillpipe to the surface. It is more instructive to introduce the new symbol $P_{pipe}(t) = p_3(t - L_c/c)$ so that the above equation can be rewritten as

$$P_{pipe}(t) - \{(A_p - A_c)/(A_p + A_c)\} \, P_{pipe}(t - 2L_c/c) =$$
$$= \{\Delta p(t - L_m/c - L_c/c) + \Delta p(t + L_m/c - L_c/c)\}/(A_p/A_c + 1) \qquad (5.1.4)$$

In the above form, the indexes on the right-side of the equation contain the correct space-time dependencies, with the solution $P_{pipe}(t)$ applying to $x = L_c$. The resulting P_{pipe}, of course, will decrease in amplitude as it travels to the surface, on account of attenuation (at the surface, the above P_{pipe} signal is reduced by the attenuation factor $e^{-\alpha x}$ where α depends on sound speed, kinematic viscosity, drillpipe radius and frequency – more on this in Chapter 6). The amount of attenuation depends on frequency, and fluid rheology, density and kinematic viscosity.

Finally, for our third application, note that we can shift arguments in Equation 5.1.2 by a different amount to obtain, instead of Equation 5.1.3, the following

$$p_3(t - L/c) - \{(A_p - A_c)/(A_p + A_c)\} \, p_3(t - L/c - 2L_c/c) =$$
$$= \{\Delta p(t - L_m/c - L/c) + \Delta p(t + L_m/c - L/c)\}/(A_p/A_c + 1) \qquad (5.1.5)$$

where L is a length such that $L > L_c$ (L is the distance from the origin $x = 0$). The term $p_3(t - L/c)$ is the drillpipe pressure at any $x = L$ location uphole, e.g., very far away at the surface standpipe or any other intermediate position along the drillpipe (this only accounts for the wave originating from downhole and not any downgoing reflections). If we denote $p_3(t - L/c) = p_{surface}(t)$, then Equation 5.1.5 becomes

$$p_{surface}(t) - \{(A_p - A_c)/(A_p + A_c)\} \, p_{surface}(t - 2L_c/c) =$$
$$= \{\Delta p(t - L_m/c - L/c) + \Delta p(t + L_m/c - L/c)\}/(A_p/A_c + 1) \qquad (5.1.6)$$

Attenuation effects along $x \gg 0$, that is, the long path in Section "3," are easily modeled. We had assumed a pure upgoing wave $u_3(x,t) = q(t - x/c)$ without attenuation only for the purposes of studying local wave interactions in the relatively short bottomhole assembly. Once the signal leaves the drill collar and enters the drillpipe, it is subject to a decay factor $e^{-\alpha x}$ where $\alpha > 0$ depends on frequency, fluid kinematic viscosity and pipe radius. Over large distances, pressure takes the form $p_3(x,t) = -e^{-\alpha x} B \, \partial u_3(x,t)/\partial x = +(B/c) \, e^{-\alpha x} q'(t - x/c)$. The "x" in this exponential actually refers to the distance from the source located at $x = L_m$, however, since L_m is very small compared to the surface location $x = L$, we can use $e^{-\alpha L}$ with very little error. In summary, if in Equation 5.1.6, L is very far away from the source, then the corresponding pressure is obtained by multiplying the undamped solution of Equation 5.1.6 by $e^{-\alpha L}$ to give $e^{-\alpha L} p_{surface}(t)$.

Analogously, the reverse applies at the surface for Δp determination. We can use our surface reflection cancellation filters to separate waves originating from downhole from surface reflections traveling downhole. The waves originating from downhole would consist of everything from downhole (the intended signal, plus reflections at the drillbit) and would, in fact, be of the form $e^{-\alpha L} p_{surface}(t)$. Instead of solving Equation 5.1.1 to recover $\Delta p(t)$, which contains all the well logging information, one would solve

$$\Delta p(t) + \Delta p(t - 2L_m/c) =$$
$$= e^{-\alpha L} \{(A_p/A_c +1)\, p_3(t - L_m/c) - (A_p/A_c - 1)\, p_3(t - L_m/c - 2L_c/c)\} \quad (5.1.7)$$

However, this is really not necessary, since the exponential is simply a constant factor applicable to all parts of the signal that obviously results in a proportionately reduced Δp. In other words, the shape of the Δp versus time curve is the same, and this is all that is important to retrieving our downhole well logging information – this is very fortunate, since we typically will not measure α. Similar remarks apply to Method 5-2 and will not be repeated there.

Below, we run collar-pipe-open-16 for four different scenarios. The "delta-p" functions are listed immediately before the graphical results. Additional parameters are required and are presently entered directly in the source code. We have not recorded these values as they are unimportant to demonstrating the fundamental concepts and software capabilities. The basic runs considered are obtained from the simulation menu as –

```
MWD dipole source models available ...
(1)  12 Hz PSK, sampling time DT = 0.0010 sec
(2)  24 Hz PSK, sampling time DT = 0.0001 sec
(3)  96 Hz PSK, sampling time DT = 0.0010 sec
(4)  Short rectangular pulse, DT = 0.0010 sec
```

In the source code, the additional inputs are entered directly by editing the values below. Descriptions are supplied within the parentheses.

```
C       Hardcode above parameters for testing.
        LM = 40.      (distance, feet, pulser from drillbit)
        LC = 50.      (length, feet, of MWD drill collar)
        C = 3600.     (sound speed, ft/sec)
        AP = 1.       (pipe cross-sectional area, sq feet)
        AC = 1.5      (collar cross-sectional area, sq feet)
C       We wish to calculate the drillpipe pressure at a distance
C       x = L > LC from the drillbit. At this distance, original
C       drillpipe signal leaving the drill collar will have decayed
C       exponentially due to attenuation (this satisfies a separate
C       model). Assume that DECAY is the fraction of the original
C       signal that remains. If no attenuation, DECAY = 1. Be sure
C       to try L = 100.*LC for NCASE=2 for difficult test!
        L = 5.*LC     (location > LC, feet, where we want computed
                       values of P3 pressure)
        DECAY = 0.75 (assumed decay fraction, 1.0 if no decay)
C
```

In Runs 1, 2, 3 and 4 below, green, red and black curves are shown, respectively, from bottom to top. The green curve is the assumed input $\Delta p(t)$ function; red is the pressure traveling up the drillpipe that consists of the upgoing wave from the pulser and the downward wave that reflects at the drillbit. Again, these waves reverberate within the drill collar because the cross-sectional area changes abruptly between drill collar and drillpipe. The black curve uses red data only and represents the recovered Δp(t). In all cases, the recovery is excellent. Note that, as a programming check, various choices for DECAY, namely, 1.0, 0.1, 0.5 and 0.75 were run, and as expected, the software simply rescaled the green curve.

5.1.3   Run 1. Phase-shift-keying, 12 Hz carrier wave.

```
C    CASE 1.  High data rate PSK phase shift keying.
     IF(NCASE.EQ.1) THEN
C    F is frequency in Hz, AMP is delta-p amplitude in lbf/ft^2 (PSF),
C    sampling time DT in seconds.
C    2*PI*F cycles in 2*PI secs, or F cycles per sec, F is Hz.
C    One cycle requires time PERIOD = 1/F.
     IF(T.LE.0.0) DELTAP = 0.
     IF(T.GE.0.0) THEN
     PI = 3.1415926
     AMP = 0.5
     F = 12.
     PERIOD = 1./F
     IF(T.GE.0.0.AND.T.LE.2.*PERIOD) DELTAP =
1       +AMP*SIN(2.*PI*F*T)
     IF(T.GE.2.*PERIOD) DELTAP = abs(-AMP*SIN(2.*PI*F*T))
     ENDIF
     ENDIF
```

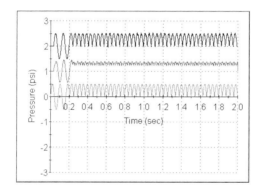

Figure 5.1b. Phase-shift-keying, 12 Hz carrier wave.

5.1.4   Run 2. Phase-shift-keying, 24 Hz carrier wave.

In this higher frequency run, notice that the red curve is very different and highly distorted from the green curve. The recovered black curve is very good.

```
C    CASE 2.  High data rate PSK phase shift keying.
     IF(NCASE.EQ.2) THEN
C    F is frequency in Hz, AMP is delta-p amplitude in lbf/ft^2 (PSF),
C    sampling time DT in seconds.
C    2*PI*F cycles in 2*PI secs, or F cycles per sec, F is Hz.
C    One cycle requires time PERIOD = 1/F.
     IF(T.LE.0.0) DELTAP = 0.
```

```
IF(T.GT.0.0) THEN
PI = 3.1415926
AMP = 0.5
F = 24.
PERIOD = 1./F
IF(T.GE.0.0.AND.T.LE.PERIOD) DELTAP =
1   +AMP*SIN(2.*PI*F*T)
IF(T.GE.PERIOD.AND.T.LE.2.*PERIOD) DELTAP =
1   -AMP*SIN(2.*PI*F*T)
IF(T.GE.2.*PERIOD) DELTAP = 0.
ENDIF
ENDIF
```

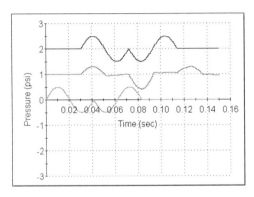

Figure 5.1c. Phase-shift-keying, 24 Hz carrier wave.

5.1.5   Run 3. Phase-shift-keying, 96 Hz carrier wave.

In this higher frequency 96 Hz run, the red curve is very different and highly distorted from the green curve. The recovered black curve is very good.

```
C     CASE 3.  Standard phase shift keying
      IF(NCASE.EQ.3) THEN
C   F is frequency in Hz, AMP is delta-p amplitude in lbf/ft^2 (PSF),
C   sampling time DT in seconds.
C   2*PI*F cycles in 2*PI secs, or F cycles per sec, F is Hz.
C   One cycle requires time PERIOD = 1/F.
      IF(T.LE.0.0) DELTAP = 0.
      IF(T.GT.0.0) THEN
      PI = 3.1415926
      AMP = 0.5
      F = 96.
      PERIOD = 1./F
      IF(T.GE.0.0.AND.T.LE.PERIOD) DELTAP =
1     +AMP*SIN(2.*PI*F*T)
      IF(T.GE.PERIOD.AND.T.LE.2.*PERIOD) DELTAP =
1     -AMP*SIN(2.*PI*F*T)
      IF(T.GE.2.*PERIOD) DELTAP = 0.
      ENDIF
      ENDIF
```

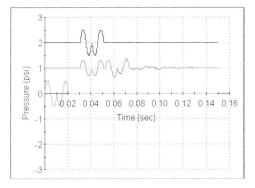

Figure 5.1d. Phase-shift-keying, 96 Hz carrier wave.

5.1.6   Run 4. Short rectangular pulse with rounded edges.

For this 0.5 sec pulse, the black curve is successfully recovered and looks like the green assumed curve. The red curve clearly shows distortions at the front and tail end of the pulse. These are due to multiple reverberations inside the collar due to area impedance mismatch at the collar and drillpipe junction.

```
C       CASE 5.  Short rectangular pulse with rounded edges.
        IF(NCASE.EQ.5) THEN
        IF(T.LE.0.0) DELTAP = 0.
        IF(T.GT.0.0) THEN
        AMP = 0.4
        R = 200.0
        DELTAP = AMP*(TANH(R*(T-0.01))-TANH(R*TANH(T-0.51)))
        ENDIF
        ENDIF
```

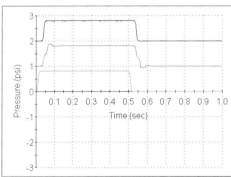

Figure 5.1e. Short rectangular pulse with rounded edges.

5.2 Method 5-2. Problems with collar-drillpipe area discontinuity, with drillbit assumed as closed end, solid drillbit reflector (software reference, collar-pipe-closed-*.for).

5.2.1 Theory.

Figure 5.1a from Method 5-1 applies to this problem, however, the drillbit is assumed as a solid reflector in the present model. If the drillbit can be modeled as a solid reflector, e.g., if nozzle areas are truly small or if very hard rock is being drilled, then the acoustic displacement satisfies $u_1(0,t) = 0$ at the bit. This requires that $u_1(x,t) = h(t - x/c) - h(t + x/c)$ which vanishes at $x = 0$. Minor changes to the analysis of Method 5-1 lead to the three applications formulas below –

$$\Delta p(t) - \Delta p(t - 2L_m/c) =$$
$$= (A_p/A_c +1)\, p_3(t - L_m/c) + (A_p/A_c - 1)\, p_3(t - L_m/c - 2L_c/c) \qquad (5.2.1)$$

$$p_{pipe}(t) + \{(A_p - A_c)/(A_p + A_c)\}\, P_{pipe}(t - 2L_c/c) =$$
$$= \{- \Delta p(t - L_m/c - L_c/c) + \Delta p(t + L_m/c - L_c/c)\}/(A_p/A_c +1) \qquad (5.2.2)$$

$$p_{surface}(t) + \{(A_p - A_c)/(A_p + A_c)\}\, p_{surface}(t - 2L_c/c) =$$
$$= \{- \Delta p(t - L_m/c - L/c) + \Delta p(t + L_m/c - L/c)\}/(A_p/A_c +1) \qquad (5.2.3)$$

Identical cases to those performed in Method 5-1 were run. Refer to Method 5-1 for a description of the $\Delta p(t)$ functions assumed. Excellent recovery is achieved in all instances, that is, black and green curves identical.

5.2.2   Run 1. Phase-shift-keying, 12 Hz carrier wave.

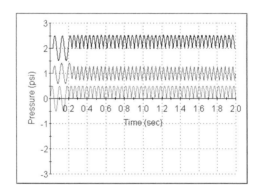

Figure 5.2a. Phase-shift-keying, 12 Hz carrier wave.

5.2.3   Run 2. Phase-shift-keying, 24 Hz carrier wave.

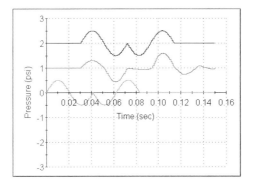

Figure 5.2b. Phase-shift-keying, 24 Hz carrier wave.

5.2.4   Run 3. Phase-shift-keying, 96 Hz carrier wave.

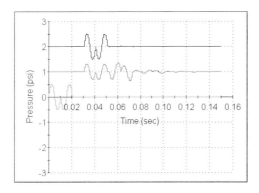

Figure 5.2c. Phase-shift-keying, 96 Hz carrier wave.

5.2.5   Run 4. Short rectangular pulse with rounded edges.

This run is quite interesting. As noted, the recovery is excellent – the green $\Delta p(t)$ curve is a well-defined rectangle and the computed recovered black curve is identical.   The red curve is completely different, bearing little resemblance to the green, due to the assumed solid reflector condition at the drillbit which reverses the sign of the wave and causes destructive interference. MWD engineers cannot explain this, claiming only that "the drilling channel differentiates the signal." It is true that the spikes resemble spatial derivatives at the edges of the rectangular pulse, but the conventional explanation is wrong.

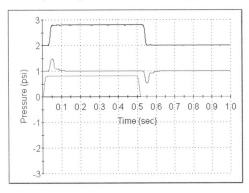

Figure 5.2d. Short rectangular pulse with rounded edges.

This concludes our discussion on surface and downhole reflection processing. Reflections are not "forgiving" in one-dimensional MWD waveguides which, unlike those in three dimensions, "have no where to go." Additional filters are needed for signal processor development and are briefly described in Chapter 6.

5.3 References

Oppenheim, A.V. and Schafer, R.W., *Digital Signal Processing*, Prentice-Hall, New Jersey, 1975.

Oppenheim, A.V. and Schafer, R.W., *Discrete-Time Signal Processing*, Prentice-Hall, New Jersey, 1989.

6

Signal Processor Design
and Additional Noise Models

The MWD signal processor must accommodate numerous noise sources, where both type and relative contribution may vary with bottomhole assembly, bit design, drilling mud, rock type, drilling rig, surface piping arrangement, transmission frequency, and so on. It is important for MWD designers to accurately characterize the noise environment through field and laboratory tests. This chapter is not intended to provide an exhaustive treatment on noise and filtering. Quite the opposite, we only introduce the reader to basic ideas, directing him to detailed literature elsewhere. Only physical mechanisms related to fluid and wave processes, the author's areas of specialty, are developed from first principles. While we discuss noise components separately, for the purposes of presentation, we emphasize that in the design of the signal processor, filter operations are not commutative, that is, the process of applying "A, then B" may not yield results identical to "B, then A." Some experimentation is therefore necessary, especially in determining the best sequencing for our reflection filters and others.

Contents

1. Desurger Distortion
2. Downhole Drilling Noise (Positive Displacement Motors, Turbodrills, Drillstring Vibrations)
3. Attenuation Mechanisms
4. Drillpipe Attenuation and Mudpump Reflection
5. Applications to Negative Pulser Design in Fluid Flows and to Elastic Wave Telemetry Analysis in Drillpipe Systems
6. LMS Adaptive and Savitzky-Golay Smoothing Filters
7. Low Pass Butterworth, Low Pass FFT and Notch Filters
8. Typical Frequency Spectra and MWD Source Properties

6.1 Desurger Distortion

The desurger, for example, as shown in Figure 6.1a, used at the rigsite during drilling operations, suppresses strong fluid transients arising from mudpump actions and is important to rig safety. Its role in MWD signal processing, however, has been a source of confusion. Some engineers will claim that it severely distorts the MWD signal, so much that signal detection is sometimes impossible; others, however, never find problems with desurgers and in fact embrace them. This chapter explains the discrepancies. Analytical solutions for the desurger-distorted signal are derived which assist in developing signal processing algorithms.

Figure 6.1a. Desurger.

The Lagrangian fluid displacement u(x,t) in a one-dimensional acoustic field satisfies the classical wave equation $\partial^2 u/\partial t^2 - c^2\, \partial^2 u/\partial x^2 = 0$ where c is the speed of sound. When a signal (wave) travels uphole and reflects at the mudpump piston (assuming that there is no desurger), the pressure doubles at the piston face, then turns around and travels downward. The sign of the pressure does not change at a solid reflector – this is evident from measurements in very long wind tunnels, but this can also be proven mathematically from a more complicated three-dimensional solution. Again, at the solid reflector, the signal reflects with the same sign, but importantly, the shape of the signal does not change and remains the same. The author has, in fact, instrumented pressure transducers near to the mudpump to demonstrate improved signal recovery.

When a desurger is present, however, events can be unpredictable. A desurger is typically used in the drilling setup at the rigsite to absorb dangerous pressure transients from the mudpump. It is essentially a thick rubber membrane within a metal housing that is "charged" with gas that is responsible for energy absorption. However, the desurger can also affect the MWD signal. When communicating with MWD engineers, one finds that different engineers from different companies offer very different experiences that may be contradictory and confusing.

For example, engineers operating higher data rate siren tools will explain that desurgers are never problematic, that is, desurgers never distort signals and in fact are an essential part of the MWD surface setup. On the other hand, operators of lower data rate tools will describe how desurgers distort signals so badly that the original signal is sometime unidentifiable. These remarks are not hearsay: these comments are based on the author's first-hand experiences. It turns out that, on returning to first principles, these discrepancies can be explained logically. Again, we turn to the modeling principles established earlier in Chapters 2-5. The one-dimensional wave model used applies so long as a typical wavelength greatly exceeds the cross-dimensions of the drillpipe, a condition easily met in all mud pulse operations.

The Lagrangian displacement variable $u(x,t)$ is the distance that a fluid particle moves as the wave travels past a given location "x." The corresponding velocity is $\partial u/\partial t$ while the local acoustic pressure p is $p(x,t) = -B \ \partial u/\partial x$, where B is the bulk modulus. If ρ is the fluid density, then $c^2 = B/\rho$. The general solution to the above wave equation is $u(x,t) = f(ct-x) + g(ct+x)$, as is well known from partial differential equations. The "f" represents a right-going wave and the "g" represents a left-going wave. For our convention, let us assume that a reflector is located at $x = 0$. The wave originating from downhole is "g" and the reflected wave is "f." If the reflector is a solid reflector like a mudpump piston, then the fluid displacement $u = 0$ at $x = 0$. Thus, $u(0,t) = f(ct) + g(ct) = 0$ at $x = 0$ so that $f = -g$. If we take x-derivatives of the general solution, then $\partial u(x,t)/\partial x = -f'(ct-x) + g'(ct+x) = -f'(ct) + g'(ct) = 2g'(ct)$, thus proving that at the piston, the original pressure $g'(ct)$ doubles – an elementary derivation indeed and a correspondingly exact result.

This result, interestingly, has been used to develop the simplest analogue signal amplifier – the "hundred feet hose" attachment shown in Figure 6.1b and discussed in detail in the author's U.S. Patent Nos. 5,515,336 and 5,535,177. Assume that an amplitude "A" is measured at a pressure transducer installed on the standpipe. If this transducer is removed and attached to one end of a long hose, with the other end then installed to the standpipe, the measured pressure is obtained as "2A." This application has been validated in field tests. Simply installing the transducer ahead of a pump piston will also double the signal. This is not usually done because the pump is viewed as a noise source.

Figure 6.1b. "One-hundred feet hose" analogue signal amplifier.

An extension of the above analysis can be used to model desurger effects on acoustic signal reflection. Now, the rubber membrane within a desurger essentially acts like a large spring as opposed to a solid piston. The spring will move with a distance "u" and the force exerted by the membrane would be ku(0,t) where k is the spring constant. This force must equal that exerted by the acoustic wave. If the acoustic pressure is -B $\partial u/\partial x$ and $\pi D^2/4$ is the area of the membrane, D being the diameter of the circular membrane, then the boundary condition is ku(0,t) = T $\partial u(0,t)/\partial x$ where T = $B\pi D^2/4$. Note the following two physical limits: the mud pump piston possesses "infinite k" so that u = 0, while a centrifugal pump satisfies $\partial u(0,t)/\partial x = 0$ instead.

6.1.1   Low-frequency positive pulsers.

Consider the situation at the top of Figure 6.1c. A rectangular pulse travels toward the desurger location x = 0 at time t = 0 – but what does a rectangular "u" pulse physically mean? Again, pressure is the spatial derivative or slope of "u." The head of the wave shows a large positive slope while the tail has a large negative one. The flat portion of the rectangular has zero slope and is therefore not associated with any compression or expansion. Thus, the initial t = 0 schematic in Figure 6.1c represents a sudden valve opening followed by a sudden closure.

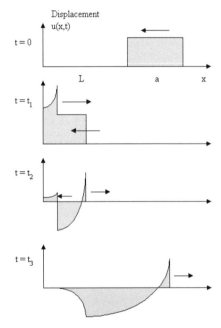

Figure 6.1c. MWD signal distortion for low data rate pulsers
(confirmed experimentally in mud loop tests).

The two pressure spikes, which possess opposite signs, are characteristic of dipole sources; and since they are separated in time, the dipole source shown applies to positive pulser valves – which are also known for large pressure amplitudes. Physically, the acoustic force exerted on the desurger membrane is "large" and acts over "long enough" times to be dynamically significant (these are characterized dimensionlessly later). If the wave equation for displacement is solved subject to the above spring boundary condition, we can derive the following exact solution for desurger response to the rectangular wave, namely,

$$f(ct-x) = \quad g(ct-x) - 2[1-\exp\{-k(ct-x-L)/T\}] \, H(ct-x-L)$$
$$+ \, 2[1-\exp\{-k(ct-x-L-a)/T\}] \, H(ct-x-L-a) \qquad (6.1a)$$

where H is the Heaviside step function. Note that unlike the simple piston solution previously obtained for mudpump piston reflections in which $f = - g$ without changes in shape, Equation 6.1a is complicated and contains exponential distortions to the rectangular pulse. This is the shape distortion observed for positive pulsers associated with longer duration with high pressure amplitude. An expression not too different from Equation 6.1a should apply to negative pulsers. Unless their downward motions are filtered by the multiple transducer techniques of Chapter 4, their presence will lead to serious surface signal processing and interpretation problems (these effects are also likely to be found with negative pressure MWD signals).

6.1.2   Higher frequency mud sirens.

Figure 6.1c does not apply to higher frequency mud siren operations, which employ periodic frequencies, plus signal amplitudes that are typically much lower than those of positive pulsers. Changes in phase or frequency are used to telemeter 0's and 1's to the surface. When frequencies are high, the wave literally "does not have time" to act on the desurger membrane, and vice-versa – this inaction is, to be sure, more so, given the lower pressure amplitudes involved. Thus, one does not expect desurgers to affect signals adversely. For higher frequency mud siren applications, desurgers will do what they are suppose to: remove dangerous transients and eliminate mudpump noise. These statements can be demonstrated mathematically.

Again, the partial differential equation governing mud pulse acoustics is the classical wave equation $\rho_{mud} u_{tt} - B u_{xx} = 0$, and its general solution is $u(x,t) = f(ct-x) + g(ct+x)$, where f and g represent right and left-going waves respectively. This solution need not include damping because only the desurger nearfield is considered. Over short distances, damping in the pipeline itself is unimportant. Our objective is simple: what happens to a wave of a given shape upon reflection? In this acoustic study, we will assume that the functional form of the incident wave $g(ct+x)$ transmitted by the pulser and impinging at the desurger $x = 0$ is given. The problem consists in solving for the reflected wave $f(ct-x)$, and then, the complete superposition solution $u(x,t)$, so that the acoustic response is fully determined everywhere and at all instances in time.

To accomplish this, we now model the desurger more precisely.   We envision the desurger as a mass-spring-damper system with a mass M (e.g., the bladder mass plus a time-averaged fluid mass of the partially filled volume), a spring constant k (due to the elasticity of the bladder and the charge pressure of the compressed gas, as compared to the standpipe pressure), and an attenuation factor γ (due to orifice losses and internal friction).   The rubber membrane is excited by the acoustic pressure $p_a = - Bu_x(x,t)$, so that the ordinary differential equation satisfied by the end value $u(0,t)$ at the boundary $x = 0$ now takes the form $Mu_{tt} + \gamma\, u_t + ku = (B\pi D^2/4)u_x$ (in this right-side term, having units of force, D is an effective bladder diameter).   Away from the desurger, both incoming and outgoing waves exist, and radiation conditions do not apply.

The general solution to this boundary value problem is difficult if we attack the partial differential equation directly.   Instead, we will solve the desurger ordinary differential equation at $x = 0$ exactly, and use the general solution $u(x,t) = f(ct-x) + g(ct+x)$, plus the Convolution Theorem, to analytically continue the solution into the domain $x > 0$.   This equally rigorous approach allows us to construct the exact general solution as

$$u(x,t) = g(ct+x) - g(ct-x) \qquad\qquad (6.1b)$$

$$+ \{2(B\pi D^2/4)/[Mc^2(a-b)]\} \int_0^{ct-x} g(\sigma)\, [ae^{a(ct-x-\sigma)} - be^{b(ct-x-\sigma)}]\, d\sigma$$

It is important to recognize that M, γ, k, B, D, c and ρ_{mud} do not appear individually in the complete solution.   Rather, they appear implicitly through the lumped parameters

$$a = \{- (\gamma c+(B\pi D^2/4)) - \sqrt{[(\gamma c+(B\pi D^2/4))^2 - 4kMc^2]}\,\}/2Mc^2 \qquad (6.1c)$$

$$b = \{- (\gamma c+(B\pi D^2/4)) + \sqrt{[(\gamma c+(B\pi D^2/4))^2 - 4kMc^2]}\,\}/2Mc^2 \qquad (6.1d)$$

In the general solution for the displacement "u," the "g(ct+x)" term represents the known incident waveform, whereas the term second term -g(ct-x) represents the reflection at a rigid interface, e.g., the piston faces of a positive displacement mud pump if the desurger were not functioning (again, u = 0 at piston faces). The last term in Equation 6.1b represents the distortion of signal due to reflection at the desurger; this distortion consists of a phase shift and a shape change that is again exponential in nature.   In order to determine its effects quantitatively, we give exact solutions for a class of important incident signals.

Let us consider the incident upcoming displacement wave taking the form "A sin $\omega(ct+x)/c$."   This function is particularly important to MWD, since a general transient signal can always be written in terms of its harmonic Fourier components. The exact solution corresponding to this assumption is

$$u(x,t) = A \sin \{\omega(ct+x)/c\} - A \sin \{\omega(ct-x)/c\} \tag{6.1e}$$

$$+ 2A(B\pi D^2/4)b\{b \sin (\omega(ct-x)/c) + (\omega/c) \cos (\omega(ct-x)/c)\}/\{Mc^2(a-b)(b^2 +\omega^2/c^2)\}$$

$$- 2A(B\pi D^2/4)a\{a \sin (\omega(ct-x)/c) + (\omega/c) \cos (\omega(ct-x)/c)\}/\{Mc^2(a-b)(a^2 +\omega^2/c^2)\}$$

$$+\{2A(B\pi D^2/4)\omega/[Mc^3(a-b)]\} \{ae^{a(ct-x)}/(a^2 +\omega^2/c^2) - be^{b(ct-x)}/(b^2 +\omega^2/c^2)\}$$

The first line, again, represents the incoming wave and its solid wall reflection, where A is the signal amplitude corresponding to the fluid displacement "A sin $\{\omega(ct+x)/c\}$" leaving the pulser. The second and third lines represent the phase distortion (of the sinusoidal signal) introduced by the desurger, while the fourth line shows that the desurger produces a non-sinusoidal distortion that will contain exponential smearing.

A more detailed examination of Equation 6.1e indicates that these distortion effects vanish in the limit of high frequency ω, because "the waves do not have sufficient time" to act on the system, especially in the limit of a "heavy desurger" with large M or a "small amplitude A." How high a frequency is needed so that desurger distortion is not important? Apparently, the carrier frequency of 12 Hz used in present mud siren operations is high enough – thus higher frequency operations should be safer insofar as signal distortions are concerned. Of course, the mathematical structure of the terms in Equation 6.1e reveals the complicated nature of the dimensionless parameters controlling the physical problem.

In electric engineering, the resistance, capacitance, and inductance values of the circuit elements determine the non-dimensional time scales associated with the particular circuit. Analogies exist here. The physical time scales "ac" and "bc" inferred from the exponential terms above define two of the time scales important to the problem; the third is the period $1/\omega$. The mud density ρ_{mud} does not explicitly appear in our solutions, but the dependence can be easily recovered if we note that $c^2 = B/\rho_{mud}$. This leads to $B/c^2 = \rho_{mud}$. If the B/c^2 coefficient multiplying the integral is replaced instead by ρ_{mud}, the distortion is seen to be proportional to mud density, or more precisely, a ratio that depends on ρ_{mud}/M, among other quantities. The dependence is complicated by the appearance of the same variables in "a" and "b." As noted above, the distortion appears to increase with increasing mud weight. All quantities being equal, the distortion is largest for low frequencies ω.

Interestingly, in recent mud flow loop experiments, measured pressure signal levels and shapes in the 1-2 Hz range were particularly affected by the desurger, whereas for 5 Hz and above, the effect of the desurger was insignificant. In wind tunnel experiments designed to understand near-static pressure response, manometer-based results were noticeably affected by fluid column oscillations at 1-5 Hz but not at 9 Hz and above. Finally, the exact solution emphasizes that pressures will depend on the location x and the time t. Depending on the telemetering method used, standing wave patterns created temporarily can be identified with systems of nodes and antinodes. Interpretations for measured data using single transducers may not be useful and it may be more meaningful to measure changes between multiple transducers. Nodal positions will change with frequency from rigsite to rigsite, and desurger charges will vary just as unpredictably. Since it is impossible to characterize the complete telemetry channel in practice, the most likely solution to eliminating signal processing problems is through the use of multiple transducer methods. Software is not provided for the low and high frequency models derived here because exact, closed form, analytical solutions are available above.

6.2   Downhole Drilling Noise

Noise originating from downhole near the bit is problematic in the sense that it cannot be filtered using the multiple transducer methods in Chapter 4 for surface signal processing – Methods 4-3 and 4-4 remove only downgoing noise. Frequency-based filtering may be an option, providing such noise falls outside the MWD signaling range. MWD signals traveling downward, impinging on rubber drill motor rotors and reflecting to travel upwards, however, may be difficult to remove. Multiple varieties of noise are present and special filters are required for each type. Developing these requires not only an understanding of the physics, but detailed calibration with "clean" data – that is, measurements not contaminated by other possibly coupled noise sources.

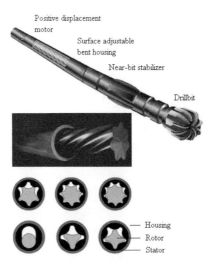

Figure 6.2a. Positive displacement motor (see noise spectra in Figure 6.8a).

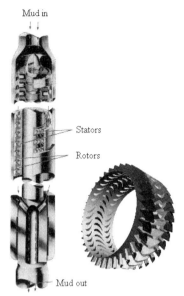

Figure 6.2b. Turbodrill motors.

6.2.1   Positive displacement motors.

Downhole drilling motors, drawing hydraulic power from the flowing mud, are used to turn the drillbit without turning the entire drillstring. Two types are available, namely, the positive displacement drilling motors and turbodrills represented, respectively, in Figures 6.2a and Figure 6.2b. In Figure 6.2a, downward flowing mud is forced through the cross-sectional space between the metal (gray) rotor and the rubber (black) stator. This rotates the spiraled rotor shaft which in turn drives the drillbit.

The MWD source, not shown in these figures, is positioned above the drilling motor. When the pulser opens and shuts, it creates "intended signals" that travel uphole (which embed the 0's and 1's position-encoded by valve action), but as noted earlier in this book, it creates equally strong signals of opposite sign that travel downward. Chapter 4 assumes that the drillbit can be modeled as a solid reflector or an acoustic open-end; these simple models predict phases that are 180° apart, but in either case, signal shape remains undistorted. Chapter 2 more generally treats the drillbit as one segment of a six-segment waveguide, providing the needed transition between the two simpler limits of Chapter 4. On the other hand, the previous section on desurger noise demonstrates that not all reflections are so simple: those associated with elastic boundary conditions at rubber interfaces may significantly distort MWD signals.

While the rubber reflector in our desurger is conveniently located at a single point "x = 0," the reflection in Figure 6.2a is distributed along the entire length of the positive displacement motor. How MWD signals reflect will depend on this length, the stiffness of the stator rubber, the cross-section of the actual rotor-stator geometry (typical shapes are given in Figure 6.2a), wave amplitude and frequency, and so on. An acoustic path from the top of the motor to the drillbit definitely exists, since mud does flow through the entire length and out the nozzles – whether the reflector can be treated as a solid or open end, and whether or not any anticipated smearing is significant, remains to be determined. Signal processing problems can be difficult to remedy since multiple transducer methods that eliminate downward noise cannot used. Frequency-based filters may be of limited usefulness because incident and reflected signals possess similar frequency content. Controlled lab tests would be useful in developing models that can potentially eliminate such upgoing noise sources.

6.2.2 Turbodrill motors.

Drilling motors like those represented in Figure 6.2b are known as turbodrills. They are made up in multiple "stages," each stage consisting of one stator and one rotor. The stator, which is fixed to the housing, deflects flow tangentially; this tangential flow imparts additional turning momentum to the rotor, which is fixed to a rotating shaft. As shown, rotors and stators are made completely of metal; there are no rubber reflectors that produce signal distortions. In fact, the stage "see through" area is approximately 50% or more and downgoing waves have no trouble reflecting at the bit, as modeled in earlier chapters. In addition, the noise associated with rotor-stator interactions does not propagate to the surface, since it is associated with very small wavelengths that are not plane wave in nature. The acoustic passage associated with turbodrill motors can be conveniently modeled as one of the segments provided for in the six-segment waveguide of Chapter 2.

6.2.3 Drillstring vibrations.

Practical consequences associated with dangerous drillstring vibrations, e.g., damaged MWD well logging tools, borehole instability, formation damage, well control, and so on, are well known. These are of three main categories: axial, torsional and lateral (or bending) vibrations, although other motions, for instance, whirling, also exist. These are described in detail in the lead author's book Wave Propagation in Petroleum Engineering, *with Applications to Drillstring Vibrations, Measurement-While-Drilling, Swab-Surge and Geophysics* (Gulf Publishing, 1994) and in Chin (1988) explaining vibration subtleties near the neutral point. In general, all three modes are coupled and do not act independently as is usually assumed.

Such vibrations invariably affect the drillpipe and are important in "drillpipe telemetry" where MWD signals are transmitted through metal. Vibration effects appear in the mud which can be filtered out using standard frequency-based methods. The most significant problem is "bit bounce," that is, the low-frequency bouncing of the drillbit that occurs when unstable drillbit and formation interactions are encountered. These nonlinear effects have not been studied in the literature due to their complexity. Typically, in conventional models, individual partial differential equations for axial, torsional and lateral vibrations are solved subject to standard boundary conditions, e.g., sinusoidal displacements are prescribed and resonant conditions are obtained.

There are serious limitations associated with such approaches. Assuming that bit displacement is sinusoidal does *not* imply that a Fourier component of the general transient problem is being considered – in fact, by assuming that bit motion is always sinusoidal, one importantly precludes the modeling of highly nonlinear events like "bit bounce." In the author's book, a general boundary condition related to rock-bit interactions is used, and the transient "up and down" displacement effect of the rotating bit is in fact modeled using "accordion-like, displacement sources" like those in earthquake engineering. When the general initial value problem is solved subject to specified starting conditions and bottomhole geometrical constraints, the resulting bit motion will be sinusoidal – or, erratically bouncing, if the rock-bit interaction dictates.

Other common fallacies exist that the book addresses which may prove important to MWD signal processing. For example, one can state that all resonances are dangerous; however, not all dangerous events arise from resonances. The well known fact that many drillstring failures occur at the neutral point in the drill collar while undergoing strong lateral vibrations is one consequence that cannot be predicted by traditional resonance-based models. The neutral point is simply the location within the drillstring where *axial* forces change from tension to compression, that is, axial stresses vanish at the neutral point. But what do axial forces have to do with lateral transverse vibrations?

It turns out that axial and transverse vibrations are dynamically coupled. An exact, closed form solution based on "group velocity" methods in theoretical physics shows that axial vibration energy tends to coalesce and trap near the neutral point and instigates high-cycle bending fatigue, e.g., see Chin (1988). Mathematically, the solution represents a "singularity" in the governing partial differential equations, popularly termed a "black hole." This implies serious consequences in developing MWD drillpipe telemetry technology based on transmissions through metal itself. Severe lateral vibrations at the drillbit may affect the rock-bit interaction boundary condition enforced for axial vibrations, resulting in axial bit bounce. Because lower frequencies are typically involved, direct bit bounce effects can be removed using high-pass filters; unfortunately, bit bounce is associated with significant fluctuations in mud flow rate through the nozzles, and hence, with large changes to siren torque, control system response, signal strength and phase, and so on, that will require both better designed tools and very robust signal processing.

It is well known that free vibrations of axial, torsional and lateral bending modes are each associated with different characteristic frequencies. Signals correlated with these frequencies, which are typically different from those of the MWD signal, can be removed by low-pass, high-pass or notch filters (see

Section 7 below). Ideally, surface signal processors would determine these frequencies for the bottomhole assembly under consideration (using standard mechanical engineering formulas) and automatically remove their effects before application of our echo cancellation and other filters.

6.3. Attenuation Mechanisms (software reference, Alpha2, Alpha3, MWDFreq, datarate).

Numerous models are available for wave attenuation modeling in the engineering literature, however, they are developed for ultrasonic applications where wavelengths are very, very small and interactions with particles are considered in detail. In MWD applications, wavelengths are typically hundreds of feet and reliable measurements are not available due to mudpump transients. These are typically made in wells with flowing non-Newtonian mud, or in very long flow loops with multiple twists, turns and (undocumented) area changes. Vortex flows, density segregation, secondary flows, and so on, may be present. The action of positive displacement mud pumps and their unsteady pistons renders fluid flows highly transient; moreover, the effects of constructive and destructive interference, and those associated with nodes and antinodes for standing wave patterns, are usually not separated out. Good data is difficult to find. Also, given the complexity of the mathematical problem, and the fact that field situations are rarely controlled, approximate methods are appropriate and are therefore considered here.

6.3.1 Newtonian model.

Almost all oil service companies use a classic formula during job planning for sound wave attenuation developed for steady laminar Newtonian flow in a circular pipe, e.g., see Kinsler *et al* (2000). It appears to be reliable if used properly and cautiously, and is derived from rigorously formulated fluid-dynamic models. If ω is circular frequency, μ is viscosity, ρ is mass density, c is sound speed and R is pipe radius, then the pressure P corresponding to an initial signal P_0 is determined from

$$P = P_0 \, e^{-\alpha x} \tag{6.3a}$$

where x is the distance traveled by the wave and α is the attenuation

$$\alpha = (Rc)^{-1} \sqrt{\{(\mu\omega)/(2\rho)\}} \tag{6.3b}$$

In other words, the damping rate varies as the square root of frequency and depends on density and viscosity only through the "kinematic viscosity" μ/ρ. The software model ALPHA2.EXE, emphasizing "Newtonian" in Figure 6.3a, performs the required calculations. Once the input data are entered in the white text boxes, clicking on "Find" will give the value of α and the pressure ratio P/P_0 as shown in Figure 6.3b.

6.3.2   Non-Newtonian fluids.

When the flow is non-Newtonian, few published models for very low frequency waves are available. In order to obtain rough attenuation estimates, we adopt the following procedure. Non-Newtonian drilling fluids are typically modeled by a simple power law rheology model with "n" and "K" coefficients. The apparent viscosity is variable throughout the pipe cross-section of the pipe and depends on the radius R and the volume flow rate Q. The volume flow rate Q in steady flow is exactly

$$Q = \{\pi R^3/(3 + 1/n)\} \; \{R/(2K)\}^{\,1/n} \; (dp/dz)^{\,1/n} \tag{6.3c}$$

from which we can calculate the pressure gradient dp/dz. What is the effective Newtonian viscosity $\mu_{\text{effective}}$ that will give the same flow rate Q for identical parameters n, K and R? For this, we use the well known pipe flow Hagen-Poiseuille formula rewritten in the form

$$\mu_{\text{effective}} = \{\pi R^4/(8Q)\} \; dp/dz \tag{6.3d}$$

and substitute the pressure gradient known from the first calculation. This viscosity value can then be used with the calculator of Figures 6.3a and 6.3b. Again, this procedure gives an approximate "engineering solution" that is roughly correct and useful for MWD job planning purposes.

Figure 6.3a. Newtonian model input form.

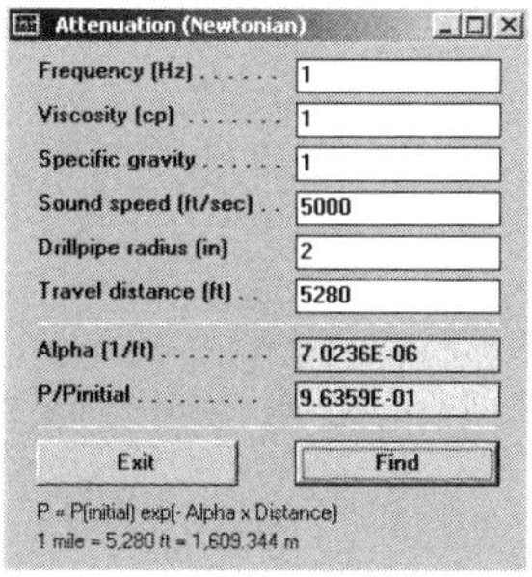

Figure 6.3b. Newtonian attenuation solutions.

The non-Newtonian model is implemented in ALPHA3.EXE. Clicking on this filename produces a message box reminder and three programs as shown in Figure 6.3c. The message notes that n and K values must be entered into the bottom attenuation calculator before "Find" can be executed. Two methods are available to determine n and K. The first assumes that Fann dial readings are available and the second assumes that viscosities and shear rates are available. For example, let us use the Fann dial readings shown in the above left calculator and click "Calculate." Then, n and K will appear in the bottom boxes of that calculator. These values should be copied and pasted into the attenuation calculator (which emphasizes "non-Newtonian" at the top). Then, click "Find" to obtain the results in Figure 6.3d.

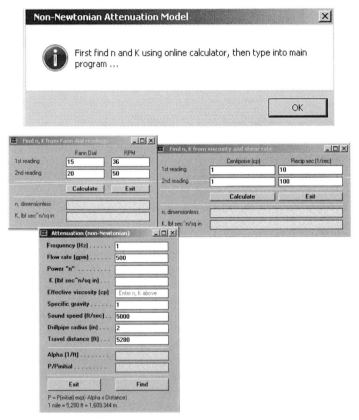

Figure 6.3c. Non-Newtonian flow menu.

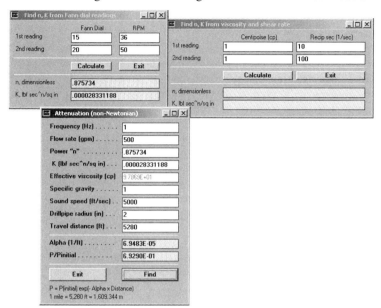

Figure 6.3d. Example non-Newtonian attenuation results.

In Figure 6.3d, the bottom box indicates that 69.29% of the signal will remain after 5,280 feet (1 mile). Also, for information purposes, the effective viscosity in centipoise is shown in the shaded box. If further calculations are required in either Newtonian or non-Newtonian modes, simply edit the data in the white text boxes and click "Find" again.

Finally note that a complementary wave attenuation model has been developed that predicts the "critical frequency" above which MWD signals are damped, given fluid properties, drillpipe geometry and length, source signal strength and transducer sensitivity. Refer to the discussion for Figure 10-7 on the math model underlying MWDFreq.vbp and datarate*.for.

6.4   Drillpipe Attenuation and Mudpump Reflection (software reference, PSURF-1.FOR).

This model assumes that the signal entering the drillpipe at the top of the MWD drill collar – that is, the complicated waveform containing initially upgoing waves at the pulser, downgoing waves that reflect at the drillbit and travel upward, plus the complete reverberant field associated with an acoustic impedance mismatch at the drill collar and drillpipe junction – is known from the models developed in Chapters 2, 3, 4 and 5. The Newtonian or non-Newtonian attenuation models in the foregoing section are applied to this signal as it travels along the length of the very long drillpipe.

When the diminished wave signal reaches the surface, it reflects at the mudpump, which may be a solid or open-end reflector (we assume that, for high enough carrier wave frequencies, the effects of desurger distortion are unimportant), and attenuates downward. The present model computes the pressure signal that is obtained at any point along the drillpipe – the main positions of interest, of course, are those along the standpipe. It is assumed that, because of attenuation, multiple reflections do not occur in the drilling channel.

6.4.1 Low-data-rate physics.

To understand the significance of the model in this section, let us consider first a relatively long rectangular pulse, namely, the 0.5 sec duration rectangular waveform considered previously in Method 4-1, Run 1. In this example, the reflection occurs at the mudpump modeled as a solid reflector and pump noise is omitted for clarity. This waveform nominally represents a 1 bit/sec data rate, a low rate typical of present MWD tools.

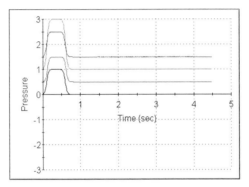

Figure 6.4a. Wide signal – low data rate.

Figure 6.4a shows pressure measurements at a standpipe transducer. The incident upgoing assumed signal (black) is a broad pulse with a width of about 0.5 sec. The red curve is the reflection obtained at a solid reflector with no attenuation assumed; there is very little shifting of the red curve relative to the black curve, since the total travel distance to the piston is very short. The transducer will measure the superposition of incident and reflected signals which broadly overlap. This superposition appears in the green curve – it is about twice the incident signal due to constructive wave interference – moreover, it does not cause any problems and actually enhances signal detection. The blue curve is the signal extracted from data using only the green curve and the algorithm of Method 4-1. Colors above are, respectively, black, red, green and blue, starting from the bottom curve. The key conclusion from Figure 6.4a is obvious: *at low data rates, constructive wave interference is always found at positive displacement mudpumps and always enhances signal detection.*

6.4.2   High data rate effects.

At higher data rates, that is, those achieved with rapidly varying sinusoidal or periodic waveforms, this is not the case. Signals do not necessarily interfere constructively because of phase differences.

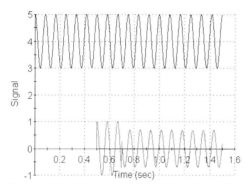

Figure 6.4b. Pressure versus "t" at given "x."

In Figure 6.4b, model PSURF*.FOR is used to illustrate our point. The clean, constant frequency, blue signal enters the drillpipe and the red signal is measured uphole (zero attenuation is assumed in this example). At first, silence (zero pressure) is found because the signal has not reached the standpipe transducer. Then the wave arrives with an amplitude initially identical to that of the blue trace. This signal proceeds to the mudpump and reflects. In this case, it diminishes the pressure measured at the standpipe due to destructive interference arising from phasing effects. Short wavelengths make signal processing more difficult. In this example, the (upper) blue line represents the intended signal, but it is the (lower) red trace that is actually recorded at the standpipe transducer. Note the kink in the (bottom) red curve at t = 0.7 sec. The multiple transducer methods of Chapter 4, of course, will remove the foregoing destructive interference effects to recover the blue trace.

6.5 Applications to Negative Pulser Design in Fluid Flows and to Elastic Wave Telemetry Analysis in Drillpipe Systems

We have studied MWD signal analysis and reflection cancellation assuming dipole sources, that is, signal generators such as positive pulsers and mud sirens which create disturbance pressures whose polarities are antisymmetric with respect to source position, and additionally, assuming fluid flow systems. We emphasize that the methods are equally applicable, with some re-interpretation of variable names, to negative pulser design, and then, to both fluid and elastic systems. These notions and applications are developed next.

The ideas behind "conjugate harmonic functions" are well documented in the theory of elliptic partial differential equations, the best known and simplest application being that for Laplace's equation. In short, whenever the equation $U_{xx} + U_{yy} = 0$ holds, there exists a complementary model $V_{xx} + V_{yy} = 0$ describing a related physical problem; these are connected by the so-called Cauchy-Riemann conditions $U_x = V_y$ and $U_y = -V_x$. These relationships lie at the heart of the theory of complex variables.

For example, the "velocity potential" describing ideal, inviscid, irrotational flow past a two-dimensional airfoil satisfies Laplace's equation; the complementary model for the "streamfunction" describes the streamline pattern about the same airfoil. These ideas have been used by this author to study steady-state pressure distributions, torque characteristics and erosion tendencies associated with MWD mud sirens (Chin, 2004), while detailed applications to Darcy flows in petroleum reservoirs are developed in Chin (1993, 2002). The siren application is developed in detail in Chapter 7 of this book.

The conjugate function approach, though, has never been applied to hyperbolic equations. However, simple extensions for the classical wave equation used in this book to model the acoustic displacement function $u(x,t)$ lead to powerful practical implications. A rigorous derivation is easily given. Recall that for long waves, $u(x,t)$ satisfies $\partial^2 u/\partial t^2 - c^2 \partial^2 u/\partial x^2 = 0$, which can be rewritten in the form $\partial(\partial u/\partial t)/\partial t - c^2 \partial(\partial u/\partial x)/\partial x = 0$ (again, x is the direction of propagation, t is time and c is the speed of sound). We introduce, without loss of generality, a function $\phi(x,t)$ defined by $\partial\phi/\partial x = \partial u/\partial t$ and $\partial\phi/\partial t = c^2 \partial u/\partial x$. This merely restates the identity $\partial^2\phi/\partial t\partial x = \partial^2\phi/\partial x\partial t$. However, the definition importantly implies that $\phi(x,t)$ satisfies $\partial^2\phi/\partial t^2 - c^2 \partial^2\phi/\partial x^2 = 0$. In other words, the function ϕ likewise satisfies the wave equation. But what do mathematical boundary value problems similar to those for $u(x,t)$, for which we have already developed numerical solvers, model in engineering practice?

Consider first the formulation $\partial^2 u/\partial t^2 - c^2 \partial^2 u/\partial x^2 = 0$, with the jump or discontinuity $[\partial u/\partial x]$ specified through the source position (that is, a "delta-p" function of time is prescribed at the pulser) and then $\partial u/\partial x = 0$ at the drillbit (a uniform pipe is assumed and outgoing wave conditions are taken at $x = \infty$). This is the dipole formulation previously addressed, assuming an opened acoustic end, for which we already have a solution algorithm. Now we ask, "What problem is physically solved if we simply replace 'u' by 'φ' in the formulation?" Thus, we wish to interpret the formulation $\partial^2\phi/\partial t^2 - c^2\ \partial^2\phi/\partial x^2 = 0$, with the jump $[\partial\phi/\partial x]$ specified through the source position and $\partial\phi/\partial x = 0$ at the drillbit. At the drillbit, $\partial\phi/\partial x = 0$ implies that $\partial u/\partial t = 0$, that is, a solid reflector. From our definitions, a jump in $[\partial\phi/\partial x]$ is simply a jump in the velocity $\partial u/\partial t$. In other words, we have a discontinuity in axial velocity, as one might envision for accordion motion or for a pulsating balloon, which models a negative pulser. Once the solution for $\phi(x,t)$ is available, local pressures would be calculated by evaluating $\partial\phi/\partial t$ and then re-expressing the result in terms of $\partial u/\partial x$, which is, of course, proportional to the acoustic pressure. That is, our displacement dipole formulation for open drillbits is identical to that for monopoles with solid reflectors – the latter solution is a "free" byproduct of the first*!*

To understand this, in perspective, recall that a dipole source (that is, positive pulser or mud siren) is associated with antisymmetric disturbance pressure fields and velocities continuous through the source point, while a monopole source (or negative pulser) is associated with a symmetric disturbance pressure fields and velocities discontinuous through the source. Contrary to popular engineering notions, it is not necessary to have a nonzero "delta-p" in order to have MWD signal generation; in fact, the "delta-p" associated with negative pulser applications is identically zero whatever the valve motion.

The slightly different formulation $\partial^2 u/\partial t^2 - c^2\ \partial^2 u/\partial x^2 = 0$, with the jump $[\partial u/\partial x]$ specified through the source position (again, a "delta-p" function of time prescribed at the pulser) and $\partial u/\partial t = 0$ at the drillbit (that is, a solid reflector assumption) has also been addressed previously and a numerical solution algorithm is already available. If we replace 'u' by 'φ,' what does the resulting boundary value problem solve? The only difference from the foregoing formulation is '$\partial\phi/\partial t = 0$' at the bit. Now, $\partial\phi/\partial t = 0$ implies, per our definitions, that $\partial u/\partial x = 0$. In other words, the formulation for φ solves for the pressures associated with negative pulsers when the drillbit satisfies a zero acoustic pressure, open-ended assumption. Pressures are obtained as before.

The above paragraphs demonstrate how solutions to our dipole source formulations (for positive pulsers and sirens) under a Lagrangian displacement description provide "free" solutions for monopole formulations for negative pulser problems without additional work other than minor re-interpretation of the pertinent math symbols. However, the extension can be interpreted much more broadly even outside the context of mud pulse telemetry. The above approaches and results also apply directly to MWD telemetry applications where the transmission mechanism involves axial elastic wave propagation through drillpipe steel. A dipole source would model, say, piezoelectric plates "oscillating back and forth," while a monopole source might model piezoelectric transducers stacked so that they "breathe in and out, much like pulsating balloons." For elastic wave applications, the above analogies can be used to develop models for complicated bottomhole assemblies with multiple changes in bottomhole assembly and borehole geometry and acoustic impedance along the path of signal propagation – models analogous to the six-segment waveguide formulation in Chapter 2 and the two-part waveguide approaches in Chapters 3 and 5 are easily developed for elastic systems. This is obvious because the only formulation differences are changes required at acoustic impedance discontinuities located at drillpipe and drill collar junctions.

For the fluid flows considered earlier, we assumed that acoustic pressure and volume velocity were continuous, that is, $(\partial u/\partial x)_{collar} = (\partial u/\partial x)_{drillpipe}$ and $A_{collar} (\partial u/\partial t)_{collar} = A_{drillpipe} (\partial u/\partial t)_{drillpipe}$ hold. These would be replaced by continuity of displacement and net force, that is, $(\partial u/\partial t)_{collar} = (\partial u/\partial t)_{drillpipe}$ and $A_{collar} (\partial u/\partial x)_{collar} = A_{drillpipe} (\partial u/\partial x)_{drillpipe}$. In terms of $\phi(x,t)$, we would have $(\partial \phi/\partial x)_{collar} = (\partial \phi/\partial x)_{drillpipe}$ and $A_{collar} (\partial \phi/\partial t)_{collar} = A_{drillpipe} (\partial \phi/\partial t)_{drillpipe}$. In other words, the models and numerical solutions developed in Chapters 2-5 can be used without modification provided the dependent variables are interpreted differently! Our discussion, for simplicity, assumes that the moduli of elasticity for drillpipe and drill collar are identical, but extensions to handle differences are easily constructed. The wave equation transforms used here to develop our physical analogies were originally introduced by the author in Chin (1994). It is important to note that, while "drillpipe acoustic" methods work in vertical wells, they perform poorly in deviated and horizontal wells where rubbing of the drillstring with the formation is commonplace.

6.6   LMS Adaptive and Savitzky-Golay Smoothing Filters
(software reference, all of the filters in Sections 6 and 7 are found in C:\MWD-06)

As explained in our chapter objective, our aim is not an exhaustive treatment of standard signal processing, but rather, a concise one which directs readers to more detailed publications, e.g., Stearns and David (1993) and Press *al et* (2007). Exceptions are Sections 1-5 above, which explain relevant downhole concepts in detail. Of the more complicated methods available, LMS (least mean squares) adaptive filters provide some degree of flexibility in applications with slowly varying properties. Figure 6.6a shows a raw unprocessed wave signal with random noise and a propagating wave, while the LMS processed waveform remarkably appears in Figure 6.6b. Smoothing filters may need to be applied to data such as that in Figure 6.6b, or to noisy datasets such as the one in Figure 6.6c. The Savitzky-Golay smoothing filter, for instance, removes high-frequency noise in the top curve to produce the lower frequency red line shown at the bottom.

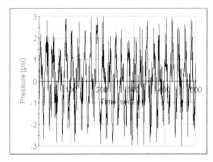

Figure 6.6a. Raw unprocessed wave signal
with random noise and propagating wave.

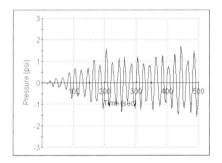

Figure 6.6b. LMS processed signal, with random noise
removed by adaptive filtering.

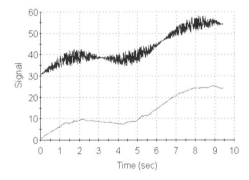

Figure 6.6c. Savitzky-Golay smoothing filter.

6.7 Low Pass Butterworth, Low Pass FFT and Notch Filters

The capabilities shown in this final section are standard and available with numerous software tools. We provide typical results without further comment.

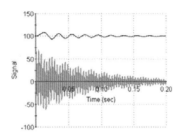

Figure 6.7a. Low-pass Butterworth filter.

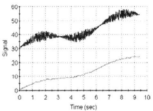

Figure 6.7b. Low-pass FFT filter.

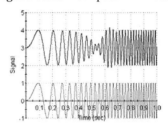

Figure 6.7c. Notch filter.

Software reference. Fortran source code is available for the five filters in Sections 6 and 7 in C:\MWD-06 under obvious filenames. These, together with the algorithm codes in Chapters 2-5, provide the basic tools for high-data-rate signal processor testing and design. It goes without saying, of course, that field data is all-important, and should be collected with a "wave orientation." In other words, careful attention should be paid to bottomhole assembly details, borehole annular geometry, drillbit type, surface setup and mud sound speed, noting that the latter may vary along the drillstring as pressure and temperature conditions change.

6.8 Typical Frequency Spectra and MWD Signal Strength Properties

We have discussed the physical properties of typical noise components in the drilling channel.   It is of interest, from an experimental perspective, to examine typical MWD signals and how they might appear together with drilling anomalies.   Figure 6.8a displays "clean" frequency spectra for a siren pulser operating at a constant frequency f at both low and high flow rates.   Because the sound generation mechanism and fluid-dynamical equations are nonlinear, harmonics at $2f$, $3f$, $4f$ and so on, will also be found, as shown.   At the present time, there is no clear method for their prediction and removal.   Higher harmonics are associated with inefficient signal generation and also complicate surface signal processing.   It is speculated that "swept" siren rotors might minimize harmonic generation, but further study is required.   The results in Figure 6.8a were obtained in a wind tunnel and are "clean" in that they are free of real noise.

On the other hand, the sketch in Figure 6.8b (from an unreferenced Schlumberger source) shows a siren operating at 12 Hz in mud under typical conditions.   Note the existence of pump noise and mud motor noise.   Because mudpump noise propagates in a direction opposite to the upgoing MWD signal, it can be effectively removed using directional multiple-transducer surface signal processing techniques, e.g., as shown in Figures 4.4b,c.   These would eliminate the pump spectra in Figure 6.8b.   On the other hand, mud motor noise, which travels in the same direction as the upgoing MWD signal, would require frequency-based filtering, effective only if the siren frequency were different from that of the mud motor.   In practice, MWD frequencies should always be chosen so that they differ from that of the motor to facilitate noise removal.

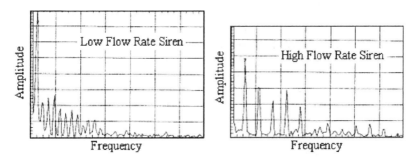

Figure 6.8a. Signal strength harmonic distribution.

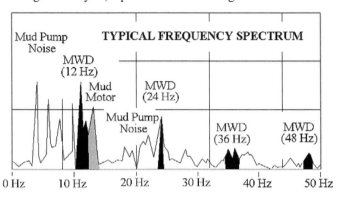

Figure 6.8b. Typical frequency spectrum for 12 Hz carrier signal.

The physical properties of the MWD source signal Δp are also of interest. There is lack of clarity on signal strength as a function of frequency at the present time. For example, Montaron, Hache and Voisin (1993) and Martin *et al* (1994) suggest that the amplitude of the pressure pulse created is roughly independent of frequency at higher frequencies. On the other hand, Su *et al* (2011) shows experimentally that Δp decreases with increasing frequency when the flow rate is fixed. Additional investigation is needed to determine the exact dependence of signal strength on frequency and flow rate. This empirical knowledge would supplement the wave interaction models developed in Chapters 2-5.

Other sources of "noise," or more accurately, uncertainty, arise in signal processing. Boundary condition differences associated with "hard versus soft rock" may be responsible for "closed versus open" downhole end reflections that confuse analysis. At other times, reports do not distinguish between positive versus negative pulsers, which create completely different reflection and wave propagation patterns. And still, some rig operators have indicated "loss of signal" at higher depths that can be attributed to mechanical malfunctions under high pressure – effects that have nothing to do with signal processing. All of these are valid concerns, but because of their diverse nature, illustrate why the design of a general signal processor very challenging.

6.9   References

Oppenheim, A.V. and Schafer, R.W., *Digital Signal Processing*, Prentice-Hall, New Jersey, 1975.

Oppenheim, A.V. and Schafer, R.W., *Discrete-Time Signal Processing*, Prentice-Hall, New Jersey, 1989.

7

Mud Siren Torque and Erosion Analysis

Three-dimensional flowfields related to the mud sirens used in Measurement-While-Drilling are studied using a comprehensive inviscid fluid-dynamic formulation that models the effects of key geometric design variables on rotor torque. The importance of low torque on high-data-rate telemetry and operational success is discussed. Well known field problems are reviewed and aerodynamically based solutions are explained in detail. Both problems and solutions are then studied numerically and the computer model – developed using flow concepts known from aerospace engineering – is shown to replicate the main physical features observed empirically. In particular, the analysis focuses on geometries that ensure fast "stable-opened" rotary movements in order to support fast data transmissions for modern drilling and logging operations. This chapter, which extends work first presented in Chin (2004), also addresses erosion problems, velocity fields, and streamline patterns in the steady, constant density flow limit. Studies related to drillpipe mud acoustics, signal propagation and telemetry, where transients and fluid compressibility are important, have been presented earlier in this book.

7.1   The Physical Problem

In drilling longer and deeper wells through unknown, hostile, and high-cost offshore prospects, the demand for real-time directional and formation evaluation information continues to escalate. Because wireline logging cannot provide real-time information on pre-invaded formations from deviated, horizontal, and multilateral wells, Measurement-While-Drilling and Logging-While-Drilling tools are now routinely used instead.

177

Possible transmission methods are several in variety, e.g., mud pulse, electromagnetic, and drillpipe acoustic. These methods send encoded information obtained from near-bit sensors to the surface without the operational difficulties associated with wireline methods. In this chapter, we introduce and solve special problems associated with the most popular, namely, mud pulse telemetry, which presently experiences wide commercial application. In particular, we consider special design issues important to "mud siren" signal sources, which unlike competing positive and negative pulsers, enable the highest data rate and the potential for even faster transmissions.

A vintage 1980s mud siren is shown in Figure 7.1 with its key elements housed in the tool drill collar. Water hammer dominated pressure pulses are created by the moving *upstream rotor* while rotating about the stationary *downstream stator*, as it interrupts the downward flow of drilling mud. In the schematic, the resulting pressure wave is shown propagating upward, although, of course, an equally strong signal of opposite sign propagates downward. If the rotation continues at constant rate without change, the only wave form created is a periodic one, which obviously contains no useful information.

Encoded '0' and '1' data is transmitted by non-periodic siren (also referred to as "valve") movements using any number of telemetry schemes, e.g., "phase shift keying," "frequency shift keying," "pulse width modulation," "pulse amplitude modulation," and so on. This chapter does not deal with telemetry issues and modulation schemes, which have been addressed previously. Instead, it focuses on the fundamental problem of efficient mechanical mud siren source design. Here, the basic design problems and their solutions, and fast and efficient computational methods needed to assess the practical viability of new siren pulsers, are considered in detail.

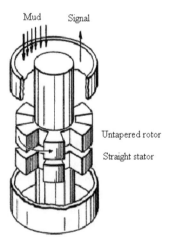

Figure 7.1. Early-to-late 1980s "stable-closed" design.

7.1.1   Stable-closed designs.

The model in Figure 7.1, which is "obvious" in concept, causes significant operational problems. Rock particles and other debris present in the drilling fluid tend to lodge between the axial clearance or gap between the rotor and the stator, thus impeding rotary motion. Usable gap distances are almost always small, say much less than 1/8 inch, since larger spacings do not create detectable surface signals. This "1/8 inch rule" is known from simple valve tests, e.g., larger gaps produce very small water hammer pressures (strong areal blockage is required to produce the plane waves that ultimately reach the surface). When rotary motion is stopped, the rotor surprisingly stops in the closed position: the solid lobes of the upstream rotor completely block the opened ports of the downstream stator, with this closed position completely stable and resistant to any attempts to re-open the valve.

This stable-closed behavior has several undesirable operational consequences. (1) High pressures developed in the drillpipe induce the mudpump seals to break, causing possible pump damage and introducing a rig floor safety hazard. (2) Excessive pressures may fracture or damage the formation, and most definitely, increase unwanted invasion. (3) High pressures will damage the MWD tool, while high flow rates through the narrow gap will severely erode the rotor and stator. (4) The need to remove the tool, especially from deep wells, means a loss of expensive offshore rig time, not to mention additional formation invasion while tool retrieval is in progress.

7.1.2   Previous solutions.

In the 1970s, several methods were developed to solve this problem. The jamming problem was first discussed in detail in U.S. Patent No. 3,770,006. The inventors noted that ". . . in logging-while-drilling tools of this type, the signal generating valve normally develops certain hydraulic torque characteristics as a function of the flow rate through the valve which tend to force the valve to its closed position. This creates problems as drilling mud is pumped down the drill string and through the valve before the tool begins operation and the motor begins to power the valve." This is crucial when power is produced by a mud turbine downstream of the valve, of course, and less crucial for battery powered tools.

A simple mechanical solution is given. Essentially, "a spring means is included in the tool which has sufficient force to bias the rotor upwardly away from the stator when a low rate of flow is passing down the drill string." Recognizing that the "gap in the tool must be relatively small during operation of the tool in order for the signal generated by the valve to have sufficient strength to reach the surface," the authors proceed to note how as "the flow rate increases, the pressure drop across the rotor also increases. When the pressure drop exceeds the force of the spring means, the rotor moves downward toward the stator which establishes the gap necessary for satisfactory operation of the tool." Finally, "by allowing the gap to be large during the time the tool is not operating, the passages available for flow through the valve during this time are increased and the problem of plugging the valve is substantially decreased."

Again, the plugging or jamming is a fluid-dynamic or aerodynamic characteristic of the mud siren. It is not necessarily caused by debris or lost circulation material in the mud, as the above discussion emphasizes – *jamming of the valve is possible even in clean water!* U.S. Patent 3,792,429 provides an alternative solution. "The tool includes a means for both biasing the valve toward an open position and holding it there when the tool is not operating and for canceling or substantially reducing the torque loads applied to the drive train of the tool when the tool is operating. This means comprises a magnetic unit which develops a magnetic torque characteristic which opposes the normal hydraulic torque characteristics of the valve." The authors point out additional problems associated with this jamming. "Due to the composition of standard drilling mud, solid material is normally present therein which tends to strain out of the mud as it is forced through restricted passages in the valve which are present when the valve is in its closed position. This solid material may continue to collect in the valve and does present a real problem in that it may plug the valve to such a extent that the valve cannot be opened by the motor when operation of the tool is commenced."

The proposed solution is a means that "is comprised of a magnetic unit which has a magnet attached to the tool housing and a cooperating magnetic element attached to the drive train of the tool. The unit develops magnetic torque characteristics which are greater than the hydraulic torque characteristic of the valve when the tool is not operating so that the valve will be biased toward and held in its open position when the tool is not operating." It is noted that "the hydraulic torque characteristic of the valve is an increasing function of the flow rate through the valve. Since the maximum flow rate will normally occur during operation of the tool, the torque characteristic of the unit is designed to be roughly equal to the hydraulic torque at this operating flow rate. By positioning the unit so that the magnetic torque is 180° out of phase with the hydraulic torque, the resulting torque applied to the drive train at any time during operation will be negligible."

Although this solution is reasonable, it implies significant operational difficulties and added cost: magnets affect navigational measurements and requires additional shielding offered by expensive lengths of nonmagnetic drill collar. U.S. Patent No. 3,867,714 provides still another solution, namely, a torque producing turbine upstream of the rotor designed to maintain an open position using hydraulic torque drawn from the mudstream. From the Abstract, "a mud conditioning means comprising a jet and a spinner is positioned in the drill string above the valve of said tool wherein said means imparts angular motion to at least a portion of the drilling fluid in such a manner that the power hydraulically developed by the valve as the mud flows therethrough will be a desired function of the flowrate and/or density of said mud."

7.1.3   Stable-opened designs.

The above solution methods are "brute force" in nature because they address the symptoms and not the cause of the root problem. The primary reason for stable-closed behavior is an aerodynamic one: the rotor will naturally close by itself (when turning torque is not supplied through the shaft) even when the fluid is clean and free of debris. Of course, the presence of debris worsens the problem. In other words, the torque acting on the rotor is such that it will always move to the most stable position, which happens to be closed. In airplane design, commercial jetliners are engineered so that, despite wind gusts and turbulence, aircraft always return to a stable horizontal cruise configuration. By analogy, a safe stable-opened position is the objective of good mud siren re-design. But how is this achieved?

A completely aerodynamic solution to the stable-closed problem was developed and reported by Chin and Trevino (1988). In this re-design, the stator is located upstream while the rotor is placed downstream. As illustrated in Figure 7.2, the rotor also contains several important physical features. Importantly, (1) its sides must be "slightly" tapered, (2) the azimuthal width of its top should be "a bit" less than that of the stator bottom, and (3) rotor-stator gap distance should not be "too small." While stable-open behavior is desirable, even a stable-closed characteristic is tolerable provided opening hydraulic torques are small. Torques needed to open closed valves for sirens of the type in Figure 7.2 are much smaller than those for Figure 7.1 for the same flow rate.

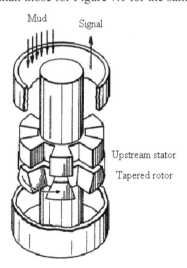

Figure 7.2. New 1990s "stable-opened" improvement.

Figure 7.3. A "stable-closed" four lobe siren used in flow loop tests.

These design considerations are validated from numerous (literally *hundreds* of) CNPC wind tunnel tests and verified in water and mud flow loops. To what extent each of (1) – (3) is necessary, and in which combinations, of course, depends on geometric details, but these three items appear to be the main relevant design parameters for the basic siren configuration in Figure 7.2. We emphasize that, despite our appreciation for the roles of these variables, the financial and time costs for a realistic test matrix are still substantial.

Having identified these parameters as pertinent from the aerodynamic standpoint – and demonstrated that their effects are repeatable from a testing perspective - we next ask if fast and efficient computational methods can be developed to quickly identify not only new design principles, but to provide engineering trends and details on closing and opening torques, optimum numbers and sizes of lobes for a given flow rate and drill collar size, pressure magnitudes, surface velocity predictions for erosion estimates, and so on.

7.1.4 Torque and its importance.

For high-data-rate telemetry, low rotor torque in addition to high signal amplitude and frequency are required. Originally the work of Chin and Trevino (1988) focused on achieving stable-opened designs, but with recent requirements for high-data-rate telemetry on the horizon, research is focusing on developing designs that not only do not jam, but which are extremely low in torque as well. Why is low torque of paramount importance? There are three principal reasons. Low torque (1) reduces jamming tendencies of debris temporarily lodged in the rotor-stator gap, (2) allows mud sirens to modulate signals faster and achieve higher data rates, and (3) implies lower power requirements and thus decreases erosion incidence in downhole turbines. Additionally, (4) low power consumption allows additional sensors to be operated, while (5) low torque provides additional flexibility in selecting optimal telemetry schemes – that is, data rate increases can also be realized from signal processing advantages and not mechanical considerations alone.

7.1.5   Numerical modeling.

The above needs spurred the development of a three-dimensional computational system that quantifies not only the relationships needed between the taper, azimuthal width and rotor-stator gap constraints summarized above, but also between new design parameters such as annular convergence and divergence in the inner drill collar wall and central hub space, both leading toward and away from the siren assembly. The matrix of experimental tests needed to validate any mechanical design can be significantly reduced by identifying important qualitative trends by computer simulation.

In very early work, the geometry shown, say in Figure 7.3, was "unwrapped" azimuthally and solved by modeling the two-dimensional planar flow past a row of periodic, block-like, rotor-stator cascades. While this approach is standard and very successful in aircraft turbine and compressor design, the method led to only limited success because the three-dimensional character of the radial coordinate was ignored. In aircraft applications, the radial extent of a typical airfoil blade is very small compared to the radius of the central hub, e.g., much less than 10% and even smaller. In downhole tools, geometrical constraints and mechanical packaging requirements increase this ratio to approximately one-half, making the above "unwrapping" questionable at best: centrifugal effects cannot be ignored. Thus, the method that is described in this chapter was required to be accurate physically, as well as computationally fast and efficient, while retaining full consistency with the experimental observations identified in Chin and Trevino (1988). Here, the motivation, mathematical model and numerical solution, together with detailed computed results for streamline fields, torques, pressures, and velocities, are described.

7.2   Mathematical Approach

Flows past mud sirens, even stationary ones, are extremely complicated. Present are separated downstream flows and viscous wakes even when the bluntly shaped lobes are fully open and not rotating. Such effects cannot be modeled without empirical information. Flow prediction in downstream base regions is extremely challenging, e.g., the viscous flow behind a simple slender cone defies rigorous prediction even after decades of sophisticated missile research in the aerospace industry. Some physical insight into certain useful properties is gained from classical inviscid airfoil theory (Ashley and Landahl, 1965). In calculating lift (the force perpendicular to the oncoming flow), viscosity can be neglected provided the flow does not separate over the surface of the airfoil. This assumption applies at small flow inclinations. However, viscous drag (parallel to the flow) and separation effects cannot be modeled without using the full equations; at small angles, of course, boundary layer theory is used to estimate resistance arising from surface shear stresses. Theoretical versus experimental lift results for the NACA 4412 airfoil in Figure 7.4, for example, show excellent agreement prior to aerodynamic stall.

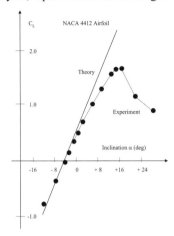

Figure 7.4. Typical inviscid lift coefficient versus inclination.

By analogy, we expect that the torque acting on siren rotors and stators – controlled principally by the upstream attached flow – can be calculated accurately using inviscid theory. (As in airfoil analysis, torque is perpendicular to the direction of the oncoming flow.) This is motivated by laboratory and field experience: the stable-closed or stable-open character of any particular siren design is independent of flow rate, viscosity, mud weight, or water versus oil mud type, i.e., it is largely unaffected by rheology. This is borne out by qualitative and quantitative experimental comparisons to be discussed. Here, because the work focuses on torque at low oncoming speeds, or more precisely, low Mach numbers, fluid compressibility is neglected. Compressibility, of course, is important to signal generation and propagation, subjects already treated in this book. To simplify torque analysis, we restrict ourselves to steady flow and examine its "static stability" as stationary siren rotor and stator sections are altered with varying degrees of closure. This philosophy is adopted from the classical approach used in airplane stability analysis and design.

We emphasize that separated viscous downstream flows and streamwise pressure drops are not described adequately in this approach. Highly empirical methods are instead needed. These flows are affected by rheology, and pressure drops at high flow rates can range in the hundreds of psi's (numbers quoted are qualitative and intended to convey "ballpark" estimates only). Engineers new to mud siren design often equate large pressure drops or "delta-p's" with strong MWD signals. This is not the case. A localized static pressure drop does not propagate and transmit information. Only dynamic, acoustic components of time-dependent pressures – water hammer signals, for instance – are useful in data transmission. For positive and negative pulsers, this propagating signal can exceed 200-300 psi at high flow rates, although the power needed to generate such signals are enormous. Sirens typically create acoustic Δp's or peak-to-peak pressures below 100 psi at the source, which is not necessarily bad from a telemetry viewpoint. This is so because continuous wave signaling schemes permit lower probabilities of bit error and more efficient data transmission.

7.2.1   Inviscid aerodynamic model.

If viscous stresses are ignored, fluid motions are governed by Euler's equation $D\mathbf{q}/Dt = -1/\rho \nabla p$, where D/Dt is the convective derivative, $\mathbf{q}$ is the Eulerian velocity vector, ρ is the constant mass density, and p is the static pressure. This applies to all coordinate systems. In practical applications, physical quantities related to directions perpendicular to the oncoming flow, e.g., lift on airfoils, radial forces on engine nacelles, and torque on turbine and compressor blades – and torque on siren stator and rotor stages – can be modeled and successfully predicted using inviscid models.

Parallel forces associated with viscous shear, however, require separate "boundary layer flow" analyses where inviscid pressures are impressed across thin viscous zones. Such are the approaches used in calculating drag when flows are streamlined. Even when flows are strongly separated, fluid characteristics upstream of the separation point can be qualitatively studied by inviscid flow models. Of course, details related to the separated region must be examined by alternative, often empirically-based methods, not considered here.

When viscosity is neglected and the far upstream flow is uniform, the fluid motion is said to be irrotational and satisfies the kinematic constraint $\nabla \times \mathbf{q} = 0$. This allows $\mathbf{q}$ to be represented as the gradient of a total velocity potential ϕ,

$$\mathbf{q} = \nabla \phi \qquad (7.2.1)$$

so that

$$\nabla^2 \phi = 0 \qquad (7.2.2)$$

by virtue of mass conservation, that is, $\nabla \cdot \mathbf{q} = 0$. A consequence of Euler's equation is "Bernoulli's pressure integral," which takes the general vector form

$$p + \tfrac{1}{2}\rho \,|\nabla\phi|^2 = p_0 \qquad (7.2.3)$$

where p_0 is the stagnation pressure determined completely from upstream conditions.

Again, these equations apply three-dimensionally to all coordinate systems. In inviscid single-airfoil formulations, the boundary value problem associated with Equation 7.2.2 is first solved subject to geometric constraints, and corresponding surface pressures are later calculated using Equation 7.2.3. This equation, we emphasize, does not apply to the viscosity-dominated downstream wake. For further details about inviscid flow modeling, its applications and limitations, consult the classic book by Batchelor (1970).

7.2.2   Simplified boundary conditions.

Geometric complexities arising from the three-dimensional blunt body character of the siren lobes preclude exact analysis. Thus we are led to examine approximate but accurate "mean surface" approaches for fluid-dynamical modeling. One successful method used in aerodynamics is "thin airfoil theory," which is classically used to predict lift, the force perpendicular to the direction of the oncoming flow. This model, summarized in Figure 7.5, is discussed and solved in the book of Ashley and Landahl (1965), with Marten Landahl at M.I.T., this author's doctoral thesis advisor, being one of its principal developers.

In modeling flows past two-dimensional airfoils, as noted in the upper diagram, an exact tangent flow kinematic condition applies at the geometric surface itself, while a "Kutta condition" related to smooth downstream flow applies at the trailing edge (this mimics viscous "starting flow" effects). In the approximate model, shown in the lower diagram of Figure 7.5, the tangent flow condition is replaced by a simpler boundary condition (setting the ratio of vertical to horizontal velocities to the local airfoil slope), and applied along a mean surface $y = 0$ (y is the vertical coordinate perpendicular to the oncoming flow). These assumptions form the basis of thin airfoil theory, used successfully for most of the twentieth century in aerodynamic design. When planform (e.g., wing areal layout) effects are important, three-dimensional geometric boundary conditions are traditionally evaluated on a mean flat surface; this method forms the basis for classical lifting surface theory. In all these methods, the evaluation of forces perpendicular to the direction of flow is very accurate, while parallel forces (related to viscous effects) require separate boundary layer or empirical separated flow corrections.

Correspondingly, an analogous model can be designed for nacelles, which house the engine turbomachinery components installed beneath airplane wings. It is known that the presence of the engine can improve or degrade the aerodynamics of an optimally designed "wing alone" flow. Thus, one objective of nacelle design is favorable aerodynamic performance, if possible, to offset any unfavorable interference effects. In this approach, outlined in Figure 7.6 for baseline axisymmetric nacelles, exact tangent flow kinematic conditions applied on the nacelle surface are replaced by approximate conditions along a mean cylinder with constant radius $r = R$, noting that r is the radial coordinate perpendicular to the oncoming flow. Not shown, for clarity only, are the internal actuator disks used to model energy addition and pressure increase due to the presence of engine turbomachinery.

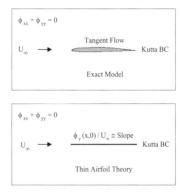

Figure 7.5. Exact model and thin airfoil theory.

The axisymmetric model was motivated by cylindrical coordinate methods used to model flows past aircraft fuselages, rocket casings and projectiles. The ideas have been extended to three-dimensional nacelle flows, solving Equation 7.3.1 below, which include the effects of lower "chin inlets" containing gear boxes and azimuthal pressure variations accounting for wing-induced effects. A schematic showing how a sub-grid nacelle model is embedded within the framework of the complete rectangular-coordinate airplane model is given in Figure 7.7. Iterations are performed between sub and major grid systems until the computer modeling converges. The successful approaches in Figures 7.6 and 7.7 were developed by this author for Pratt and Whitney Aircraft Group, United Technologies Corporation, in the late 1970s, and are described in the aerodynamics literature; for example, see Chin *et al* (1980, 1982).

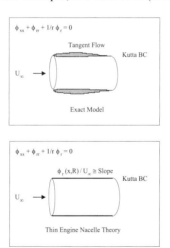

Figure 7.6. Exact model and approximate thin engine nacelle theory.

Figure 7.7. Three-dimensional mean surface approach.

7.3 Mud Siren Formulation

Aerospace engineering experience shows that small geometric details – say, contours selected for airfoils and ailerons – greatly affect aerodynamic performance, e.g., changes to quantities like moment, pressure distributions, lift, flow separation point, pitch stability, and so on. End results are not apparent to the naked eye and must be modeled rigorously. For instance, the radically altered behavior going from Figure 7.1 to Figure 7.2 arose principally from the effect of small streamwise rotor tapers, whose end effects could not have been anticipated a priori. One can only surmise the effects of different combinations of tapers in stators, rotors, and upstream and downstream annular passages.

Practical computational concerns require geometric simplification, but these cannot be made at the expense of incorrect physical modeling. In this section, we adopt the philosophy suggested in Section 2, namely, that geometric boundary conditions can be successfully modeled along mean lines and surfaces while retaining the three-dimensional partial differential equation in its entirety, as suggested in Figure 7.7. With this perspective and philosophical orientation, we correspondingly assume a cylindrical coordinate system for analysis, that is, the ones implied by Figures 7.8 and 7.9.

From aerospace analogies, we expect that the torque acting on the siren lobe (again, perpendicular to the direction of flow) can be accurately predicted since it is inviscidly dominated and is largely independent of shearing and rheological effects. This is particularly so because strong areal convergence at the lobes precludes local separation. On the other hand, viscous pressure drops in the streamwise direction and downstream separated flows cannot be modeled using inviscid theory. With these limitations in mind, we proceed with a comprehensive three-dimensional formulation.

7.3.1 Differential equation.

In cylindrical radial coordinates, Laplace's equation $\nabla^2\phi = 0$ takes the form given by

$$\phi_{xx} + \phi_{rr} + 1/r\,\phi_r + 1/r^2\,\phi_{\theta\theta} = 0 \qquad (7.3.1)$$

where x is the axial streamwise variable, r is the radial coordinate, and θ is the azimuthal angle. It is possible to solve this three-dimensional partial differential equation computationally, but the numerical model would require significant computer resources and not be useful for real-time engineering design.

Thus we ask if a simple approach embodying three-dimensionality can be developed with the convergence speed of faster two-dimensional methods. The key is an integral approach not unlike the integral methods and momentum models used years ago to solve the viscous equations for aerodynamic drag, which is fast yet rigorous mathematically. The procedure is straightforward. We multiply Equation 7.3.1 by $2\pi r$, integrate $\int \ldots dr$ radially over the radial limits $R_i < r < R_o$ and introduce the area-averaged velocity potential

$$\Phi(x,\theta) = A^{-1} \int \phi(x,r,\theta) \, 2\pi r \, dr \qquad (7.3.2)$$

Here, R_i is the inner hub radius at the bottom of a siren lobe, R_o is the inner drill collar radius at the top of the lobe, and A is a reference area to be defined. Then

$$\Phi_{xx} + 2\pi A^{-1} (r\phi_r) \Big|_{R_i}^{R_o} + A^{-1} \int_{R_i}^{R_o} 1/r^2 \, \phi_{\theta\theta} \, 2\pi r \, dr = 0 \qquad (7.3.3)$$

Now approximate the r in $1/r^2$ by the mean value $R_m = \tfrac{1}{2}(R_i + R_o)$ to obtain

$$\Phi_{xx} + 1/R_m^2 \, \Phi_{\theta\theta} \cong - 2\pi A^{-1} (r\phi_r) \Big|_{R_i}^{R_o} \cong 2\pi A^{-1} \{R_i \, \phi_r(x,R_i,\theta) - R_o \, \phi_r(x,R_o,\theta)\} \qquad (7.3.4)$$

Next observe that $\phi_r(x,R_i,\theta)/\phi_x(x,R_i,\theta)$ represents the ratio of radial to streamwise velocities at the inner surface; kinematically, it must equal the geometric slope $\sigma_i(x,\theta)$. Thus, $\phi_r(x,R_i,\theta) = \sigma_i \, \phi_x(x,R_i,\theta)$ at the inner radial surface. Similarly, $\phi_r(x,R_o,\theta) = \sigma_o \, \phi_x(x,R_o,\theta)$ for the outer surface, so that

$$\Phi_{xx} + 1/R_m^2 \, \Phi_{\theta\theta} \cong 2\pi A^{-1} \{R_i \, \sigma_i \, \phi_x(x,R_i,\theta) - R_o \, \sigma_o \, \phi_x(x,R_o,\theta)\} \qquad (7.3.5a)$$

From thin airfoil theory, the complete horizontal speed ϕ_x is approximated by the oncoming flow speed U_∞, which is permissible away from the siren lobes themselves where flow blockage is significant, so that, in the absence of azimuthal variations, Equation 7.3.5a becomes

$$\Phi_{xx} + 1/R_m^2 \, \Phi_{\theta\theta} \cong 2\pi A^{-1} U_\infty \{R_i \, \sigma_i - R_o \, \sigma_o \} \qquad (7.3.5b)$$

Note that the right side represents a non-vanishing distributed source term when general annular convergence or divergence is allowed. In spaces occupied by solid lobes, the flow speed above is increased by the ratio of total annular area to total "see through" port area, as will be explained in greater detail later.

7.3.2 Pressure integral.

Bernoulli's equation applies in the absence of viscous losses and arises as an exact integral for inviscid irrotational flow. In cylindrical radial coordinates, Equation 7.2.3 takes the form

$$p(x,r,\theta) + \tfrac{1}{2}\rho\,(\phi_x{}^2 + \phi_r{}^2 + 1/r^2\,\phi_\theta{}^2) = p_0 \qquad (7.3.6)$$

Since Equation 7.3.6 is nonlinear in $\phi(x,r,\theta)$, a simple formula connecting Φ and p cannot be obtained. Special treatment is needed so that the radially-averaged potential variable $\Phi(x,\theta)$ can be used. First, we expand $\phi = \phi_0 + \phi_1$ where $|\phi_0| \gg |\phi_1|$. Here "0" represents the uniform oncoming flow and "1" the disturbance flow induced by the presence of the siren. Neglecting higher order terms,

$$p(x,r,\theta) + \tfrac{1}{2}\rho\,\phi_{0x}{}^2 + \rho\,\phi_{0x}\phi_{1x} + \ldots = p_0 \qquad (7.3.7)$$

As before, multiply throughout by $2\pi r$ and integrate over (R_i,R_o) to obtain

$$\int p(x,r,\theta)\,2\pi r\,dr + \tfrac{1}{2}\rho\,\phi_{0x}{}^2\int 2\pi r\,dr + \rho\,\phi_{0x}\int\phi_{1x}\,2\pi r\,dr \cong p_0\int 2\pi r\,dr$$

$$(7.3.8a)$$

noting that ϕ_{0x} and p_0 are constants. Introducing

$$\int p(x,r,\theta)\,2\pi r\,dr = A\,p_{avg}(x,\theta) \qquad (7.3.8b)$$

$$\int \phi_{1x}\,2\pi r\,dr = A\Phi_{1x}(x,\theta) \qquad (7.3.8c)$$

$$\int 2\pi r\,dr = \pi\,(R_0{}^2 - R_i{}^2) \qquad (7.3.8d)$$

we find that Equation 7.3.8a becomes

$$A\,p_{avg}(x,\theta) + \tfrac{1}{2}\rho\,\phi_{0x}{}^2\,\pi\,(R_0{}^2 - R_i{}^2) + \rho\,\phi_{0x}\,A\Phi_{1x}(x,\theta) \cong p_0\,\pi\,(R_0{}^2 - R_i{}^2)$$

$$(7.3.9)$$

Without loss of generality, we now set the reference area to $A = \pi\,(R_0{}^2 - R_i{}^2)$ so that Equations 7.3.5 and 7.3.9 simplify as follows,

$$\Phi_{xx} + 1/R_m{}^2\,\Phi_{\theta\theta} \cong 2U_\infty\,\{R_i\,\sigma_i - R_o\,\sigma_o\}/(R_0{}^2 - R_i{}^2) \qquad (7.3.10)$$

$$p_{avg}(x,\theta) + \tfrac{1}{2}\rho\,\phi_{0x}{}^2 + \rho\,\phi_{0x}\,\Phi_{1x}(x,\theta) \cong p_0 \qquad (7.3.11)$$

Equations 7.3.10 and 7.3.11 are the final governing partial differential equations solved. We next discuss auxiliary conditions used to obtain specific solutions. The normalization used for A was selected so that Equation 7.3.11 takes the form of Bernoulli's equation linearized about the mean speed ϕ_{0x}. The quadratic terms neglected in this approximation can be added back to Equation 7.3.11 without formally incurring error to this order. If this is done, the modified formulation is advantageous since it is exact for planar cascade flow.

7.3.3 Upstream and annular boundary condition.

Auxiliary conditions are key in solving the two-dimensional elliptic Poisson equation prescribed by Equation 7.3.10. Despite the reduced order of the partial differential equation, geometric complexities can render fast solutions difficult. However, motivated by the "thin airfoil" and "thin engine nacelle" approaches offered in Section 2, in which exact flow tangency conditions are approximated along mean lines, planes, and cylindrical surfaces, we adopt a similar approach but for three-dimensional mud sirens. Before describing the details of the method, we introduce three additional design variables not discussed in Section 1.

First consider the siren configuration shown in Figure 7.8. This diagram indicates still another design variable – a swirling upstream flow that can be induced by the presence of deflection vanes located just upstream of the lobed pair, as suggested in U.S. Patent No. 3,867,714. The degree of imposed swirl can be introduced by prescribing the value of $1/R_m \Phi_\theta$ as a boundary condition. Figure 7.9 illustrates the downstream central hub, which is always present in existing designs; it also shows an upstream hub, which may or may not be present. Although it is natural to design the associated annular passages with surfaces that are completely aligned with the direction of the oncoming flow, this is not necessary or recommended. In fact, the right side of Equation 7.3.10 shows that local geometric curvatures will affect computed torques, velocity and pressure fields, although their consequences are not immediately evident.

Possible choices for the annular passages leading up to the siren and away from it are shown in Figure 7.10. By no means are these the most general. Streamwise surface slopes shown in the individual diagrams here only increase or decrease monotonically, but they can increase and decrease, decrease and increase, or for that matter, take on the "wavy wall" form studied in aerospace literature. The sign of $\{R_i \sigma_i - R_o \sigma_o\}$ in Equation 7.3.10 is seen to be an important design parameter. For instance, referring to Figure 7.10.1, it is positive for the converging case and negative for the diverging case. Figure 7.10.2 shows one possibility (of several) tested for its influence on torque and signal. We emphasize again that the three-dimensional character of typically small hub radii is modeled in the integral approach leading to Equation 7.3.10.

Figure 7.8. Basic mud siren with upstream swirl.

Figure 7.9. Three-dimensional mud siren with annular passages.

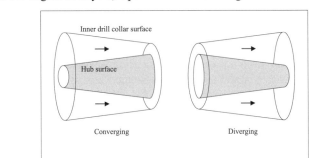

Figure 7.10.1. Annular passage flow examples
(refer to the "inlet cones" tested for signal influence).

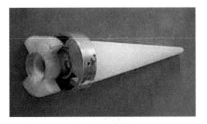

Figure 7.10.2. Example for guided flow device.

7.3.4   Radial variations.

Now we address the modeling of the rotor and stator rows themselves. To properly motivate the problem, we turn to classical aerodynamic modeling methods for turbine and compressor blade rows, and consider the "unwrapped" periodic blade row shown in Figure 7.11 used in aerospace engineering. We have remarked that a naive geometrical unwrapping of the siren lobes in (say) Figure 7.3 is incorrect because important centrifugal flow changes in the radial direction are ignored. Computed results did not agree with experimental observations. But Equation 7.3.10, which is averaged in the "r" direction, shows how radial effects can be incorporated.

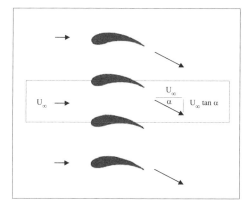

Figure 7.11. Aerodynamic cascade problem.

In our approach, streamwise curvatures of upstream and downstream annular spaces appear directly on the right-side of the equation. By contrast, the mean R_m on the left side contains the effects of hub radius, while individual boundary conditions at R_i and R_o (to be described) implicitly include the effects of the "siren lobe height," that is, $R_o - R_i$. Hence, the two necessary length scales R_m and $R_o - R_i$ appear as required in our three-dimensional formulation. Although Figure 7.11 shows only single airfoil configurations, we emphasize that standard formulations (outlined below) apply to "multi-element" combinations as well, e.g., wings with trailing flaps, ailerons with control structures, and so on.

7.3.5   Downstream flow deflection.

The theory underlying cascade analysis is summarized in the classic turbomachinery book of Hawthorne (1964) or in the comprehensive aerodynamics book of Oates (1978). Both give detailed derivations of fundamental equations. The most relevant flow characteristic in studying single and multi-element cascade flows is downstream streamline deflection. In general, if a nonzero lift is exerted on the blade row, then the far downstream flow must exhibit an exit angle deflection that is consistent with the momentum theorem. For airfoil cascades, the deflection angle α in Figure 7.11 satisfies

$$\tan \alpha = \tfrac{1}{2}\, s^{-1} \int (\, |\nabla\phi_l|^2 - |\nabla\phi_u|^2 \,)/U_\infty^2 \; dx \qquad (7.3.12a)$$

following the nomenclature in Figure 7.4, where s is the vertical blade separation between neighboring airfoils, and l and u denotes lower and upper blade surfaces. The integral is taken from the upstream leading edge to the downstream trailing edge. The deflection is independent of the density ρ and the oncoming speed U_∞. The deflection of the downstream flow is associated with a pressure drop

$$\Delta p = \tfrac{1}{2}\, \rho U_\infty^2 \tan^2 \alpha \qquad (7.3.12b)$$

arising purely from inviscid considerations. This loss is small compared to the loss that would be realized in the actual viscous flow, one that should be estimated from wake models or orifice formulas. We emphasize that Equation 7.3.12b is not to be used in computing pressure drops through the downstream wake. Finally, periodic velocities are assumed at the top and bottom of the problem domain in Figure 7.11. It is computationally important that single-valued velocities are modeled, noting that velocity potentials themselves are multivalued if torques are nonzero and lead to programming complexities.

7.3.6 Lobe tangency conditions.

Equation 7.3.5a for the averaged potential is solved with flow tangency conditions on lobe surfaces. These are easily derived. For example, the ratio of the vertical to the horizontal velocity is kinematically $R_m^{-1}\phi_\theta /\phi_x = f_x(\theta,x,r)$ where $f(\theta,x,r)$ represents the surface locus of points and the subscript x is the streamwise derivative. We can rewrite this as $R_m^{-1}\phi_\theta = f_x(\theta,x,r) \phi_x$, multiply through by $2\pi r$, carry out the former integration, and introduce our definitions for Φ. If $f_x(\theta,x,r)$ is approximated by $f_x(\theta,x,R_m)$, we obtain $R_m^{-1}\Phi_\theta /\Phi_x = f_x(\theta,x,R_m)$ to leading order, where Φ_x is roughly U_∞.

One significant modification to $\phi_y(x,0)/U_\infty \cong$ "slope" in Figure 7.5 must be made. In airfoil cascade analysis, as in thin airfoil theory, flow blockage is very minimal, and the left side of the approximate tangency condition expresses the ratio of the vertical to the horizontal velocity, taken to leading order as the freestream speed itself. This treatment does not apply to mud sirens because flow blockage in the neighborhood of port spaces is significant, nominally amounting to half of the flow area. Thus, U_∞ in the boundary condition must be replaced by U_{hole}, which can take on different values for rotor and stator. At x locations not occupied by solid siren lobes, the velocity ϕ_x can be approximated by U_∞. But at locations occupied by the siren, it is convenient to introduce a "see through" area A_{hole}. Mass flow continuity requires that $A_{hole}U_{hole} = A_{total}U_\infty$, where $A_{total} = \pi (R_o^2 - R_i^2)$. Thus, the value of U_{hole} is completely determined everywhere along the streamwise direction.

7.3.7 Numerical solution.

The classical aerodynamic cascade problem, in summary, solves Poisson's equation subject to (1) uniform flow far upstream, (2) approximate tangency conditions evaluated on a mean line, (3) periodic velocities at the top and bottom of the computational box in Figure 7.11, and finally, (4) selection of a deflection angle α far downstream that is consistent with momentum conservation. Although isolated closed form analytical solutions are available for simple classical airfoil problems, the boundary value problem developed here is highly nonlinear, owing to Equation 7.3.12a, and must be solved numerically by an iterative method. The integral in our Equation 7.3.12a is evaluated over both solid surfaces.

With the aerodynamic formulation stated, emphasizing that our "unwrapping" includes radial effects, we develop a boundary value problem model appropriate to cascades of siren rotors and stators. We begin by considering a portion of the unwrapped lobe structure, as shown in Figure 7.12. Since the velocity field is periodic going from one set of rotor-stator lobes to another, a smaller computational box can used in the flow domain at whose upper and lower boundaries we invoke periodic disturbance velocities.

To be consistent with aerodynamic analysis, we focus on a primary blade pair (arbitrarily) located along a (bottom) horizontal box boundary, as shown in Figure 7.13. If the slopes associated with streamwise annular curvatures are small, as they generally are in actual mud sirens, then Equation 7.3.12 applies to leading order. If we note that setting $R_m\theta$ to y transforms Equation 7.3.10 to classical form (e.g., see Figure 7.5), that is,

$$\Phi_{xx} + \Phi_{yy} \cong 2\{R_i\sigma_i\,\phi_x(x,R_i,\theta) - R_o\sigma_o\,\phi_x(x,R_o,\theta)\}/(R_0^2 - R_i^2)\ (3\text{-}13a)$$

$$\cong 2U_\infty\{R_i\sigma_i - R_o\sigma_o\}/(R_0^2 - R_i^2) \qquad (7.3.13b)$$

but modified by a non-zero right side Poisson term, it is clear that computational methods developed for aerospace problems can be applied with minor modification to mud siren torque analysis problems.

In our software, the source code and data structure of Chin (1978) was modified to solve the foregoing problem. The flow in Figure 7.13 is solved by a finite difference column relaxation procedure. The solution is initialized to an appropriate guess for the flowfield, which can be taken as an approximate analytical cascade solution or the flowfield to a slightly different siren problem whose solution has been stored. The mesh is discretized into, say, 100 streamwise grids and 50 vertical grids. The solution along each column is obtained, starting from the left and proceeding to the right, with one such sweep constituting one iteration through the flowfield. Latest values of the potential are always used to update all tridiagonal matrix coefficients. At the end of each sweep, the deflection angle is updated based along the far right vertical boundary. For the quoted 100 × 50 grid, as many as 10,000 sweeps may be required for absolute convergence; this requires approximately three seconds on typical computers for the efficiently coded velocity potential solution, and an additional two seconds for the streamfunction streamline tracing solution. For any particular siren design, torque characteristics are of interest for several degrees of relative rotor-stator closure. If six or seven equally spaced azimuthal positions are studied in order to define the torque versus closure curve, a complete solution can be obtained in about one minute.

7.3.8   Interpreting torque computations.

How might computed torques be interpreted in terms of stable-opened and stable-closed performance? Consider first the left upstream lobe in Figure 7.13. If the indicated force vector is positive, the black lobe will tend to move upward and close the valve, should that be chosen as the rotor. Next consider the right downstream lobe. If the indicated force is positive, it will tend to open the valve if it were chosen as the rotor; if it is negative, on the other hand, the lobe will tend to close the valve. If $F_{single-lobe}$ is the force acting on one lobe, as calculated by numerical integration, the torque associated with that lobe is given by

$$T_{single\ lobe} = R_m F_{single-lobe} \qquad (7.3.14)$$

The square and trapezoidal shapes shown are for illustrative purposes only – in the computer program, different lobe aspect ratios and taper angles may be assigned. In addition, lobe taper slopes at top and bottom need not be equal and opposite; they can take arbitrary values and hold identical signs. In more general engineering designs, when identical signs are taken, the resulting rotors can be stable-closed, stable-open, or self-rotating, drawing upon the kinetic energy of the oncoming mudstream; see, e.g., refer to the patent descriptions in Chin and Ritter (1996, 1998) or U.S. Patents 5,586,083 and 5,740,126. One analogy to airfoil analysis should be mentioned. In Figure 7.4, both positive and negative lifts can be obtained depending on the angle-of-attack. Similarly, positive and negative forces on siren lobes are in principle possible. However, experimentally and numerically, it has never be possible to achieve a stable open design for upstream rotors. Downstream rotors, depending on the taper chosen, may be stable open, stable closed, neutrally stable in both positions, and also stable in the partially-open position. Also, unlike the flow past turbine blade rows (with upward lift), for which the flow deflection is downward, computations show that the downstream deflection can be downward or upward, accordingly as the total force on both rotor and stator is upwards or down.

7.3.9 Streamline tracing.

The streamline pattern assumed by any particular flow sheds insight on locations prone to flow separation and those likely to induce surface erosion. It can be constructed by tracing velocity vectors, by integrating "dy/dx = (vertical velocity)/(horizontal velocity)," but this procedure is very inaccurate. For example, particle locations easily "fly off" the computational box whenever high surface speeds are encountered in the numerical integration. An alternative method applicable to weak annular convergence can be implemented. The theory is developed by first writing Equation 7.3.13a in the more concise form

$$\Phi_{xx} + \Phi_{yy} \cong \Lambda \qquad (7.3.15a)$$

where Λ denotes the right side of Equation 7.3.13a, and then re-expressing it as

$$\partial\{\Phi_x\}/\partial x + \partial\{\Phi_y - \Lambda y\}/\partial y = 0 \qquad (7.3.15b)$$

This "conservation form" implies the existence of a function Ψ satisfying

$$\Phi_x = \Psi_y \qquad (7.3.15c)$$

$$\Phi_y - \Lambda y = -\Psi_x \qquad (7.3.15d)$$

This derivation extends the classic derivation (Ashley and Landahl, 1965) for the streamfunction Ψ to problems with non-zero Λ. If we now differentiate Equation 7.3.15c with respect to y and Equation 7.3.15d with respect to x, and eliminate the velocity potential, we obtain the governing equation

$$\Psi_{xx} + \Psi_{yy} = 0 \qquad (7.3.15e)$$

This takes the same form as Equation 7.3.15a so that the same solution algorithm applies. The boundary conditions used are obtained from Equations 7.3.15c and 7.3.15d since Φ is already known. The normal Neumann derivative Ψ_y is applied along the horizontal upper and lower box boundaries, while the normal derivative Ψ_x is applied along the vertical left and right boundaries.

How is Ψ used to trace streamlines? Note that streamline slope dy/dx is kinematically equal to the velocity ratio Φ_y/Φ_x, that is,

$$dy/dx = \Phi_y/\Phi_x = -\Psi_x/\Psi_y + \Lambda y/\Psi_y \qquad (7.3.15f)$$

Then the total differential satisfies

$$d\Psi = \Psi_x\, dx + \Psi_y\, dy = \Lambda y\, dx \qquad (7.3.15g)$$

When annular convergences are small, Λ can be neglected so that $d\Psi = 0$. Thus, Ψ is constant along a streamline. In this limit, streamlines are easily constructed by using a contour plotter for the converged numerical field $\Psi(x,y)$.

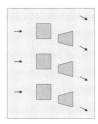

Figure 7.12. Periodic cascade of upstream and downstream lobes.

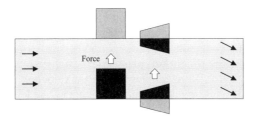

Figure 7.13. Periodic boundary value problem flow domain.

7.4   Typical Computed Results and Practical Applications

In this section we discuss typical qualitative and quantitative computed results. Figure 7.14 shows streamline patterns obtained as relative siren rotor and stator lobe positions are changed from opened to closed. The tapers clearly indicated in Figure 7.13 could not be plotted by the graphical software used to generate Figure 7.14; only shown are the mean lobe box boundaries where taper boundary conditions are applied. The streamlines correspond to the radially averaged flow obtained per Equation 7.3.2. Similar comments apply to Figure 7.15, which illustrates the absolute value of the velocity magnitude in the flow domain and at solid surfaces; to visually enhance the color scheme, its logarithm is plotted instead. Red zones in the velocity plots identify areas of high fluid and surface speed that are susceptible to sand erosion.

Erosion concerns are paramount to practical mud siren design. Very often, a poorly shaped mud siren will not survive more than several hours in heavy weight muds flowing at high speeds, e.g., 12 ppg muds with gpm's exceeding 700. Mechanical engineers new to siren design often quote a nominal 100 ft/sec as *the* dangerous critical velocity to avoid, that is, it is the velocity at which erosion is incipient. However, this rule-of-thumb is not completely accurate. Every highway driver has witnessed bug impacts on windshields: the high inertia of typical insects does not allow them to flow tangentially with the wind. Consequently, they collide into the automobile. Sand particle convection by flowing mud follows similar principles. Solids collide into the surfaces of mud sirens, not to mention turbines, strainers, and other downhole equipment. The speed of impact is, of course, important. However, the impingement angle and the impact velocity together dictate the predominant erosion mechanism, that is, whether metal removal is controlled by brittle fracture, ductile shearing, or both.

While diagrams like Figures 7.14 and 7.15 do not predict particle impact velocity vectors, they do provide a qualitative indicator that may be useful in empirically correlating field and laboratory observed erosion patterns. In the aerospace industry, particularly in jet engine design, computed results like those shown in Figures 7.14 and 7.15 are actually used in particle-hydrodynamic simulators to predict turbine blade erosion. These color plots are followed by design calculations in which torque predictions are addressed. We again emphasize that computed torque results are entirely consistent with the results of detailed experiments in which key geometric design parameters were varied systematically over different oncoming freestream velocities. Refer to Chapter 9 for descriptions of mud flow loops and wind tunnels used for validation of the computed results obtained here.

7.4.1   Detailed engineering design suite.

In this section, an illustrative set of ten siren design calculations is described, using our mathematical model and software implementation. In Figures 7.16 to 7.25, "Lobe Force" appears on the vertical axis, while "Open at left, closed at right" appears on the horizontal axis, referring to the azimuthal coordinate. All torque trends indicated qualitatively agree with experimental results. In all the runs considered, a specific gravity of 1.0 assuming pure water is taken. The inner hub radius at the siren lobe base is 1 inch, the outer lobe radius is 2 inch, so that the blade height is 1 inch. The effective moment arm is approximately 1.5 inch. Also, the volume flow rate is fixed at 1,200 gpm. Both sets of siren lobes were centered in the computational box. For brevity, detailed gridding information and tabulated torques are not discussed and we focus on fundamental physical effects instead.

Run A. The geometry in the cascade plane assumes square rotor and stator lobes having 1 in × 1 in dimensions, separated by a gap of 0.25 inches. This is considered large by MWD standards and will not generate significant signal, however, the run was performed to furnish baseline numbers for comparison. Again, neither lobe possesses tapers, so that the run corresponds to the sirens shown in Figures 7.1 and 7.3. Also, there is no streamwise annular taper. The upper (red) curve corresponds to forces obtained for the upstream lobe, while the lower (green) curve corresponds to forces on the downstream lobe. The forces peak at -19 lbf and +19 lbf for downstream and upstream lobes. The straight line behavior of the force curves versus the inclination of the lobe pair is consistent with inviscid aerodynamics, as seen in Figure 7.4 for computed lift coefficients. Note that the force varies linearly with closure and therefore acts like a linear spring. The corresponding spring constant can be used to estimate mechanical response times. To estimate the maximum torque acting on the siren system, consider the outer radius of 2.0 inch, having a circumference of about 12 inches. This fits to a six lobe system in the cascade plane (six solid lobes combined with six port spaces, each space being 1 inch). The moment arm is about 1.5 inch. Thus, the torque is approximately 6 lobes × 20 lbf/lobe × 1.5 inch or about 180 in-lbf, in rough agreement with mud loop experiments.

Run B. Here, the geometry in Run A is altered by decreasing the gap from 0.25 inches to 0.1 inches. This distance represents the small gap that might be used in siren-type tools. Note how computed forces increase approximately 50% to the –28 lbf to +28 lbf range, a trend that agrees with experiment. Also, the green curve shifts to left, indicating a decreased stable-open character.

Run C. In this simulation, we change the geometry in Run B by adding an outward taper of 10° to the downstream lobe, consistently with the shape in Figure 7.2. The –28 lbf to +28 lbf range obtained previously is significantly reduced to –21 lbf to + 18 lbf, again consistently with experiment. In addition, there is a noticeable shift of the green curve to the right, indicating improved stable open characteristics for the downstream lobe, also observed empirically.

Run D. We now repeat Run C, and increase the taper angle from 10° to 15°. The force range is narrowed, now falling in the –20 lbf to 11 lbf band.

Run E. We will retain the 15° used above, but reduce the thickness of the downstream lobe from 20 to 5 gridblocks. The 1 in thick lobe becomes 0.25 in (recent siren advertisements suggest that thin downstream rotors are used in present tools). Results show some force decrease in the downstream lobe, also consistent with experiment. Lab results for the upstream lobe are not available.

Run F. In this run, we return to the baseline Run A with square lobes and no lobe tapering, and a large 0.25 inch gap. In that run, a force range of −19 lbf to +19 lbf was computed. We now place the siren in an annulus that converges in area at both walls as the flow moves to the right, with 10° inclinations at each wall. The computed force range is −21 lbf to +16 lbf, a noticeable reduction in the upstream force.

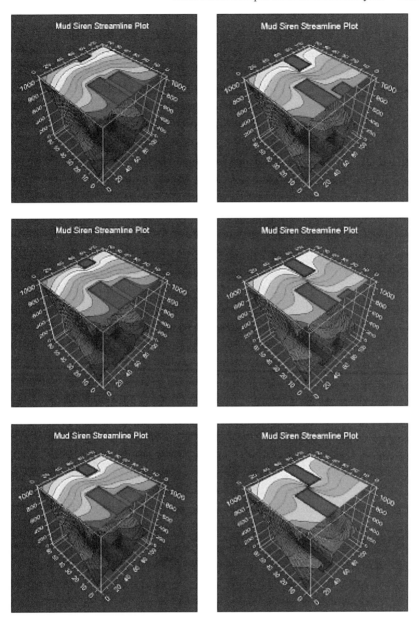

Figure 7.14.   Streamline patterns at different degrees of closure.

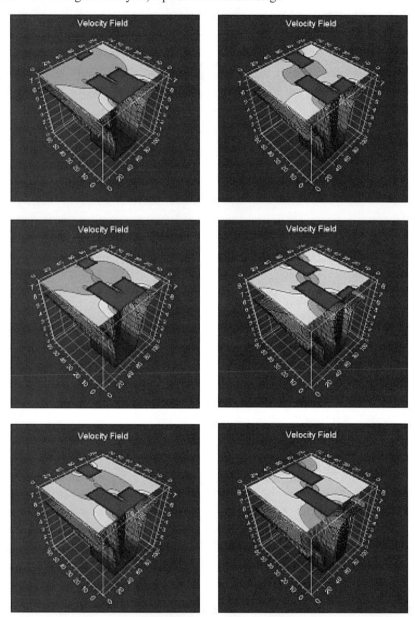

Figure 7.15. Velocities for erosion prediction at different degrees of closure.

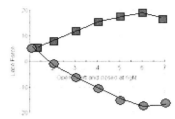

Figure 7-16. Run A siren torques.

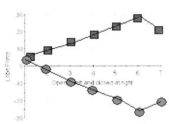

Figure 7-17. Run B siren torques.

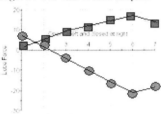

Figure 7-18. Run C siren torques.

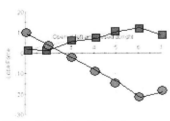

Figure 7-19. Run D siren torques.

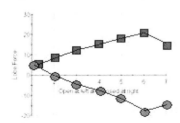

Figure 7-20. Run E siren torques.

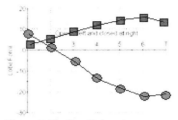

Figure 7-21. Run F siren torques.

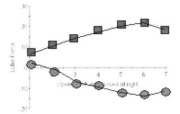

Figure 7-22. Run G siren torques.

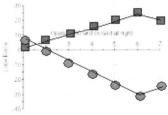

Figure 7-23. Run H siren torques.

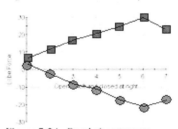

Figure 7-24. Run I siren torques.

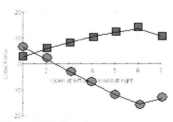

Figure 7-25. Run J siren torques.

Run G. We now reverse the annular geometry and allow the flow area to diverge instead. The computed results show a significant reduction in the force acting on the downstream lobe.

Run H. In this simulation, the gap distance is decreased significantly, as we had allowed in going from Run A to Run B. Aside from this change, the geometry is identical to that in Run B, where the force range was –28 lbf to +28 lbf. Here we allow the annular area to converge again, and as in Run F, we obtain some decrease in the upstream force.

Run I. We allow the annular area to diverge instead, and the effect is a sharply reduced force (about 25%) on the downstream lobe.

Run J. Finally, we allow both a diverging annular area and a 15° outward taper on the downstream lobes. The result is a sharply reduced force range, now falling in a –16 lbf to +14 lbf band. This combination of design tapers appears to offer the potential for fast modulations in high-data-rate MWD transmissions.

7.5 Conclusions

In this chapter, a three-dimensional inviscid flow formulation for MWD mud siren torque prediction is justified and developed. The computational model is not applicable to pressure drop determination in the downstream viscous wake. The numerical algorithm hosts a stable and rapidly converging finite difference solution of the governing fluid-dynamical equations. Column relaxation methods are used which provide diagonally dominant intermediate matrices, which allow for robust simulation and numerical convergence by practicing engineers, without intervention from specialists in numerical analysis. This combination of fluid mechanics, software design and integrated color graphics permits the depth-averaged model to be used in real-time engineering design, guiding the formulation of test matrices and the interpretation of flow data. The basic operational problems associated with mud siren design, and their implications in high-data-rate MWD telemetry, have been discussed and addressed both experimentally and numerically. Considerations related to "stable-open," "stable-closed," "self-rotating," "low torque," and "erosion" issues were in particular discussed in detail.

7.5.1   Software reference.

The Fortran simulation engine for the three-dimensional model are found in C:\MWD-01\siren-22.for while the graphics module is embedded in the Visual Basic 6.0 program sfline.vbp. Note that a complementary three-dimensional model for Δp signal generation is possible for periodic rotor turning. This model, solving a Helmholtz-type formulation derived from the more general wave equation, would require a greater degree of complexity and more computing resources.

7.6 References

Ashley, H. and Landahl, M.T., *Aerodynamics of Wings and Bodies*, Addison-Wesley, Reading, Massachusetts, 1965.

Batchelor, G.K., *An Introduction to Fluid Dynamics*, Cambridge University Press, 1970.

Chin, W.C., "Algorithm for Inviscid Flow Using the Viscous Transonic Equation," *AIAA Journal*, Aug. 1978.

Chin, W.C., "MWD Siren Pulser Fluid Mechanics," *Petrophysics*, Journal of the Society of Petrophysicists and Well Log Analysts (SPWLA), Vol. 45, No. 4, July – Aug. 2004, pp. 363-379.

Chin, W.C., Golden, D. and Barber, T., "An Axisymmetric Nacelle and Turboprop Inlet Analysis with Flow-Through and Power Simulation Capabilities," *AIAA Paper No. 82-0256, AIAA 20th Aerospace Sciences Meeting*, Orlando, FL, Jan. 1982.

Chin, W.C., Presz, W., Ives, D, Paris, D. and Golden, D., "Transonic Nacelle Inlet Analyses," *NASA Lewis Workshop on Application of Advanced Computational Methods*, Nov. 1980.

Chin, W.C. and Ritter, T., "Turbosiren Signal Generator for Measurement While Drilling Systems, U.S. Patent No. 5,586,083, Dec. 17, 1996.

Chin, W.C. and Ritter, T., "Turbosiren Signal Generator for Measurement While Drilling Systems," U.S. Patent No. 5,740,126, April 14, 1998.

Chin, W.C. and Trevino, J.A., "Pressure Pulse Generator," U.S. Patent No. 4,785,300, Nov. 15, 1988.

Hawthorne, W.R., *Aerodynamics of Turbines and Compressors*, Volume X, High Speed Aerodynamics and Jet Propulsion, Princeton University Press, 1964.

Oates, G.C., *The Aerothermodynamics of Aircraft Gas Turbine Engines*, Air Force Aero Propulsion Laboratory Report No. AFAPL-TR-78-52, July 1978.

Patton, B.J., Prior, M.J., Sexton, J.H. and Slover, V.R., "Logging-While-Drilling Tool," U.S. Patent No. 3,792,429, Feb. 12, 1974.

Patton, B.J., "Torque Assist for Logging-While-Drilling Tool," U.S. Patent No. 3,867,714, Feb. 18, 1975.

Sexton, J.H., Slover, V.R., Patton, B.J. and Gravley, W., "Logging-While-Drilling Tool," U.S. Patent No. 3,770,006, Nov. 6, 1973.

In this chapter, a three-dimensional inviscid flow formulation for MWD

8

Downhole Turbine Design and
Short Wind Tunnel Testing

As an experienced professional embarking on a career in MWD in 1981, having worked as Research Aerodynamicist at Boeing and Turbomachinery Manager at Pratt & Whitney Aircraft, the jet engine manufacturer, and having earned graduate engineering degrees from Caltech and M.I.T., I had envisioned MWD turbine design as a low-tech "slam dunk" affair. Nothing could have been farther from the truth. In fact, nothing taught in classical aerodynamics applied that could have been reasonably used for downhole tool design. To understand why turbine design is frustrating, one needs only to compare the contrasting operating environments seen by MWD versus jet engine turbines. We will do this in the following section; fortunately, it turns out that good turbine design can be approached systematically using basic physical principles.

8.1   Turbine Design Issues

A standard MWD turbine is shown in Figure 8.1 where the mud flows from left to right. A single "stage" consists of an upstream "stator," which does not rotate, and a downstream "rotor" which does. The shaft and alternator combination to which the rotor is connected is not shown. The turbine transforms the kinetic energy of the mud into electrical energy used to power both logging sensors and siren pulser. The demands on the turbine are nontrivial. Up to several horsepower, e.g., 2-3 HP, may be required in a modern high-data-rate tool, which is significant in view of mechanical packaging constraints and severe drill collar space limitations.   As power demands increase, erosion accelerates and life spans decrease rapidly. But very often, generating the required power is not the problem. For an oncoming axial speed U, it is known that torque varies with U^2 while power varies like U^3. Since positive displacement mud pumps rarely pump with constant speed, a $\pm 10\%$ speed fluctuation easily translates into a $\pm 30\%$ variance in power that must be regulated electrically. The designer must, of course, err on the side of excess power, since insufficient power renders an MWD tool useless.

In designing an MWD turbine, one naturally turns, at first, to the wide body of literature available in aerospace and mechanical engineering for practical guidelines. However, MWD turbines are a special breed. To see why, we examine typical aircraft turbines as shown in Figure 8.2. Since longitudinal space constraints are not severe, such turbines are built with numerous stator-rotor pairs. Thus, power generation is shared by multiple stages, and blade pitch angles (required for torque and power production) need not be as highly inclined as those for MWD – implying that inefficiencies due to flow separation are avoided. To further enhance flow effectiveness, distances between blades are close. And finally, tip-to-housing clearances are vanishingly small – the left diagram in Figure 8.3 shows that pressure loadings across the span of the rotor blade are almost uniform, so that every part of the blade is effective in creating torque and power.

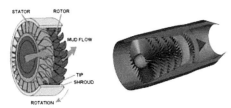

Figures 8.1 and 8.2. Single-stage MWD turbine versus multistage jet engine.

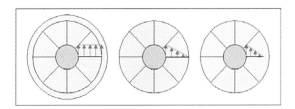

Figure 8.3. Spanwise pressure loading as function of rotor tip clearance
(rotor tips do not support transverse pressure loadings).

As noted, MWD turbines must develop all the required power in a single stage occupying just inches in the drill collar. Thus, blade pitch angles must be high so that the flowing mud "pushes" as hard as possible. To make matters worse, blade-to-blade separations cannot be small since debris entrapment may lead to high localized erosion zones and even complete plugging of the tool. These two effects significantly decrease turbine efficiencies since massive flow separation is the rule. As if this were not enough, tip-to-housing clearances cannot be small, since the slightest bend or transverse vibration in the drill collar would jam the rotor and curtail power production. Sticking due to mud gelling and debris entrapment is also a concern.

The middle diagram in Figure 8.3 shows how, since pressure loadings necessarily vanish at the rotor tip, torque creation is least at large radii, an unfortunate situation. As highly pitched blades are associated with vortex shedding from the tips and rapid local erosion, power generation worsens, as suggested at the right of Figure 8.3 (this effect is explained later). From a fluid mechanics perspective, we have highly separated, three-dimensional, unsteady flows which cannot be analyzed with rigor. Given the time, cost and labor constraints associated with typical engineering projects, the outlook for any turbine design, let alone a good one, at first appeared pessimistic. Figure 8.4 summarizes the major differences between aircraft and MWD turbines.

	Aircraft	**MWD**
Number of stages	Many	Single (small volume size constraint)
Blade separation	Close	Wide (avoid debris jamming)
Tip-to-shroud clearance	Tight	Large (prevent jamming from vibration, doglegs, gelled mud - means low torque)
Jamming	No	Yes (vibration, doglegs, debris)
Erosion	Normal	High rotor tip wear (sand recirculation by vortex flow), constant stator wear (sand abrasion from rock entrapment)
Shock and vibration	None	High cycle fatigue, bit bouncing, strong transverse loads
Efficiency	High	Low (high blade pitch angles and large tip clearances, flow separation)

Figure 8.4. Qualitative comparison, aircraft versus MWD turbines.

8.2   Why Wind Tunnels Work

Designing an MWD turbine is expensive, time-consuming and labor-intensive. Once torque and power requirements at a given volume flow rate are specified, the turbine geometry is to be determined. "Geometry" includes many parameters: annular inner and outer radii, number of blades, cross-section contour of a blade, pitch of the blades, rotor tip clearance, stator-rotor separation, and so on. Usually, because of downhole mechanical packaging constraints, only one stage can be accommodated, that is, one stator and one rotor. Still, the number of possible configurations is vast, perhaps hundreds. Testing of metal models in flowing mud is inconvenient, with theoretical analysis being equally difficult. Aircraft companies rightly deal with potential flow analyses such as that presented in Chapter 7. Commercial software packages are often less rigorous and should be carefully evaluated.

In MWD applications, very high pitch angles are needed because all of the power desired must come from a single stage. This, together with large blade-to-blade distances means that the flow will separate. Clearances between rotor tips and housing imply high three-dimensionality, inefficiencies in torque creation, massive vortex shedding, and so on. Rotor-stator flow interactions render fluid motions unsteady. Any one of these conditions means that analysis is impossible. However, simple observations have led to accurate means for MWD turbine design which provide almost perfect results: wind tunnel analysis.

Although wind tunnels have been used extensively in the petroleum industry to model unsteady loads associated with vortex shedding from offshore platforms in water, its application in downhole tool design was apparently not well known until the publication of Gavignet, Bradbury and Quetier (1985) which examined flow in tricone drillbits. These investigators, at the lead author's suggestion, used "air as a flowing fluid," noting that the "substitution of fluids is justified by the highly turbulent nature of the flow." In the past decade, the lead author has developed the approach more extensively. As we will see, the cited reason represents only part of the justification: air, which is convenient, free, clean and providing of quick turn-around, is optimal for other reasons.

Fluid mechanics books typically introduce the subject by developing ideas in "dynamic similarity." In the context of the aerodynamic design of turbines (and mud sirens too), consider a fluid with density ρ, viscosity μ and speed U, and a geometry with a characteristic surface area S and a characteristic radius R. The dimensional torque T will be a function of (i) the geometry or shape of the turbine and the shape of the blades, (ii) the dimensionless Reynolds number Rey = $\rho UR/\mu$, and finally, (iii) the dimensional quantity $\rho U^2 SR$. If two different situations are such that (i) and (ii) are identical, then the two are physically equivalent even if the torque values themselves are not.

Now consider the possibility of using a wind tunnel. This would mean inexpensive and fast tests since the models can be made of, say, balsa wood, constructed using simple wood-working tools. Hundreds can be tested in a matter of days. Because downhole turbines (and sirens) are relatively small, we will test them "full scale" with identical size and shape. Thus, condition (i) is satisfied. Next, consider Reynolds number. Since the test fixture is small and turbines do not substantially block the flow, we can test at the actual downhole speed U using simple squirrel cage blowers.

For *constant density, incompressible* laminar flow, we need only require that μ/ρ, known as the "kinematic viscosity," be identical. Let us consider the kinematic viscosity versus temperature relationships in Figure 8.5 for various fluids. Surprisingly, it turns out that the kinematic viscosity of a typical drilling mud is not that of water, as one might surmise, but that of two gaseous fluids, air and methane. Methane is dangerous because it is explosive. Air at room temperature and pressure is free and abundant. By using air, our Reynolds numbers are very close, satisfying condition (ii). If *Rey* is such that turbulent flow is found, then the argument of Gavignet *et al* applies. We do emphasize that, in the laminar case, Reynolds numbers need not be too close; in fluid dynamics, Reynolds numbers that differ by, say, a factor of ten, may be close enough. For both turbine and siren testing, we are actually more than fortunate – the effects of viscosity (that is, Reynolds number) on torque are secondary, as viscosity primarily affects only the thrust acting in the direction of flow.

An unanticipated benefit is the use of wind tunnel analysis in *compressible* flow, that is, in MWD sound transmission modeling. While high turbine rotation speeds will result in very short wavelength sound which will not travel to the surface, the opposite is true of siren valves which typically operate at low frequencies. Although this chapter deals with turbine flows, it is important to digress temporarily to study siren acoustic modeling which, after all, forms the main subject of this book. In a downhole situation, mud sound speeds vary from 3,000-5,000 ft/sec. In the most optimistic case, consider a carrier frequency of 100 Hz. The associated wavelength is 30-50 ft, which greatly exceeds a typical drillpipe diameter; thus, our waves are acoustically long.

Now consider sound wave propagation in a very long wind tunnel. The sound speed is approximately 1,000 ft/sec. For the same 100 Hz, the wavelength is now 10 ft, which still greatly exceeds a typical pipe diameter. Since MWD transmissions in air are also long waves, the wind tunnel can be used to study important problems in transmission, reflection, and constructive and destructive interference, provided results are properly scaled and interpreted. Interestingly, thermodynamic attenuation is also amenable to such modeling. As we have discussed elsewhere in this book, acoustic pressure decays like $P_0 e^{-\alpha x}$ where P_0 is an initial value, x is the distance traveled by the wave and α is the attenuation rate $\alpha = (Rc)^{-1} \sqrt{\{(\mu\omega)/(2\rho)\}} = (Rc)^{-1} \sqrt{\{(\nu\omega)/2\}}$. The effects of viscosity appear *only* through the kinematic viscosity and not viscosity or density individually – and since kinematic viscosity values of air and mud are comparable, air as a working medium is again justified.

We will develop the foundations underlying wind tunnel analysis thoroughly in Chapter 9, where we introduce techniques for short, intermediate and long wind tunnels in the context of siren design. Again, short wind tunnels are used to evaluate torque and erosion; intermediate wind tunnels are used to determine siren Δp, while long wind tunnels are used to develop telemetry concepts. In this chapter, we work with the short wind tunnel exclusively for turbine design. In the next section, we assume that turbine "stall torque" and "no-load rotation rate" are both available from short wind tunnel modeling, that is, from constant density, incompressible air flow measurements about the very complicated geometries described early in this chapter. We then demonstrate how performance curves can be developed for muds of arbitrary density at any downhole flow rate using simple wind tunnel data and aerodynamic conversion formulas. For more detailed discussions on turbine design, the reader is referred to Hawthorne (1964) and Oates (1978).

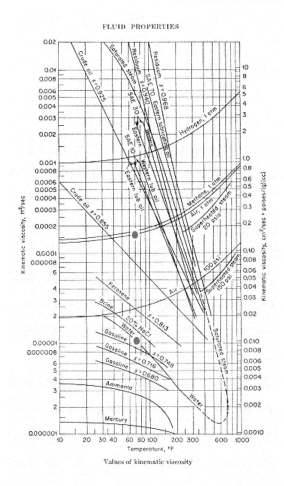

Figure 8.5. Kinematic viscosity versus temperature for various fluids.

8.3   Turbine Model Development

Having justified the use of air as a working fluid, we now address the details needed to apply wind tunnel analysis to turbine design. We ask, "What are the important flow parameters?" Also, "How are wind measurements converted to those for actual downhole flow?" Now, recall from our above discussion that with conditions (i) and (ii) satisfied, condition (iii) implies that the torque T depends on $\frac{1}{2}\rho U^2 SR$ and a dimensionless number C_T which depends on geometric shape only (the "1/2," added for convenience, is customary in aerodynamics). The C_T is our dimensionless "torque coefficient," analogous to the "lift coefficient" used in classical airfoil theory.

Without loss of generality, we assume $T = \frac{1}{2}\rho U^2 S R C_T$. It is important to remember that the torque coefficient is a function of geometry only and not fluid properties, at least to first order. Thus, once we have tested a wood model in the wind tunnel, it can be determined from the formula

$$C_T = T_{air}/(\frac{1}{2}\rho_{air}U_{air}^2 SR) \qquad (8.1)$$

Now, consider a test under field conditions with a mud of density ρ_{mud} and downhole flow speed U_{mud}. This yields a torque T_{mud}. Its torque coefficient would be $C_T = T_{mud}/(\frac{1}{2}\rho_{mud}U_{mud}^2 SR)$. But since the two torque coefficients must be the same, we have $T_{mud}/(\frac{1}{2}\rho_{mud}U_{mud}^2 SR) = T_{air}/(\frac{1}{2}\rho_{air}U_{air}^2 SR)$ or

$$T_{mud} = (\rho_{mud}/\rho_{air})(U_{mud}/U_{air})^2\, T_{air} \qquad (8.2)$$

That is, the torque in mud is linearly proportional to the ratio of mud densities and varies quadratically with the ratio of oncoming speeds (and hence, the volume flow rates).

Next, focus on the wind tunnel test. Whether we perform a test using a wind tunnel, a mud or water test loop, it is essential to use bearings that are low in friction; otherwise, the torques needed to overcome bearing friction may cause significant error in interpreting true fluid-dynamic properties. Because air densities are typically 700-800 times smaller than those in mud, it is essential that the highest quality low-friction bearings be used. Many of these are sealed so that contaminants do not enter. It is also preferable to test at higher air speeds in order to minimize bearing errors associated with torque measurement.

A simple method is available to determine if torques are measured correctly. We assume that a manometer system has been set up to measure the flow speed U. Since the equation

$$T = \frac{1}{2}\rho U^2 S R C_T \qquad (8.3)$$

holds, one should measure torque at several flow speeds U, increasing U from low to high speed. The equation shows a quadratic dependence on U. Thus, if the plot of T versus U is not parabolic, measurements for U, T, or both, may be incorrect. This provides a simple error-checking procedure.

We now turn to turbine performance and experimental details. If the turbine is installed in a wind tunnel and held still so that it does not move while wind of speed U_{air} is blowing past it, the torque that is measured can be denoted as the "stall torque, air," that is, the air turbine "stalls" and does not move, and the torque is given the symbol $T_{s,air}$. This torque can be measured by drilling a small hole through the wind tunnel wall and inserting a linear force gauge – the torque is simply the product of the measured force and the moment arm. At the opposite extreme, let the turbine turn freely at its maximum or "no-load rotation speed" (when this occurs, there is no load across blade upper and lower surfaces and the torque is zero). We denote this rotation speed by $\omega_{NL,air}$. Then, it is true in linear theory (that is, for small flow angles relative to blade pitch under rotating conditions), but also experimentally observed when the angles are large, that

$$T_{air} = T_{s,air} (1 - \omega/\omega_{NL,air})$$ (8.4)

where ω will vary from 0 to $\omega_{NL,air}$. In other words, the relationship between dynamic torque and turbine rotation speed is linear. This is only true of turbines and not generally applicable to sirens (again, sirens may not even move!). Now, turbine power P_{air} is simply the product of T_{air} and ω_{air}, or,

$$P_{air} = T_{s,air} \omega_{air} (1 - \omega/\omega_{NL,air})$$ (8.5)

That is, turbine power is a quadratic function of ω. It vanishes under stall conditions ($\omega = 0$) and no-load ($\omega = \omega_{NL,air}$) conditions. Whether we deal with air or mud, for every desired power level P^* in practice, two different values of ω will give that power because Equation 8.5 is a quadratic equation in ω – the rotation rate chosen in practice will depend on considerations other than turbine aerodynamics. The two values of rotation speed needed to provide P^* are given by the solution to Equation 8.5 as a quadratic equation

$$\omega_{1,2} = [\, T_s \pm \sqrt{\{T_s^2 - 4\, T_s\, P^*/\omega_{NL}\}}\,] / (2T_s/\omega_{NL})$$ (8.6)

In engineering design, the appropriate rotation rate may be dictated by the possibility of shaft vibrations, mechanical packaging constraints, dynamic seal performance, electrical alternator efficiency, and so on. The maximum power possible from this turbine has the value

$$P_{max} = T_s \omega_{NL}/4 \ldots \text{ at } \omega = \tfrac{1}{2} \omega_{NL}$$ (8.7)

Figure 8.6 shows typical turbine properties. Also note from the plot of speed U versus ω_{NL} that this should be a linear relationship. This should be demonstrated experimentally during any test. If it is not obtained, there are measurement errors that must be corrected, e.g., excessive bearing friction, errors in calculating U from manometer measurements, and so on.

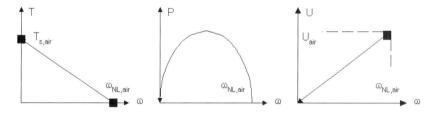

■   Three wind tunnel data points needed to completely
    characterize turbine properties for all muds and flow rates

Figure 8.6. Torque and power versus ω. Also, ω_{NL} versus axial speed U.

It is important to emphasize an additional property of turbines with respect to no-load rotation speeds. It is known from aerodynamics, that under most conditions, the no-load speed is linearly proportional to the speed of the oncoming flow as seen in the far right diagram of Figure 8.6. That is,

$$\omega_{NL} = \text{alpha} \times U \qquad (8.8)$$

where "alpha" is a constant of proportionality that does not depend on fluid properties (it depends only on the geometry of the turbine – the same "alpha" is obtained whether we test in air or in wind and at any speed U). The easiest and most accurate way to determine it is to use a fast air speed in a wind tunnel (to minimize bearing friction effects) and to measure the corresponding no-load rotation speed. Note that we have only needed to obtain two other data points, the stall torque and the no-load speed – as we will show, this all that is required to determine all turbine performance for any mud flowing at any speed.

In summarizing, we have a wind tunnel setup with a free-standing turbine installed *without* any drive motors. We measure the stall torque and the no-load rotation speed. A high oncoming speed should be used to allow accurate measurement of high stall torque $T_{s,air}$ since low torques may be degraded by bearing friction effects – similar considerations apply to no-load speed. The wind tunnel plots for Equations 8.4 and 8.5 are easily created and appear as shown in Figure 8.6.

We now ask, how do we extrapolate these results to any mud of any density flowing at any speed? To do this, we observe that we have assumed that the same geometries (that is, flow patterns when the angle of the oncoming flows are considered under rotating conditions) for both air and mud tests. This requires that the ratios of the transverse velocity to axial velocities be identical, that is, $\omega_{air}/U_{air} = \omega_{mud}/U_{mud}$ so that

$$\omega_{mud} = (U_{mud}/U_{air})\,\omega_{air} \qquad (8.9)$$

This simple relationship also follows from Equation 8.8. Since "alpha" can be computed using air or mud conditions, it follows that we again have $\omega_{air}/U_{air} = \omega_{mud}/U_{mud}$ for which Equation 8.9 follows. Now, let us combine Equations 8.2 and 8.4 to give

$$T_{mud} = (\rho_{mud}/\rho_{air})(U_{mud}/U_{air})^2\,T_{s,air}\,(1 - \omega/\omega_{NL,air}) \qquad (8.10)$$

We multiply Equation 8.10 by Equation 8.9 to give

$$T_{mud}\omega_{mud} = (\rho_{mud}/\rho_{air})(U_{mud}/U_{air})^3\,T_{s,air}\,\omega_{air}\,(1 - \omega/\omega_{NL,air}) \qquad (8.11)$$

Now, the right side of Equation 8.11 can be simplified using Equation 8.5 which states that $P_{air} = T_{s,air}\,\omega_{air}\,(1 - \omega/\omega_{NL,air})$. The left side of Equation 8.11 is the power in mud denoted by P_{mud}. Thus,

$$P_{mud} = (\rho_{mud}/\rho_{air})(U_{mud}/U_{air})^3\,P_{air} \qquad (8.12)$$

For computing purposes, we can write Equation 8.12 in the form

$$P_{mud} = (\rho_{mud}/\rho_{air})(U_{mud}/U_{air})^3 \, T_{s,air} \, \omega_{air} \, (1 - \omega/\omega_{NL}) \qquad (8.13)$$

This relationship states that, under the same dimensionless conditions, the power in mud increases by the ratio of mass densities (or specific gravities) and varies with the third power of velocity (or volume flow rate) ratio. Here, the ratio ω/ω_{NL} refers to rotation speeds for the oncoming speed U obtained for the mud test. If $U_{mud} > U_{air}$, then the corresponding no-load speed is higher as determined from Equation 8.9. If torques and powers are plotted versus rotation rate, where rotation rate appears on the horizontal axis, the range of rotation values for mud will be larger than that for air.

8.4   Software Reference

To understand physically what these equations imply, we perform calculations using the software "turbine.exe," which embodies all of the above theory. Clicking on this filename brings up the application in Figure 8.7. The numbers in the text input boxes are chosen for illustrative purposes and do not represent any real test data. They are selected to illustrate relationships between physical variables. Note, from Figure 8.7, that the two radii are needed so that the program can calculate axial speed from volume flow rate and also determine torque; these inner and outer radii are apparent from, say, Figure 8.17.

The wind tunnel data shown contains the no-load rpm, the stall torque, the air specific gravity and the volume flow rate (it is also important, for repeatability testing, to record temperature and humidity, which affect air density). The computer program will plot the 'torque versus ω' and 'power versus ω' curves for the wind tunnel first. Now, note in Figure 8.7 that we have, for simplicity, assumed the same flow rate for the mud, but a density that is 2,000 times higher. We desire predictions for this situation. (Before running the program the very first time, be sure to click "Install Graphics.") Now click "Simulate." We obtain a sequence of four graphs, as shown below – each graphics window must be closed before the next appears.

Note that the 1 in-lbf in Figure 8.7 is 0.08333 ft-lbf since there are 12 inches in one foot. The 1,000 rpm is 104.7 rad/sec since 1 rpm = 2π rad/60 sec, where "rpm" is "revolutions per minute." Next, the program calculates results for the mud test. Note in Figure 8.8c that the 104.7 rotation speed value is unchanged because in Figure 8.7 we kept the same volume flow rate. But because the density has now increased 2,000 times, the stall torque for the mud test becomes (2,000)(0.08333) or 166.7 ft-lbf as shown in Figure 8.8c. The peak power for the wind tunnel test from Figure 8.8b is about "2.2." For the mud test, this also increases by a factor of 2,000. It is now (2,000)(2.2) or about "4,400" as shown in Figure 8.8d.

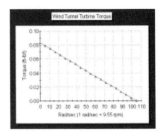

Figure 8-8a. Wind tunnel torque (foot-lbf)

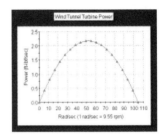

Figure 8-7. Wind vs mud performance utility. **Figure 8-8b.** Wind tunnel power (note units).

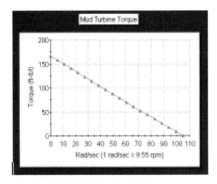

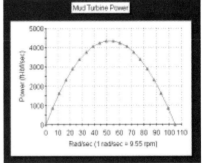

Figure 8.8c. Torque in mud test.   **Figure 8.8d.** Power in mud test.

The above results assume that density alone changes. Now, we keep the mud used, but double the mud flow rate from 200 gpm to 400 gpm. This affects the results in Figures 8.8c and 8.8d as follows. The torque will increase by 2^2 or 4 times, while the power will increase by 2^3 or 8 times. Thus, the torque number becomes 166.7 × 4 or 667 while the power number becomes 4,200 × 8 or 33,600. Because the flow rate has doubled, the turbine no load speed will also double; the 104.7 now becomes 209.4. The screen in Figure 8.7 becomes –

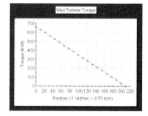

Figure 8-10a. Mud torque with flow rate doubled.

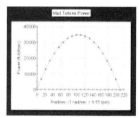

Figure 8-9. Doubling volume flow rate.   **Figure 8-10b**. Mud power with flow rate doubled.

In summary, we have described relationships for physical variables important to turbine design and demonstrated their implementation in "turbine.exe" (the Fortran engine is contained in calc-5.for). As noted, once geometric parameters and wind tunnel test results are entered, the program plots wind tunnel performance curves; also, for mud densities and mud flow rates chosen by the user, the program will plot the corresponding torque and power curves versus rotation rate.

Formulas for the two turbine rotation rates that correspond to a fixed level of desired power are given in Equation 8.6 while the maximum power available is given in Equation 8.7. Again, the numbers used to demonstrate the software do not represent real tests, but were selected for illustrative purposes only. We emphasize that it is not important to run wind tunnel tests at any particular speed, so long as that speed can be measured accurately; our conversion routines provide mud results whatever wind speed is used in the testing. Our fast calculations are almost instantaneous and do not involve iterative methods.

Finally, we give examples from actual MWD turbine hardware tests in mud, to augment our discussions on wind tunnel measurement in the laboratory. Figure 8.11 plots stall torque versus flow rate and correctly shows a parabolic dependence on speed. Figure 8.12 plots no-load rotation rate versus volume flow rate – measurements for a steel turbine in mud and a plastic mockup in air correctly fall on the same straight line. Also, the power curves in Figure 8.13 clearly follow the inverted parabolic trend as do measured data. Torque results in Figure 8.14 for mud and air agree with predicted straight-line trends and correctly scale with "ρU^2" as is demonstrated from theory.

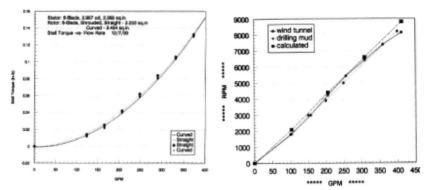

Figure 8.11. Stall torque versus flowrate. **Figure 8.12.** No-load rpm versus speed.

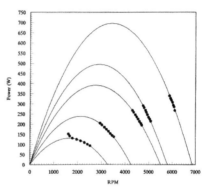

Figure 8.13. Power versus rpm (experimental data indicated).

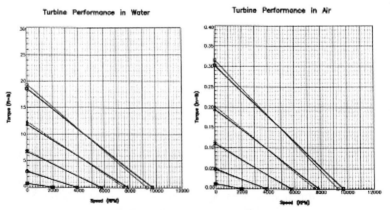

Figure 8.14. Torque results.

8.5   Erosion and Power Evaluation

The results cited here apply to turbines in MWD tools and turbodrills across a range of manufacturers. Turbine erosion and subsequent failure are detrimental to MWD operations. There are two main erosion mechanisms, namely, somewhat tolerable stator trailing edge erosion and more catastrophic rotor tip erosion. The mud flow shown in Figure 8.1 first passes through the stator row, which deflects oncoming fluid and allows it to impact rotor vanes with greater force – this enhances torque and power creation. Since stator vanes are attached to the shroud or housing, their tips cannot erode. The right photograph in Figure 8.15 shows, however, that stator trailing edges can and do erode; this occurs when small trapped rocks and debris induce local high velocity jets that literally cut away metal where the blade joins the central housing. Otherwise, stators undergo a sandpapering effect and erode slowly; because they serve only to redirect flow, stator erosion is somewhat tolerable.

Figure 8.15. Rotor tip erosion (left), stator trailing edge erosion (right).

Rotor tip clearances are required because drill collars bend and flex during drilling operations. These clearances can be large, say, 0.1 inch, in order to prevent the unacceptable jamming (and power loss) that arises from such motions; sizeable clearances also reduce the likelihood of jamming by debris entrapment and mud gelling. Fluid movement from the high-pressure undersides of rotor blades to the low-pressure upper-sides induces "tip vortex" flows similar to those at airplane wing tips, as shown in Figure 8.16. These angular momentum sources, by virtue of conservation laws well known in physics, are difficult to dissipate.

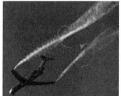  

Figure 8.16. Tip vortex flows, high lift aircraft (left), computer simulation (middle), and flow visualization (right).

In mud flows, tip vortices entrain sand; the continuous high-speed rubbing that results leads to rapid metal loss at rotor tips as shown at the left side of Figure 8.15. It is known that the greater the pitch angle, the greater the strength of the vortex. Since blade angles are high for MWD turbines, this erosion is severe. Reductions in effective radius due to erosion are responsible for large losses in torque and power. Tip vortices have been studied by aerodynamicists. Figure 8.16 also shows results for computer simulations and flow visualization.

There are, however, effective ways to deal with rotor tip erosion and power loss; the remedies used in any particular situation are selected with due attention to trade-offs and compromises. One simple solution, suggested by the aerospace examples in Figure 8.17a, is the wrap-around shroud in Figure 8.17b, a downhole mud turbine prototype for a time-tested aerodynamic concept. The shroud completely eliminates tip erosion. Also, since loadings no longer vanish at rotor tips, the shrouded turbine is more effective in generating torque and power, so that shorter blades are possible. One concern, however, is "sticking" due to mud gelling when drilling operations are interrupted. The relatively large surface shroud areas may form strong adhesive bonds that are difficult to break.

Figure 8.17a. Aerospace blade rows with shrouds.

Figure 8.17b. Turbine prototypes with and without shroud.

An alternative to shrouded turbines is twisted blade design. Figure 8.17c shows aerospace examples commonly used, with three-dimensional velocity and pressure variations highlighted in color. Sometimes, flexible blades are employed that deform at high flow rates. In either case, the design objective is reduced flow efficiency at large angles of attack (induced by flow separation) so that excessive power is not created. An example of a downhole twisted blade concept is given in Figure 8.17d.

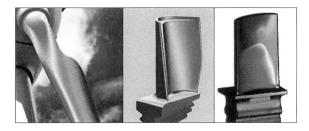

Figure 8.17c. Blade twist, aerospace designs
(pressure variations in spanwise direction clearly evident).

Figure 8.17d. Blade twist, downhole turbine concept.

Finally we emphasize that turbine jamming due to lost circulation material and other downhole debris is a serious concern in operations. This can be assessed from short wind tunnel analysis as described in the comments for Figure 9.3. We note that a viable solution is offered in Gilbert and Tomek (1997) and the reader is referred to their U.S. Patent 5,626,200 for further details.

8.6 Simplified Testing

The role played by wind tunnel analysis in MWD engineering, which is further developed in Chapter 9, is very significant. Most striking is the simplicity and low cost behind the evaluation techniques used. For example, performance trends for our plastic unshrouded and shrouded turbines were efficiently determined in the wind tunnel using the setup suggested in Figure 8.18, which provides a back view of the short wind tunnels described in Chapter 9 (here, wind is blowing out of the page). Recall that only two parameters, namely, no-load rpm and stall torque, are required to determine both 'torque versus rotation rate' and 'power versus rotation rate' performance maps applicable to all fluids at all flow rates. To determine the no-load speed, a piece of reflective adhesive tape is attached to the rearward face of the rotor; an inexpensive optical tachometer is used to measure free-spinning rpm directly. To measure stall torque, the arm of a linear force gauge is inserted through a small hole drilled into the wind tunnel wall. The required stall torque is simply the product of the measured force F and the moment arm R shown. Expensive torque meters are unnecessary. Metal prototypes for mud loop evaluation followed successful concept validation and results were described previously.

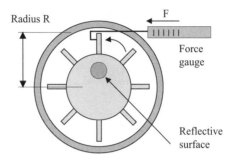

Figure 8.18. Turbine test setup schematic.

A second example demonstrating how wind tunnels can enhance physical understanding is found in the area of three-dimensional effects. The tip vortices in Figure 8.16 can be visualized using smoke, neutrally-buoyant, helium-filled soap bubbles, and small, light-weight, gas-filled beads (the author has evaluated all three). Simple flexible airfoils that can be bent and deformed from run-to-run were used for convenient testing and vortex strength can be assessed visually – the air stream will literally blow the vortex downstream into the room where all can observe its destructive tendencies. These flexible airfoils are constructed by embedding flexible skeletons made from copper wire in rubber skins. The wire skeletons are actually placed in a mold into which liquefied rubber is poured – the curing process requires several hours. Figure 8.19 shows one such flexible airfoil, with the hole revealing a broken solder joint after numerous twists and turns from wind tunnel tests.

Figure 8.19. Flexible rubber airfoil with copper skeleton.

Figure 8.20. Wind tunnel concepts for preliminary evaluation.

Figure 8.20 shows how simple aerodynamic concepts can be evaluated using "shim stock" turbines that are easily fabricated. Since MWD blade pitch angles range easily exceed 50 degrees, resulting in massive flow separation, the details of the airfoil cross-section are secondary (high angles result from the use of single, as opposed to multiple, turbine stages, a limitation imposed by mechanical packaging constraints – without large pitch angles, the required power cannot be achieved). Thus, simple blades cut from shim-stock (that is, soft rolled-metal shown at the right) suffice for qualitative evaluation purposes. Blades are easily trimmed, inserted and glued into balsa wood hubs; smooth outer circular traces are obtained by turning down diameters to the desired size in a lathe. The plastic tubular fixture shown is all that is required for stall torque and no-load rotation rate measurement – Figure 8.18 gives the view from the bottom of the test section in Figure 8.20. Once the effects of twist, blade number, span and chord are determined for a particular configuration, more precisely defined models can be constructed using computer-aided-design (CAD) methods and 3D printing prototypes. This chapter offers only an introduction to wind tunnel testing in MWD design. More sophisticated techniques are given in Chapter 9 in the context of siren analysis, using modern wind tunnels such as that shown in Figures 9-2a to 9-2h.

8.7 References

Hawthorne, W.R., *Aerodynamics of Turbines and Compressors*, Volume X, High Speed Aerodynamics and Jet Propulsion, Princeton University Press, 1964.

Oates, G.C., *The Aerothermodynamics of Aircraft Gas Turbine Engines*, Air Force Aero Propulsion Laboratory Report No. AFAPL-TR-78-52, July 1978.

Oates, G.C., *Aerothermodynamics of Aircraft Engine Components, AIAA Education Series,* American Institute of Aeronautics and Astronautics, New York, 1985.

9

Siren Design and Evaluation in
Mud Flow Loops and Wind Tunnels

We developed "short wind tunnel" analysis methods from a turbine design perspective in Chapter 8.   In fact, we showed how from two simple measurements, namely, those for stall-torque and no-load rotation rate, each of which can be taken in minutes in inexpensive wind tunnels powered by simple blowers, the complete turbine torque and power versus rpm curves can be obtained for drilling mud of any density flowing at any speed.   Here, we describe short wind tunnel design and test techniques further, in support of earlier turbine development ideas.

We also introduce mud siren design in short wind tunnels, in particular, analysis methods for stable-closed versus stable open, static versus dynamic torque, and erosion evaluation. "Short wind tunnels" are short, say, five to ten feet, and are used for hydraulic testing of properties that are independent of compressibility, e.g., torque, power, erosion, flow visualization, approximate viscous pressure drops and losses. On the other hand, acoustic properties like signal strength, harmonic content, and constructive and destructive wave interference, require not only longer wind tunnels that allow accurate simulation of wave interactions and reflections, but more sophisticated math models for data reduction, interpretation and extrapolation.

We will develop ideas for "intermediate length wind tunnels," which are typically 100-200 feet, and also, for "long wind tunnels" which range up to 2,000 feet in length. Some of the wind tunnels and devices described here were designed very early on, but in recent years, very sophisticated, highly instrumented systems have been developed. This chapter focuses on basic principles and applications of mathematical models given earlier in this book, and consistent with this philosophy, applies newly developed ideas to simple test systems that can be constructed quickly and inexpensively.

Sirens require testing using multiple wind tunnels to support detailed evaluation of design trade-offs and compromises. For instance, strong acoustic signal strength is useless unless turning torques are low and erosion is minimal. On the other hand, one might select a low-torque siren shape that by itself creates weaker signals, but which provides strong signals by virtue of a telemetry scheme which takes advantage of constructive wave interference.

Or, possibly, one might consider how two sirens placed in series might be phased in order to develop a single coherent reinforced signal. Multiple sirens address pressing issues confronting existing single-siren systems. Single-sirens require very small rotor-stator gaps for signal generation which are associated with high erosion and large torques. Multiple sirens, while complicating mechanical design, offer increased lifespan and reduced turbine power demand.

As an additional example for long wind tunnel analysis, consider turbine-siren interactions. In existing siren-based MWD tools, the siren is placed above the turbine – the usual reason offered, namely that an upstream turbine would block the MWD signal, is fallacious – simple tests in a long wind tunnel have shown that turbines do not reflect the long acoustic waves important to signal transmission. This indicates that the siren can be placed closer to the bottom of the MWD collar, and nearer to the drillbit, without incurring any penalties. This shortened distance, obviously, implies that constructive interference based on drillbit reflections can occur in less time, and thus, data rates will be higher.

The list of potential important applications goes on and on. For this reason, it is important to understand the concepts behind wind tunnel methodologies fully. The need for short wind tunnels to evaluate siren torque, power and erosion is clear, as is the need to use long wind tunnels to study wave interference absent of the complicating reflections plaguing shorter tunnel lengths. Wind tunnel turbine design was addressed in Chapter 8 because it supports mud siren integration – modern MWD systems cannot function flexibly unless they can effectively generate the power needed.

9.1   Early Wind Tunnel and Modern Test Facilities

We consider wind tunnel analysis in greater detail and review basic principles explaining why wind tunnels work. According to the principle of "dynamic similarity," applicable to all scientific disciplines including fluid mechanics, the dimensionless parameters governing two separate experiments must be close in order that they describe identical physical phenomena. For example, this is how aerospace companies perform experiments using more convenient and cheaper models. In commercial jetliner design, inexpensive small-scale models are tested in small wind tunnels powered by low-speed blowers, thus eliminating the need for jet engines and full-scale prototypes. Then, flight characteristics are extrapolated to large size aircraft flying at faster speeds or at altitudes with different air densities.

The test engineer's judgement is important when more than one dimensionless parameter governs the problem and it is not possible to match all of the parameters. Wind tunnel results for lift, i.e., the upward force keeping the aircraft in the air, almost always are corrected for "displacement thickness" effects because Reynolds numbers do not exactly match. Wind tunnel usage in the petroleum industry is not completely new. Forces and unsteady loads (due to vortex shedding) acting on offshore platforms are also routinely obtained from wind tunnel measurements extrapolated to hydrodynamic environments. Here, dimensionless Reynolds and Strouhal numbers enter the picture. In any event, which parameters are more important and how are results to be interpreted in any given application constitute difficult questions. Answering these is often more of an art form than a precise science.

9.1.1   Basic ideas.

In low speed viscous flow, typical of many downhole petroleum applications, the dimensionless "Reynolds number" Rey of the model test should be almost that of the full-scale problem. Let U be the oncoming speed of the flow, and L be a characteristic length of the model, say, the diameter; also, let v denote the kinematic viscosity, which is simply the quotient "viscosity/density." Then, the Reynolds number is given by $Rey = UL/v$. If we keep the mud and wind speeds U identical, and the scales L for test and full-scale geometries the same, both of which are easily accommodated for most downhole tools in wind tunnel analysis, then dynamic similarity is guaranteed if the kinematic viscosities are close.

In Figure 8.5, the kinematic viscosity is plotted versus temperature for different types of fluids. In the chart, the red dots show the kinematic viscosities of air (for wind tunnel application) and then water for reference. It may seem, at least superficially, that water should model drilling mud accurately, since they are equally dense and just as wet! The kinematic viscosity of water, indicated by the lower red dot, is "0.00001." But water is not drilling mud. Drilling mud is about 20-30 times more viscous and approximately twice as dense. Thus, the kinematic viscosity is about fifteen times more than that of water. If the lower red dot is followed upward vertically (along the 70° F room temperature line), it is seen to occupy the "0.00015" position occupied by the red dot for air, exactly fifteen times as much, as required!

In fluid mechanics, "close" Reynolds numbers need not be too close – a factor of five to ten difference might suffice for closeness. But the closeness in Reynolds number just demonstrated – and the fact that wind tunnel testing is just as applicable when both wind and mud scenarios are turbulent – are significant from the testing point of view. These suggest that wind tunnel tests are the best substitute for drilling mud – and the fact that air is convenient, clean, free and safe, and testing is fast, does not hurt. Only one other fluid works, namely, methane at room temperature and pressure; methane, of course, is difficult to work with, being explosive and toxic, not to mention its higher cost. These reasons explain why clever use of the wind tunnel analysis for hydraulics and acoustics enables rapid progress to be made.

While dynamic similarity is achievable, testing errors are possible, which are fortunately avoidable. Because air is about one thousand times less dense than mud, force, torque and power are also one thousand time less for the same model and flow speed. If low wind speeds are used to study higher mud flow rates, which is common in practice, the differences increase further. To reduce experimental error, siren and turbine models must be installed using low friction bearings. Also, the highest possible wind speeds should be used, meaning that large blowers are best. Volume flow rate ratings stated by HVAC manufacturers may not be not meaningful – these apply to calibrated flows in rectangular ducts where resistance is small compared to ducts containing large sirens with high blockage. In wind tunnel applications, flow resistance will generally remain high – and for the same blower settings, flow rates will vary with the particular siren or turbine in the test section. So three simple rules apply: low-friction bearings; powerful electric blowers for high test speeds and, importantly, accurate measurement of volume flow rate from test to test.

9.1.2   Three types of wind tunnels.

There are three types of wind tunnels useful to MWD testing, namely, "short," "intermediate," and "long." Short tunnels are used to study physical properties associated with the constant density, incompressible nature of the fluid, e.g., torque, power, streamline pattern, erosion, drag, viscous flow separation, free-spinning turbine speeds, and so on. Longer wind tunnels are used to study acoustic wave propagation in high-data-rate applications, where the wavelengths are conveniently shorter than those in mud flows, but still long when compared to the wind tunnel diameter of the flow cross-section. By short wavelengths, we mean several to about one hundred feet – the literature on "short wave" or "high frequency" acoustics, e.g., ultrasonics, does not apply, because wavelengths are many orders of magnitude smaller.

Long wind tunnels allow us to simulate sound reflections at the surface and downhole near the pulser, and also to measure signal strength produced by particular siren designs. They also provide a convenient laboratory for evaluating single and multiple-transducer signal processing schemes. They allow us to test different telemetry schemes under more or less ideal conditions before field tests with real mudpumps, desurgers and mud motors are present. However, acoustic data reduction, interpretation and extrapolation require more sophisticated mathematical methods than those used for turbine torque and power testing in Chapter 8. We now discuss the design and use of short wind tunnels, and later, extend our methods to long wind tunnel acoustic analysis.

9.1.3   Background, early short wind tunnel.

Whereas the "long wind tunnel" is used to evaluate pulser acoustic signal strength, wave propagation, reflections, telemetry schemes, constructive and destructive wave interference, surface signal processing schemes, and so on, the "short wind tunnel" is used to study properties like mud siren torque and power, turbine torque and power properties, strainer erosion, jamming due to debris, viscous flow separation, streamline pattern, and so on.

Because fluid compressibility and sound reflections are unimportant for these purposes at typical mud pump flow rates, the wind tunnel does not need to be long. It can be short, say ten feet or less, just long enough that three-dimensional blower inlet and outlet end effects are unimportant. The author's earliest wind tunnel is shown in Figure 9.1 for historical purposes. This 1980s vintage design met only the bare minimum requirements for "wind" and "tunnel," with a simple squirrel cage blower attached to plastic tubes mounted on a table-top – a very crude design, built for several hundred dollars literally overnight, which nonetheless proved useful for testing. Its importance, of course, was strategic, setting the groundwork for the methods described in Chapter 1 that have since been adopted by numerous organizations.

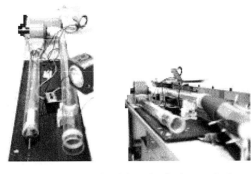

Figure 9.1. Very early (historical) short wind tunnel.

Wind tunnel usage in downhole applications was widely adopted soon after its initial usage. In "Flow Distribution in a Tricone Jet Bit Determined from Hot-Wire Anemometry Measurements," SPE Paper No. 14216, by A.A. Gavignet, L.J. Bradbury and F.P. Quetier, presented at the 1985 SPE Annual Technical Conference and Exhibition in Las Vegas, and in "Flow Distribution in a Roller Jet Bit Determined from Hot-Wire Anemometry Measurements," by A.A. Gavignet, L.J. Bradbury and F.P. Quetier, SPE Drilling Engineering, March 1987, pp. 19-26, the investigators, following ideas suggested by the lead author, showed how very detailed flow properties can be obtained using aerospace wind measurement methods. The reader is referred to these papers for descriptions of the improved set-ups. For siren and turbine test objectives, however, the three-dimensional methods developed there are not necessary.

Importantly, additional scientific justification supporting wind tunnel usage was offered in these publications, premised on the "highly turbulent nature of the flow." The investigators' arguments, in "plain English," point out that the turbulent air flow found in laboratory wind measurements is similar to the turbulent mud flow likely to be found downhole. This counter-intuitive (but correct) approach to modeling drilling muds provides a strategically important alternative to traditional testing that can reduce the cost of developing new MWD systems. Again, wind tunnel use in the petroleum industry was by no means new at the time. For instance, Norton, Heideman and Mallard (1983), with Exxon Production Research Company, among others, had published studies employing wind tunnel use in wave loading and offshore platform design.

9.1.4   Modern short and long wind tunnel system.

While the short wind tunnel in Figure 9.1 represents the author's very first rudimentary design, the wind tunnel system shown in Chapter 1 represents a state-of-the-art facility developed to handle all of the previously discussed requirements for modern mud siren and turbine testing. This MWD wind tunnel, developed at the China National Petroleum Corporation (CNPC) in Beijing, together with recent high-data-rate telemetry ideas and test results, is described by Y. Su, L. Sheng, L. Li, H. Bian, R. Shi and W.C. Chin in "High-Data-Rate Measurement-While-Drilling System for Very Deep Wells," Paper No. AADE-11-NTCE-74 presented at the American Association of Drilling Engineers' 2011 AADE National Technical Conference and Exhibition, Houston, Texas, April 12-14, 2011. Since this paper was presented, several more wind tunnel systems have been constructed for MWD applications.

Here we recapitulate CNPC highlights to show how the testing technology has evolved since the 1980s to address downhole problems and issues. The short wind tunnel system with a blower driver is shown in Figure 9.2a. After passing through the siren test section, air blowing "out of the page" turns to the right, flowing through the black tubing to the outside of the test shed, and then into the long wind tunnel in Figure 9.2b. It is important to note that abrupt turns, while affecting viscous pressure losses and pump requirements, do not reflect MWD signals because the acoustic waves are extremely long in length.

Siren tests in mud loops for torque, erosion, signal strength, harmonic content, constructive or destructive wave interference, and so on, are impractical because they are time-consuming and labor-intensive, with parts expensive to fabricate. Many geometric parameters affect the physical outcome, e.g., number of lobes, stator-rotor gap separation, rotor taper, rotor-housing clearance, flow angle into the siren, and so on – operational parameters include flow rate and rotation speed. Tests, even wind tunnel tests, must be cleverly designed, and for this reason, the computer simulation methods in Chapter 7 were developed to be used as screening tools. However, such models are not entirely accurate.

Blower with
muffler and
water coolant

Independent
power supply

Flow meter

Flow
straighteners

Siren test
section and
electric
controller

Torque meter

Drive motor

Total length,
about 60-70
feet

Observation
window to
outdoor
flowloop for
multiple
transducer and
wave
interference
testing

Data acquisition
for flow rate,
torque, multiple
and differential
pressure
transducers

Siren Δp
transducer

Connection to
outside long
wind tunnel

Figure 9.2a. Short wind tunnel test system.

Figure 9.2b. Long wind tunnel system with pressure transducers
(see Figure 1.5b.1 for additional photos).

Gridding constraints and computer resources, for example, introduce errors which do not allow the finest details of the flow to be modeled accurately; however, the computer models are useful in identifying which engineering concepts are worthy of more detailed experimental study. Properties related to torque and open-closed stability, which are weakly dependent on viscous effects, could be and were modeled quickly using the steady approaches of Chapter 7.

Various design concepts using the sirens shown in Figures 1.5d.1 and 1.5d.2 were evaluated in detailed test matrixes. These were designed to answer numerous questions, each associated with significant hardware design consequences, and perhaps, to raise even more questions. For instance, for a fixed signal frequency, is it better to spin a siren with few lobes rapidly or is it more advantageous to rotate a siren with many lobes slowly? Different choices affect torque, and hence power, and finally turbine or battery options. Tests for Δp indicate that "fewer lobe sirens" produce much stronger signals – this is the standard solution for single-siren systems. However, in high-data-rate applications, greater lobe numbers may be required; for these problems, signal data helps estimate the number of sirens to use in multiple "sirens-in-series" designs, which enhance net signal by local constructive wave interference. This motivated our re-examination of "larger lobe sirens" and, in particular, the testing of signal strength when sirens are connected in tandem.

Another interesting question is this: is it better to use two sirens with large gaps mounted on the same shaft or a single siren with an extremely small gap as is done conventionally? The latter is simpler mechanically. But here, erosion lifespan, torque and power issues are raised. And if multiple sirens are used, should they all operate simultaneously at an identical frequency, or separately at separate frequencies for very high data rate? Our point is this: many design premises that have become commonplace need to be completely reassessed without prejudice and wind tunnel analyses offer convenient, fast and accurate solutions. Wind tunnel tests and apparatus must be developed around MWD specific needs. The photographs in Chapter 1 show different components of the experimental system. Their designs are not final; they, themselves, represent improvements to earlier implementations.

As an example, the use of squirrel cage blowers as shown in Figure 9.1 is acceptable for sirens that are spinning quickly. If not, the intermittent blockage produced by the lobes induces a significant back-interaction on the blower blades that is noticeable by eye. When this occurs, it is impossible to maintain a steady flow and all experimental results are suspect. In later designs, different compressors were used to provide flow; some required lengthy charging even for short test times, and other more powerful models, more often than not, produced air discharges interdispersed with small oily droplets. Often, the choice of blower is limited by what is already available. The powerful blue blower shown in Figure 9.2a, although noisy, provided an acceptable solution. In addition, the layout of the wind tunnel system often depends on facilities and cost constraints.

Figure 9.2a shows the siren test section (within the short wind tunnel). Differential pressure transducers appear in Figure 1.5h. As noted elsewhere in this book, when a positive overpressure is found on one side of the siren, an underpressure is found on the opposite side; this difference is the so-called "Δp" signal strength associated with the valve. This strength depends on flow speed, rotation rate, geometric details and air density. The differential pressure transducer is important and useful because, provided the two taps are close, the measurement is unaffected by flow loop reflections that may be traveling back and forth (these traveling waves simply cancel). It therefore represents the siren's "true strength."

On the other hand, the pressure signal "p" that ultimately leaves the MWD collar and travels to the surface is measured by the single transducer at the bottom right of Figure 1.5m (this is located just outside the test shed). It is this "p/Δp" that is optimized in the six-segment acoustic waveguide analysis of Chapter 2. We noted that the taps of the differential transducer should be "close" so that the effects of traveling waves (due to reflections) are subtracted out identically. The required distance depends on fluid sound speed and transducer sensitivity. To determine this separation, one simple test is all that is needed. Shout "hello" down the plastic tubing: if the differential transducer does not respond, then the separation distance is acceptable. The short wind tunnel system measures the foregoing "Δp."

Chapter 1 displays other wind tunnel components, e.g, electronic flowmeter, differential pressure transducer, piezoelectric transducers, flow straighteners, motor drive, torque gauge, angular position and rpm counter assembly. Also shown are shown instruments for control and simultaneous recording of flow and rotation rate, differential pressure, single transducer pressures at four locations in the long flow loop. These measurements are taken as time progresses. Single transducer data is used to assess the degree of constructive or destructive interference due to downhole reflections, and also, to provide data for multiple transducer signal processing (noise is introduced at the opposite end of the long wind tunnel). Data relating Δp to angular position are useful in developing feedback control loops ultimately needed for telemetry. The reader interested in developing a wind tunnel facility for high-data-rate mud pulse telemetry testing should review Chapter 1. Other wind tunnel systems have since been built which improve upon the advances cited there.

9.1.5   Frequently asked questions.

Here we address common questions and issues that concern potential uses and misuses of wind tunnel modeling.

- *To which fluids does wind tunnel modeling apply?* The petroleum industry correctly emphasizes the non-Newtonian nature of drilling muds and cements versus Newtonian fluids like air and water; their rheological properties are, of course, important when studying losses in drillpipes and annuli. However, for high Reynolds number turbine and siren torque analysis, and for erosion analysis in turbines, sirens and surface strainers, rheology plays a secondary role and air is suitable as the modeling fluid. For example, in Chapter 8, actual turbine torque data collected in mud flow loops and wind tunnels fall on identical performance curves when properly normalized. In addition, direct post-mortem evaluations conducted by this author have demonstrated almost identical metal erosion patterns (for the same test hardware, e.g., turbines, sirens, flow strainers, proppant transport tools, stabilizers) whether the flowing mud is water, oil or emulsion based. Properties like pressure drop and skin friction, affected by flow parallel to the oncoming mud, are rheology-dependent, and cannot be properly modeled with air – recourse to computation or mud experiment is necessary. Again, we emphasize that turbine torque and power in air (arising from transverse flow) accurately simulates that for mud flows at any speed. This is also true of flows past sirens. Note that it may not always be possible to model foam flows because of their highly compressible, two-phase, non-Newtonian nature, and that any conclusions must be determined on a case-by-case basis.

- *Can MWD signal strength be studied using static Δp measurements in a short wind tunnel?* Several companies have characterized MWD signal strength by measuring the pressure drop (or, "Δp") across stationary pulsers and sirens, from upstream to downstream, in mud flow loops, assuming that these Δp's are representative of signal strength. Such datasets are then incorporated in telemetry planning software used at the rigsite. This procedure is completely incorrect because such Δp measurements bear no relation to the acoustic signals that travel uphole. Why? The flow past a stationary cement block, for example, is associated with a high pressure drop, but this does not mean it is useful for high-data-rate sound transmission. Of course, the block may turn slowly, causing Δp to change slowly, but this represents a slow change to hydrostatic pressure. This slow change is detectable uphole, but it is not useful for high speed telemetry. True sound transmissions can only be effected by rapid movement, e.g., the action of a piston in a pipe, the vibrations of a tuning fork in a room, the propagation of sound in acoustic mufflers and, of course, the rapid opening and closing of a mud siren against an oncoming flow. The *static* Δp method is invalid for both mud loops and wind tunnels.

- *Are Δp measurements useful at all?* Yes, they are, when performed and used properly. As we will demonstrate in our "intermediate wind tunnel" discussion, differential pressure can be used to determine signal strength for MWD signal sources where created pressures are antisymmetric with respect to source position – that is, for positive pulsers and sirens. However, a nonzero Δp is not a general requirement for MWD design. For example, the disturbance pressure field associated with a negative pulser is symmetric with respect to the source point and Δp vanishes identically whatever the signaling scheme. For negative pulsers, the correct characterization would involve "delta axial velocity" because velocities at both sides of the source are equal and opposite.

- *How accurately does the short wind tunnel model erosion?* The use of short wind tunnel modeling in evaluating and correcting turbine erosion was discussed in Chapter 8. The present chapter will develop methods for siren flows. The author, in addition to turbines and sirens, has worked extensively in two additional hardware classes, namely, (1) MWD flow strainers, placed at the surface to capture debris which would otherwise impinge upon the downhole tool, and (2) proppant transport tools used to convey sand-laden stimulation fluids downhole for hydraulic fracturing. In both applications, wind tunnel flow visualization correctly replicates areas of erosion known from field results; changes to hole, orifice and slot patterns suggested by wind tunnel analysis which would significantly reduce erosion again, in both hardware classes, led to substantial improvements to life-span. These subjects are beyond the scope of this book and will be covered separately. We emphasize that metal erosion rates cannot be estimated from wind analysis, nor can the exact surface pattern that erosion ultimately takes with time – these answers depend on the details of mud and metal interaction which can only be obtained from actual operations. The wind tunnel, however, is useful in revealing the physical mechanisms that are likely to cause trouble, and does provide means for averting such events. For instance, streamline flow patterns, which are easily visualized, provide some indication of particle impingement angles, which control ductile versus brittle fracture. Changes to these angles, possible by aerodynamic tailoring, can significantly affect life span.

- *What are potential error sources in wind tunnel measurement?* Wind tunnel analysis is useful in many ways but care must be taken to avoid incorrect conclusions. For flow visualization and erosion analysis, actual metal parts or wind tunnel prototypes of new models may be placed in the wind tunnel and evaluated directly. Simple visualization is possible using threads bonded to solid surfaces – rapid streamline convergence is an indicator of possible erosion. Observation may be facilitate by coating threads with florescent dye and photographing their trajectories in the dark. The use of lightweight beads introduced into the flowing air is very often successful, although smoke (often seen in television commercials) presents a problem since smoke filaments dissipate too rapidly. To ensure accuracy, one observes that the air emerging from the blower is generally turbulent and may contain rotating air masses associated with turns in the wind tunnel tubing. Generally, the placement of flow straighteners, e.g., as shown in Figure 9.2h, upstream of the siren or turbine would suffice for most testing purposes. In turbine and siren torque measurement, it is essential to use low friction bearings so that the torque needed to overcome bearing resistance is much less than the aerodynamic torque. If such bearings are unavailable, relative measurement error can be reduced by operating the blower at high speeds. Whether blower speeds are high or low, it is absolutely essential to be able to measure volume flow rate accurately. Digital or analog wind anemometers available from laboratory supply houses should be used with care since pre-programmed calibration data may be inapplicable. For the same blower electrical setting, different volume flow rates will be obtained depending on the blockage offered by the turbine, siren, and other appendages, such as the central hub, the electric motor used, the manner in which the model is supported, and so on. Accurate volume flow rate measurements are important because air data is used to infer performance characteristics for muds of arbitrary density flowing at general speeds – if wind measurements are inaccurate, so will our predictions. To assign numbers to these uncertainties, we note that torques measured at the same speed in water versus air are different by a factor of about eight hundred, that is, the density ratio. Thus, a small error in wind tunnel torque measurement implies a large error in drilling mud applications. If low wind speed data is extrapolated to high mud velocities, the potential for error is even greater. In a discussion below, we discuss how known properties for torque versus flow rate can be used for error-checking the experimental procedure. For example, if measured siren or turbine torques do not vary quadratically with flow rate, either the flow rate measurement is flawed, the torque determination is inaccurate, or both. Such simple tests should always be performed at the start of the day to ensure data integrity. In almost all tests with siren or turbine rotation, shaft vibration and unsteady flow will contribute to test error and uncertainty, and will invariably require some type of data smoothing.

9.2 Short wind tunnel design

Siren testing need not be complicated or expensive. The sophistication apparent in the system of Figure 9.2a is often not necessary. Short wind tunnels are easily designed and fabricated – it is their use that is subject to subtlety. We emphasize that mud sirens, turbines and strainers which are to be tested should be built to full scale, and that the inner diameter of the plastic test section should be identical to the inner diameter of the drill collar. The siren or turbine, plus the central hub it is mounted on, should replicate that used in the MWD collar. If different sections of plastic pipe are joined, care should be taken to prevent leaks at the seams (and at blower-pipe junctions) since these would lead to measurement errors for volume flow rates. The test sections should be made of clear plastic tubing with non-reflective surfaces so that digital camera pictures of test setups and flow visualization results (e.g., thread orientations) can be easily obtained. The board upon which the wind tunnel is mounted should be painted with black non-reflective paint to facilitate photographic and movie documentation. Thus a digital camera, and preferably, a digital movie camera and also a tripod, should be a permanent part of the wind tunnel system. The basic wind tunnel, in its most rudimentary form, is illustrated in Figure 9.3 and can be built for several hundred dollars.

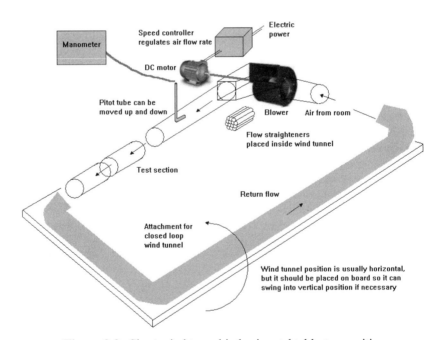

Figure 9.3. Short wind tunnel in horizontal table-top position.

The above "short wind tunnel" consists of a high-power squirrel cage blower, a DC motor and an electric controller to regulate blower speed. Volume flow rate measurements, which should be performed before each change in test conditions, should ideally be obtained from "old fashion" manometer rake procedures which apply Bernoulli's principle. Use of electronic flow meters with digital readouts, while convenient, should proceed with caution, and then, only when their results are carefully calibrated against manometer results. We will discuss a simple error-detection procedure later.

Figure 9.3 also shows flow straighteners placed ahead of the siren or turbine test section. Without flow straighteners, siren and turbine torque measurements can be highly inaccurate, as they are affected by rotating air masses that originate from the blower or which discharge from corner turns in the closed-loop mode discussed next. "Closed loop" operations are enabled by attaching secondary tubing highlighted by the gray ducting shown. This mode is useful in assessing the tendency of sirens or turbines to jam in the presence of lost circulation materials. Styrofoam pellets, glass beads, string snippets, neutrally buoyant soap bubbles and so on, can be introduced into the air stream and forced to recirculate around the wind tunnel without dispersal into the laboratory setting (sticky styrofoam pellets, e.g., from bean-bag stuffing, can be unstuck by using static removal agents like StaticGuard™ easily found in supermarkets). One effective evaluation technique includes the use of a stroboscope (with room lights turned off) with stop-action photography so that debris entrapment and escape in narrow gaps can be observed with the test apparatus rotating. More quantitative assessment is possible by measuring current fluctuations in the motor drive needed to turn at constant speed. If the particles introduced are heavy and segregate toward the bottom, the wind tunnel in Figure 9.3 can be pivoted to operate vertically, so that debris are uniformly dispersed throughout the cross-section as wind flows downward.

In some applications, e.g., determination of siren stable-open or stable-closed behavior, or turbine no-load rotation rate, the model is not driven by an electric motor. Here, a specially crafted thin-wall test section can easily slide into the wind tunnel tubing or form part of the main tubing itself (in the former case, it is secured by diametrically opposed pins). In other problems, the siren must be driven by an electric motor and operated at a given rotation speed. It is preferable if a powerful motor is used that is small enough to fit within the tubing. However, if wind speeds are large (so that torques, which vary as the square of the oncoming flow speed, are likewise large), small motors may be difficult to obtain or are expensive. The solution used in the CNPC wind tunnel is simple. The test section shown in Figure 9.2f is installed as in Figure 9.2a. The siren is rotated by a connecting shaft that extends *outside* the wind tunnel that is in turn connected to the motor. Care is taken to avoid leaks, using appropriate seals, which would compromise measurements taken in the long wind tunnel. However, if long wind tunnel measurements are not required, the short wind tunnel outlet can open directly into the room. Volume flow rate measurements are generally taken upstream of the siren and downstream of the flow straighteners.

For short wind tunnels, area discontinuities in the tubing can be tolerated. However, we emphasize that in designing a combined short and long wind tunnel system, sudden changes or mismatches in area – even small changes – will result in spurious reflections that degrade the signal traveling up the drillpipe. Consider the following "thought experiment." Suppose that the MWD collar area available for flow (that is, the area associated with the inner diameter minus the cross-sectional area of the central hub which supports the siren) and the drillpipe area are not equal. Then signals created by the siren travel toward the drillpipe, and at the collar-pipe junction, part progresses upward while part reflects downward (the latter travel to the drillbit, reflect upward to the same junction, and so on). All the time, new signals are created only to repeat this behavior. The net effect is a "blurry" signal which must be smoothed using the signal processing methods discussed earlier in this book. For this reason, one ideally matches the internal collar area to the drillpipe area if possible. Similarly, unintentional mismatches at short and long wind tunnel junctions result in noisy data. Such effects are often lost in data reduction. For example, one long flow loop used in industry MWD tests contains area discontinuities in certain buried parts of the piping that had long been forgotten.

As part of the flow rate control system, we might have the manometer and pitot probe system in Figure 9.3. The manometer and pitot probe system uses the Bernoulli principle to determine the local flow speed at the inlet of the pitot probe; this principle is explained in elementary fluid mechanics books. A single probe is shown in Figure 9.3, but ideally, in the interest of saving time during the experiments, a "rake" consisting of several probes extending completely across the diameter of the inside of the test section is preferable. It is not necessary to purchase a sophisticated system. A simple system can actually be built for a very low price. The pitot probe itself may be acquired from a laboratory supply house (many university aerospace engineering departments may supply information on vendor sources), while the pressure measurement itself may use arrays of U-tube fluid manometers than can be constructed by hand for very low cost. The U-tube manometer operates on a simple principle taught in freshman physics classes: the difference in the height of two fluid columns is converted into a pressure measurement, which is in turn translated into a velocity reading via Bernoulli's equation. This manometer-pitot tube array should be used for velocity determination. Wind anemometers may be acceptable if they actually measure local speeds and do not rely on calibration using ducts of given sizes without blockages. Whether automated or manual velocity measurements are used, it is important to set up the test procedure to record velocities at several positions in the cross-section.

From the pressure measurement for each pitot probe at a radial location "r," the oncoming axial flow velocity u(r) is calculated using Bernoulli's equation. This is not the final form of the flow rate information required. In our downhole applications, we require the integrated volume flow rate Q inside the drill collar and not the individual velocities in u(r). These quantities are related by the formula

$$Q = \int_0^R u(r)\, 2\pi r\, dr \tag{9.1}$$

which integrates u(r) across the circular area of the inside of the test section. If a pitot tube rake is used, all the required velocity inputs are available immediately and much time can be saved. Otherwise, if a single probe is used, then that probe will need to be moved across the test section and measurements are time-consuming. To make the testing procedure as efficient as possible, the integral can be calculated immediately by a personal computer that automatically accepts the input velocity data, which then computes and stores Q along with the measured velocity profile u(r). Again we emphasize the need to record Q each time the model is changed or the blower or rotation speeds are altered. To ensure accuracy in the measurement procedure, physical checks should be employed. For turbines, doubling Q should double the no-load rotor speed and quadruple the torque. For sirens, doubling Q should quadruple the opening or closing rotor torque (this measurement is described below).

Figure 9.3 also indicates "flow straighteners" (those used in the CNPC wind tunnel are displayed in Figure 9.2h). These must be placed inside the wind tunnel immediately after the air exits from the blower so that the streamlines impinging on the turbine or siren are straight. They are necessary because the blower produces very "confused" rotating air masses that do not flow in a straight manner as the flow would in a downhole situation. These flow straighteners also destroy the rotating air masses that may be formed at turns in the closed-loop system. These can be constructed very simply by gluing together a number of thin-walled plastic tubes and anchoring them in place just downstream of the blower outlet so that they cannot be blown away. The length of these tubes should be about six to eight inches.

The above discussion emphasizes the importance of clean straight flows upstream of the turbine or siren so that spurious rotating air masses do not bias erosion and torque results. When we discuss "intermediate wind tunnel" use in acoustic signal strength testing, we will explain the use of flow straighteners downstream of the rotating siren rotor as well. Essentially, strong rotating air masses are developed behind the siren rotor that are associated with swirling vortex transients. These pressures are of a hydraulic, incompressible flow nature, and are identical in frequency to the acoustic pressures generated by opening and closing the rotor relative to the stator. As compressible pressures are the ones transmitted to the surface, the hydraulic component must be eliminated to ensure accuracy in characterizing the acoustically significant Δp.

We had previously suggested the use of the gray colored "return section" shown in Figure 9.3 to provide a closed-loop system useful in lost circulation jamming evaluation. There is actually no reason why a second siren (with its own flow straightener) driven by a second motor cannot be installed in this section. If differential pressure transducers are used to measure Δp in each case, the measurements at each siren are not affected by the presence of the other siren. This test mode provides for some time and labor efficiencies since the volume flow rate Q is measured only once.

9.2.1   Siren torque testing in short wind tunnel.

Two important properties are important for high-data-rate mud pulse telemetry. First, because the siren must stop, start, increase or decrease rotation speed very often, the stopping and starting torques associated with turning should be low – or at least low enough that the driving electric or hydraulic motor does not find it problematic. Be aware that torque increases quadratically with flow rate. Mud pumps used in the field do not pump at constant speed. Thus, unplanned flow rate increases downhole may prevent a marginal siren from working at all – in fact, a "stable closed" siren will block the oncoming mud flow, prevent turbine operation, and build up dangerous pressures in the drillstring. Thus, torque measurement, and siren design for low torque, are two important objectives for short wind tunnel use. The second siren property important to high data rate is strong Δp signal strength. Because carrier frequencies need to be high, attenuation will be an important design consideration – strong Δp signals are required so that pressure waves can be detected faraway from the source with minimal error. We will discuss testing methods and telemetry design approaches in the context of "intermediate" and "long" wind tunnel testing later. As we will discuss in Chapter 10, nature may limit the maximum Δp that can be created for a given siren pulser. Fortunately, this can be increased by clever use of telemetry schemes using constructive interference, by using sirens-in-series, and so on.

In Chapter 7, we discussed the "stable open" or "stable closed" nature of many siren designs. Sirens with rotors upstream tend to be stable closed and possess restoring torques that are strong when an attempt is made to hold the rotor open – simply *try* holding open such a rotor in the wind tunnel! Downstream rotor configurations may be stable-open, stable-closed, and often, partially-open. The restoring torques which tend to keep closed rotors in the closed position tend to be much smaller than those for upstream rotor designs. All of the comments in this paragraph are based on actual empirical observation. We will focus on downstream siren rotor designs, that is, the design developed by this author in the early 1980s currently used in commercial MWD tools. Figure 9.4a provides a schematic of such a siren where, for clarity, the outer (drill collar) enclosure and the inner central hub are not shown. Here the siren is shown in the initial opened position. Let us consider the situation when wind blows past it and the rotor tends to close. If this siren is placed in a downhole tool, the closing torque will vary linearly with mud weight and quadratically with mud pump flow rate. We wish to determine what this torque would be so that a proper motor can be selected for the MWD tool.

The restoring torque is easily obtained from wind tunnel analysis. The siren is placed (without an electric motor drive) into the wind tunnel as shown in Figure 9.4b. A thin shaft extends downstream, which may be supported at one or more points to prevent flexing. A wire wheel or thin circular disk is attached to this shaft and the linear force F required to keep the rotor in the fully opened position is recorded. The product between F and the moment arm R is the restoring torque $T_{air} = FR$ for the opened position considered. If the "pulling" force is reduced, the rotor will close slightly; the resulting reduced torque is the restoring torque associated with the partially closed azimuthal position. If this process is repeated for all positions from opened to closed, a range of closing torques can be determined on a static basis.

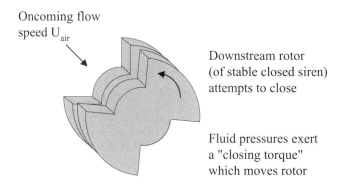

Oncoming flow speed U_{air}

Downstream rotor (of stable closed siren) attempts to close

Fluid pressures exert a "closing torque" which moves rotor

Figure 9.4a. Simple stable-closed siren.

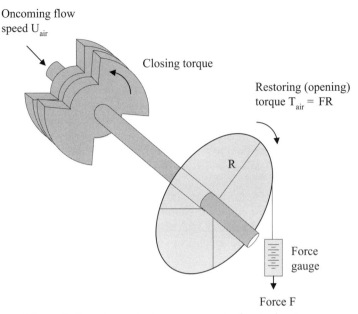

Oncoming flow
speed U_{air}

Closing torque

Restoring (opening)
torque $T_{air} = FR$

R

Force
gauge

Force F

Figure 9.4b. Conceptual measurement of restoring torque.

It is not necessary to use a linear force gauge. If a set of calibrated weights were available, restoring torques are just as easily determined versus azimuthal angle. For example, the use of a block with weight W_1 will give a torque of W_1R associated with relative rotor-stator angle A_1. Measurements are repeated for different weights W_n to give restoring torques T_n associated with different degrees of relative closure A_n. Note that we are measuring *static* torque, that is, torque when the rotor is *not* moving; the electrical drive motor is not attached to the rotor shaft and there is no resistance to rotor movement except bearing friction. This is generally the worst case torque in practice; if the motor chosen can overcome this torque, it will perform satisfactorily downhole under dynamic rotating conditions. We emphasize that, as the siren closes or opens, blower flow rates can change substantially and need to be monitored continuously. If inaccurate flow rate measurements are used, torque predictions will obviously degrade in quality. If cost were not a factor in wind tunnel construction, one would naturally turn to automated flow rate and torque measurement systems.

For the flow in Figure 9.4a, we now assume that T_{air} and Q_{air} are available from wind tunnel measurements when the air has a mass density of ρ_{air} (Q_{air} is the volume flow rate determined from u(r) as discussed). Again, good data is important; ensure that this torque data is correct by increasing Q_{air} to some higher level and verifying that the new T_{air} increases quadratically. If this quality test is passed, we would like to use their values to predict the corresponding torque T_{mud} in a flow with density ρ_{mud} and flow rate Q_{mud}. The result is simple, following derivations identical to that given in Chapter 8 for turbine torque, and given by the formula $T_{mud} = T_{air} (\rho_{mud}/\rho_{air}) (Q_{mud}/Q_{air})^2$. Self-spinning "turbosirens" such as that in Figures 9.4c,d, in the leading author's U.S. Patent 5,831,177, interestingly, do not suffer from static torque problems; they spin continuously, drawing on the power of the flow in the oncoming flow.

Figure 9.4c,d. Self-spinning "turbosiren" designed in
wind tunnel and validated in mud flow loop (U.S. Patent 5,831,177).

We next summarize the key procedural steps for siren static torque testing. These are given to guide engineers new to the test techniques developed thus far so that all necessary steps are followed.

9.2.2   Siren static torque testing procedure.

1. Record the time of day, date, temperature and humidity and describe any relevant conditions, e.g., exposure to sunshine or cold. The data are used to determine air density from a suitable physical formula.

2. Place the siren test fixture in the wind tunnel with the rotor at the downstream side and remove the electric motor from the siren assembly. Make sure that the rotor spins freely.

3. Follow the suggestions of Figure 9.4b. For example, hang a weight W_1 on the string attached to the circular wire wheel and compute the restoring torque W_1R and record the relative angle open between siren stator and rotor A_1. Repeat with different weights W_n which would correspond to torques T_n and angles A_n. Also record the flow rate Q_n obtained.

4. Ensure correct torque measurement. Remember that flow rate will change as the degree of siren closure changes. Rate needs to be monitored accurately to guarantee reliable predictions for other flow rates. Periodically increase the flow rate substantially and verify that the dependence of torque on rate (at the same stator-rotor angle) is quadratic.

5. What are the highest values of mud weight and mudpump flow rate for this mud siren in field operations? Denote these by ρ_{mud} and Q_{mud}.

6. Calculate the worst-case starting static torque using the earlier formula $T_{mud} = T_{air} (\rho_{mud}/\rho_{air}) (Q_{mud}/Q_{air})^2$. Are electric or hydraulic motors available commercially to provide this torque, given the packaging constraints known for the MWD drill collar? If not, this is an unacceptable design.

7. Calculate the moment of inertia I of the siren rotor and any attachments to the rotor shaft expected in the actual MWD tool, e.g., alternative shaft, motor parts, etc. For the torque T_{mud} and the calculated inertia I, do rotor angular accelerations appear reasonable?

8. Our discussion assumed an initially opened stable-closed rotor attempting to close. Alternatively, one might have an initially opened stable-open rotor attempting to remain open. The downward weight shown in Figure 9.4a would be placed on the opposite side and one would measure the torque attempting to close the siren. Remember, the entire range of azimuthal angles must be considered whether the siren valve is stable open or closed.

9. It is important to remember that not all siren valves are stable-open or stable-closed. Figures 9.4c and 9.4d show sirens that rotate by themselves without motor drives, drawing upon the power in the oncoming flow to aerodynamically turn the rotor without further mechanical assistance. Signal modulation for such "turbosirens" would be accomplished by braking the rotor using mechanical or electrical means.

The above discussions focus on *static torque*, that is, torques needed to open a closed stator-rotor pair or to close an opened pair. These are important because, after all, the siren rotor must stop, start, speed up or slow down in order to convey information. *Dynamic torque* is the resistive torque encountered by the rotor once it is in steady-state rotary motion. For example, if a phase-shift keying telemetry scheme is used, the rotor might turn three complete revolutions before changing phase; dynamic torque would be the torque acting on the rotor during these three rotation cycles. In general, static torques exceed dynamic torques; thus, if an electric or hydraulic motor can overcome static torques, then supplying the torque needed for steady rotation is not a problem. This fact was not apparent to early MWD designers, who went through great effort to collect dynamic torque data. Such measurements are not trivial, since expensive dynamometers must be used to test metal models under actual mud flow conditions. Testing is time-consuming and labor-intensive and, in this author's opinion, unnecessary. But interestingly, theoretical considerations could reduce test matrices substantially if dynamic results were actually deemed important.

Flowfields associated with blunt body sirens are typically separated, rotating and highly transient and, as a result, hardly amenable to analysis or computational simulation. Nonetheless, their flow properties do follow well defined rules which can be extrapolated experimentally. Thus, their complicated physical nature does not imply that hundreds of mud loop tests with metal models are necessary. For a mud siren test, the parameters that enter are the radius R, the torque T, the fluid mass density ρ, the rotation rate ω, and the flow rate Q. It is known from rotating flow turbomachinery analysis (e.g., refer to the classic books by Hawthorne (1964) and Oates (1978)) that these individual parameters by themselves are unimportant; instead, the complete physical problem depends on two parameters only, namely, the dimensionless groups $RT/\rho Q^2$ and $\pi\omega R^3/Q$. The first group measures a force-like quantity against the dynamic head; the second, also known as the "tip speed ratio," compares azimuthal to axial velocities. Plots of one versus the other will generally yield straight lines (Oates, 1982). For hundreds of data points obtained in a CNPC mud flow loop, running at different flow rates with different mud densities and viscosities, it was found that, for a given siren geometry, $RT/\rho Q^2$ is in fact a straight line function of $\pi\omega R^3/Q$. That this should have been expected is anticipated from the wide body of empirical turbomachinery literature available.

An example from mud loop tests is shown in Figure 9.5, where different flow rates, mud weights and viscosities were incorporated into the test program. Again, $RT/\rho Q^2$ and $\pi\omega R^3/Q$ are dimensionless and have no physical units. The straight line curve fit will take the form $RT/\rho Q^2$ = Slope $\times$ $\pi\omega R^3/Q$ + Vertical Intercept. Using our data, Vertical Intercept = $-$ 0.09 and Slope = (0.12 + 0.09)/1.0 = 0.21. Thus, $RT/\rho Q^2$ = 0.21 $\pi\omega R^3/Q$ $-$ 0.09. Hence, the siren torque acting at any fluid density ρ, flow rate Q and rotation rate ω is given by the equation T = (ρQ^2 /R) (0.21 $\pi\omega R^3/Q$ $-$ 0.09). This result is, of course, applicable only to the particular tapered-edge rotor design tested. But it is important to observe that dynamic torque is not a simple quadratic function of Q as in turbine flows. However, the equation applies generally to the hardware class considered and can then be used with any set of physical units desired.

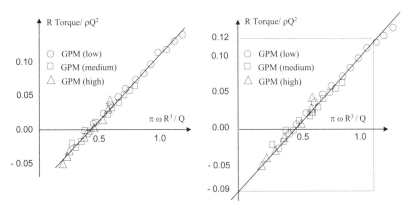

Figure 9.5. Siren torque from numerous mud tests fit to single straight line (results applicable to a research design never used commercially).

Note that the dynamic torque in Figure 9.5 does not vanish at $\omega = 0$ because asymmetric rotor tapers were evaluated in the test. The following observation from Figures 9.5 is crucial from a testing perspective. In order to determine the dimensionless torque curve, it is not necessary to use all of the data points shown. The same straight line can be constructed from a wind tunnel dynamometer test using two widely separated test points only, say those corresponding to "southwest" and "northeast." These two test points are selected as follows: fix the value of Q and perform tests at wide-apart rotation rates ω_1 and ω_2. The resulting curve, normalized as shown, applies to all flow and rotation rates and to all mud densities. These results are well known in aircraft turbine flow analysis and result from fundamental consequences of dynamic similarity. Again we emphasize that tests for dynamic torque are less important relative to those for static torque. It is also important to emphasize a physical consequence of Figure 9.5. The straight line dependence means that we can write $(R \bullet Torque)/(\rho Q^2) = \alpha \; (\omega R^3/Q) + \beta$ where α and β are constant dimensionless slope and intercept values. Thus, Torque $= (\rho/R)(\beta Q^2 + \alpha \omega R^3 Q)$. This shows that, unlike the simple quadratic dependence of force on flowrate in static problems, a dynamic correction proportional to the *first* power of flowrate and rotation speed is obtained. It is not quadratic, but the dependence on fluid density is still linear.

9.2.3   Erosion considerations.

Figure 9.6 shows eroded metal prototype parts (for a design concept not in commercial use) with obvious gouging of the metal very apparent. This is caused by intense, high speed, swirling, sand-carrying "vortex" flows which continuously remove metal. Advanced and expensive engineering measurement methods include hotwires and laser anemometers, but these are not necessary for our purposes. The "ball in cage" in Figure 9.7a can be constructed from flexible copper wires soldered in place. This cage contains a white plastic or styrofoam ball painted white on one hemisphere and black on the other. This device is introduced into the flow. A straight oncoming flow presses the ball against the back of the cage, as shown in Figure 9.7a, but a recirculating flow as in Figure 9.7b will impart an obvious spin. The higher the spin rate, the greater the implied erosion. The erosion design objective is reduction of rotation velocity or a complete elimination of the recirculation zone. "Ball in cage" test devices have been developed in different sizes, e.g., over diameters from 0.1 to 0.5 inches in order to characterize different scales of recirculating vortex motion. The author cautions against "common sense" approaches when redesigning blunt body flows. Very often, the results are unpredictable, but fortunately, reliable test results can be obtained inexpensively.

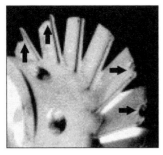 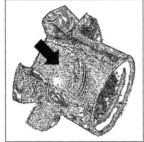

Figure 9.6. Eroded turbine rotor tips and siren lobes, both from vortex damage.

Figure 9.7a. Stationery "ball in cage" in straight flow.

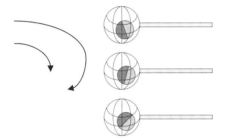

Figure 9.7b. Rotating "ball in cage" in recirculating vortex flow.

9.3 Intermediate Wind Tunnel for Signal Strength Measurement

The prior section dealt with torque measurement and erosion, two areas critical to good MWD design which can be successfully studied using the short wind tunnel. Here we consider acoustic signal strength and explain how it can be measured using an "intermediate" length wind tunnel. In general, one cannot determine signal strength by using a single transducer, say, upstream of the siren, without further signal processing; this transducer will record the created signal plus all reflections and reverberations, and the end result is difficult to decipher or interpret. A good interpretation method is thus required.

The acoustic pressure field produced by siren opening and closing is antisymmetric with respect to source position. For example, when the siren closes, an overpressure is developed upstream due to impacting fluid, while an underpressure is found downstream as fluid pulls away – these pressures are equal and opposite, a fact verifiable in both wind tunnel and mud flow loop. Because the pressure is antisymmetric, signal strength is amenable to differential pressure transducer measurement. The two piezoelectric sensors connected to differential transducers are placed upstream and downstream of the siren at equal short distances, and effectively cancel reflections, thus providing direct indicators of Δp. But such measurements are not perfect, since one sensor always resides downstream of the rotor, where strong rotating vortex flows with highly transient pressures are found; and differential pressure transducers will never characterize negative pulsers since Δp vanishes identically. Since neither single nor differential pressure measurements is ideal, one would ideally prefer both. In this section, we discuss the problems associated with each technique so that potential measurement errors could be properly understood and reduced.

When a siren opens and closes, an acoustic "water hammer" signal is created as the fluid literally crashes into a solid wall (the term "water hammer" will be retained for air flow, as it traditionally describes fluid compressibility effects). Were one to visualize nearfield flow details in all their detail, one would record a wealth of small-scale three-dimensional effects. These do not, however, propagate to the surface; only the lowest-order pressure field having an amplitude that is uniform across the cross-section travels to the surface. This is known as the "plane wave" mode. The acoustic plane wave is created by effectively stopping the oncoming fluid and is recognized as a volume effect; thus, if the rotor-stator gap is large, or the circumferential gap between the rotor and the MWD collar is not small, or both, stoppage is reduced and the created signals will be weak. Siren designs should minimize these gaps, but there may be undesirable operational consequences. For example, reducing the rotor-stator gap will increase the likelihood of debris jamming and erosion, and similarly for the circumferential rotor gap adjacent to the collar housing. An assessment related to jamming and erosion must be made using short wind tunnel flow analysis and actual mud loop or field testing.

9.3.1   Analytical acoustic model.

In Chapter 3, we introduced the use of harmonic analysis in signal modeling and considered the problems shown in Figure 9.8. In particular, we studied four scenarios, namely, Case (a), infinite system, both directions, Case (b), drillbit as a solid reflector, Case (c), drillbit as open-ended reflector, and Case (d), "finite-finite" waveguide of length 2L. It is Case (d) that is of interest in intermediate wind tunnel testing, and in this chapter, we discuss its applications.

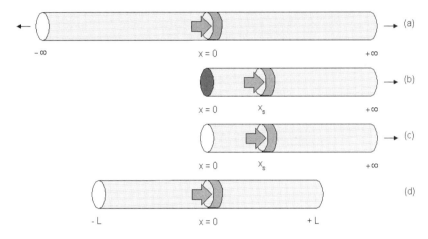

Figure 9.8. Different propagation modes (identical to Figure 3-A-1).

The sketch in Figure 9.8d shows a dipole source centered at x = 0 in a waveguide of length 2L where x is the propagation coordinate. We will assume open-ended reflectors satisfying $\partial u/\partial x = 0$ at x = ± L where u(x,t) is the fluid displacement and discuss its physical significance (both ends of the intermediate wind tunnel are opened to the atmosphere). Again, ω is the rotation frequency, c is the sound speed and t is time. Since standing waves are found at both sides of the source, linear combinations of sin ωx/c and cos ωx/c are chosen on each side to represent u(x,t). Use of Equations 3-A-8 and 3-A-9 at the source x = 0 previously gave us the acoustic pressure solutions

$$p_1(x,t) = - \{p_s/(2 \tan \omega L/c)\} [\sin \omega x/c + (\tan \omega L/c) \cos \omega x/c] e^{i\omega t}$$
$$\text{on} - L < x < 0 \qquad\qquad (3\text{-A-}14)$$

$$p_2(x,t) = - \{p_s/(2 \tan \omega L/c)\} [\sin \omega x/c - (\tan \omega L/c) \cos \omega x/c] e^{i\omega t}$$
$$\text{on } 0 < x < + L \qquad\qquad (3\text{-A-}15)$$

These solutions describe a siren source whose differential pressure Δp, or signal strength, is $p_s(t)$. As the rotor turns, equal and opposite pressure fields are created which travel in opposite directions, reflect at the open ends, and reverberate continuously to develop the solutions above. It is easily verified that $p_1 = 0$ at $x = -L$ and $p_2 = 0$ at $x = +L$. Also, the acoustic pressure is antisymmetric with respect to the $x = 0$, where the Δp is maintained. This is a very important solution for wind tunnel testing determination of Δp. It allows placement of a single (not differential) piezoelectric transducer anywhere, at any position "x" for measurement of the "p," which includes the effects of all reflections needed to set up the standing wave. In wind tunnel applications, standing waves are set up within seconds.

Then, depending on whether Equation 3-A-14 or 3-A-15 is used, we can solve for the p_s representing "delta-p" in "$p_2 - p_1 = p_s\, e^{i\omega t}$" directly. This is repeated for different values of ω. It is important to understand that different sources of experimental error and consideration will arise. In particular,

(1) We have mathematically assumed that "$x = 0$" was our dipole source. In practice, the siren stator-rotor and electric motor drive may be as much as 1 foot long and is not located at the point "$x = 0$." To minimize measurement error, the use of a longer wind tunnel is preferred, say 100 – 200 feet. It is not necessary to use the very, very long wind tunnel – a simple pipe with a blower and a piezoelectric pressure transducer suffices.

(2) If the dipole source were an electric speaker, the pressures p_1 and p_2 will be perfectly antisymmetric because there are no other sources of noise. However, when a mud siren is used, there is turbulent flow noise upstream of the stator associated with the blower, and also, turbulent flow noise plus strong pressure oscillations due to a swirling vortex motion imparted by the turning rotor in the downstream (the frequency of the downstream vortex noise will be identical to that of the acoustic signal). In order to measure acoustic Δp properly, the noise on each side of the siren may need to be filtered. The noise upstream of the stator might be eliminated using white noise or Gaussian noise filters. The noise downstream of the rotor, associated with a rotating flow, is more challenging (this rotating flow is easily observed using the flow visualization techniques described previously). However, the analog use of flow straighteners should satisfactorily filter this effect.

(3) Again, two means of acquiring Δp data are recommended although not required, namely, single pressure transducer and differential transducer methods. In either case, sensitive piezoelectric gauges should be used that are connected to oscilloscopes, signal analyzers and data recorders. These redundant measurements provide useful checks for experimental error and also provide increased physical insight which is invaluable in research. In the case of single pressure transducer testing, we recommend the use of piezoelectric pressure data upstream of the stator as it is less noisy and does not contain the swirling effects of the flow downstream of the rotor.

If the blower itself is noisy, flow straighteners may need to be installed at the blower outlet. The single transducer should not be located at the far left end because acoustic pressure vanishes at open ends; it is preferable, for example, to select the upstream location x = – ½ L in order to avoid downstream noise associated with the swirling rotor vortex flow having the same frequency. Turbulence noise may need to be filtered before one can measure the pressure. Then, p_s can be calculated from Equation 3-A-14. In the case of differential transducers, flow straighteners will be needed just downstream of the rotating rotor in order to filter out vortex noise.

9.3.2   Single transducer test using speaker source.

As these techniques are new, we describe simple tests to acquaint the engineer with wind tunnel acoustic testing. Figure 9.9a shows waves traveling within a plastic tube of the type used for our intermediate length acoustic wind tunnel. When multiple pipe lengths are joined, care should be taken to eliminate seams or small area changes since spurious reflections and noise are created as such junctions. An electric speaker (say, a low frequency woofer) is connected to a waveform generator and a very short duration signal is created which travels to the left. In the figure below, there is no blower and the signal source is the speaker alone. The single (not differential) piezoelectric transducer used will record its pressure trace as the wave travels by. Consider the top diagram, where the left end is opened. This wave will continue to travel to the opened left end and reflect with an opposite sign in pressure. This basic test should be set up so that there is no overlap of left and right-going signals, thus allowing clear identification of left and right-going signals, and the engineer should verify this reflection result. Repeat the experiment with the left end closed, e.g., attach a metal plate to the end; now, the pressure should reflect with the same sign. Once the basic electronic instrumentation, that is, wave form generators, signal analyzers and recorders, and so on, is set up, we are prepared to perform MWD measurements.

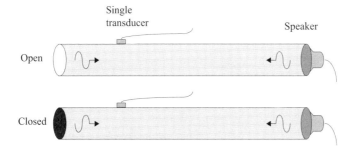

Figure 9.9a. Single transducer speaker test set-up.

9.3.3 Siren Δp procedure using single and differential transducers.

Consider the top drawing in Figure 9.9b, with the speaker in the middle of the intermediate length tunnel, again without a blower. The waveform generator should create a sinusoidal signal of frequency ω. Test frequencies should be less than 100 Hz. Then, the wavelength λ, equal to sound speed (approximately 1,000 ft/sec in air) divided by Hertzian frequency, will exceed 10 ft and remain large compared to the diameter, so that wind tunnel testing remains valid.

Two transducer types will be considered. For the differential transducer, each sensor should be placed at equal distances ahead and behind the speaker. So that local three-dimensional effects can dissipate before affecting measurements, a recommended separation is at least five tube diameters. For this test, the electronic displays should show both the signal source and the differential pressure measurement versus time on the same screen. The differential pressure measurement is straightforward. For a given siren configuration, our experimental objective is the relation $p_s = p_s(\omega)$.

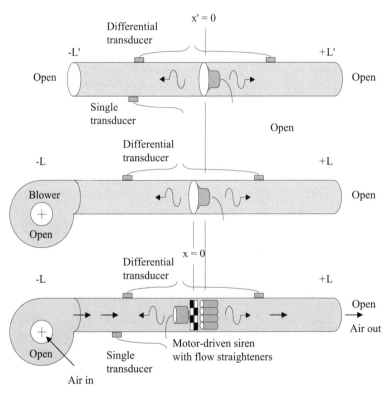

Figure 9.9b. Siren signal test set-up explained in detail.

Next consider the single transducer shown at the bottom of the tubing, again, for the top drawing of Figure 9.9b. This transducer will measure the initial signal plus all reflections and reverberations. It will be extremely difficult to understand the meaning of this measurement. However, since the single transducer location "x," say $x = -L/2$ as suggested earlier, is known, along with values for ω, L, c and the measured pressure $p_1(-L/2,t)$, it is a straightforward to calculate the source strength p_s from Equation 3-A-14. This p_s should equal the value obtained from the above differential pressure measurements.

The middle drawing in Figure 9.9b also shows a speaker test, now attached to a squirrel cage blower which is *not* blowing. Again, we have an open-open system. Note that the blower basically changes the length of the wind tunnel. Thus, if a single transducer test is performed when a blower is attached, the proper physical source location of "x = 0" must be determined before Equation 3-A-14 can be properly used. One would need to adjust the location of the speaker (by trial and error) until the p_s value determined using Equation 3-A-14 agrees with that obtained from differential pressure measurement.

In the bottom drawing in Figure 9.9b, a siren replaces the speakers used earlier. The stator-rotor gap location should be identified as the "x = 0" determined in the above paragraph. Unlike simple speaker tests, a number of fluid-dynamical complications arise. First, a swirling vortex flow will be found downstream of the rotating rotor. Thus, it will be necessary to place flow straighteners as shown, in order to remove the strong pressure oscillations associated with this flow – only then will differential pressure measurements be accurate. Single pressure transducer measurements, for this very same reason, should be obtained upstream of the siren stator in order to minimize vortex noise. Flow straighteners may also be required at the blower outlet to reduce wind noise, as in Figure 9.3. In both cases, because wind is flowing, background turbulent noise will be found. Equation 3-A-14 is used in the same manner as before. Again, redundant methods for p_s signal strength determination provide error checks, and neither method alone is likely to provide absolute accuracy.

We now assume that the engineer has developed a suitable test matrix of parameters. For example, this might include two-lobe, three-lobe and four-lobe sirens operated at different flow speeds and frequencies. In addition, each siren prototype may be evaluated with different taper angles, rotor-stator gaps, different convergence and divergence angles in the central hub, different rotor circumferential clearances with the collar housing, and so on. Each of these configurations should also be evaluated in the short wind tunnel for torque and erosion tendencies. Wind tunnel data should be extrapolated to field conditions for muds flowing under actual pump speeds, in order to determine if electric or hydraulic motors can be found for the design. A good acoustic test procedure is suggested below which addresses a number of key engineering questions which arise and summarizes the steps required to fully characterize the siren.

9.3.4   Intermediate wind tunnel test procedure.

1.   Record time, date, and air temperature and humidity, and note any unusual conditions, e.g., exposure of the room to sunlight.   Using standard engineering formulas, determine the mass density ρ of the air and the corresponding speed of sound c.

2.   Place the siren in the wind tunnel with the rotor-stator gap located at x = 0. Insert flow straighteners at the blower outlet and also just downstream of the rotor.   Install the differential pressure transducer and single pressure transducer at the upstream location x = $-\frac{1}{2}$ L ahead of the stator.

3.   The siren will be turned by an electric motor installed in the upstream or downstream location (upstream may be preferable since this mimics the role of the central hub in the MWD tool).   The torque acting on different sirens will be different, so that different electrical settings will be required to maintain the same rotation.   Again, siren torques will depend on the design geometry and also the oncoming flow rate.

4.   Install all electronic instrumentation, e.g., DC controllers for the electric motor, differential pressure and single transducer outputs to signal analyzers, simultaneous measurements for dynamic torque, and so on (refer to the "long wind tunnel" section for details).

5.   We now perform tests.   With the siren rotating at the desired frequency, determine the average volume flow rate Q as described in Chapter 8.   First, measure p_s using the differential transducer; this p_s is the one in $\Delta p = p_s e^{i\omega t}$ = p_s cos ωt in Chapter 3.   Second, perform single transducer measurements. Set x = $-\frac{1}{2}$ L in Equation 3-A-14 and introduce L, ω and c.   The result is $p_1(-L/2,t) = -\{p_s /(2 \tan \omega L/c)\}$ [$-$ sin $\omega L/2c$ + (tan $\omega L/c$) cos $\omega L/2c$] cos ωt. The transducer signal is $p_1(-L/2,t)$. Its measured peak-to-peak value equals $-\{p_s /(2 \tan \omega L/c)\}$ [$-$ sin $\omega L/2c$ + (tan $\omega L/c$) cos $\omega L/2c$] which can be solved for p_s.   Do the differential and single transducer p_s values agree?

6.   For the same frequency, repeat the above with different values of Q.   Is the dependence on Q linear, quadratic or something else?

7.   Repeat the test for different frequencies, and then, different Q's for each frequency.   What is the dependence of p_s on frequency, or better, as a function $p_s(Q,\omega)$ of both frequency ω and volume flow rate Q?

8.   Actual signals will be additionally dependent on the mud density ρ_{mud}. With all other quantities fixed, $p_{s, mud} = p_{s, air} (\rho_{mud}/\rho_{air})$.   When further extrapolated to the downhole flow rate, is the signal strong enough to overcome attenuation?

9.   Our analytical model assumes that p_s is a constant and that the transient delta-p takes the form $\Delta p = p_s e^{i\omega t}$.   With the rotor turning at a constant frequency ω, we wish to determine experimentally if p_s is, in fact, constant.

To do this, connect the differential pressure output to a spectrum analyzer. Is p_s a constant? Or does it consist of multiple peaks, as would be expected of a nonlinear system? What are the Δp's associated with the primary and higher harmonics? The results can be surprising. Amplitude spectra for the same siren operating at the same rotation rate ω, for high and low flow rates, are shown in Figure 9.10a (the vertical and horizontal scales are different). The left-most peaks represent the primary harmonic associated with ω, but the shape distributions of the higher harmonics differ owing to nonlinearity. Optimum shaping would maximize signal strength in the primary harmonic, and ideally, remove energy from higher ones which waste energy and increase signal processing complications. If nonlinear effects are important, it may be necessary to filter out higher harmonics before ascertaining relationships between Δp, flow rate and frequency.

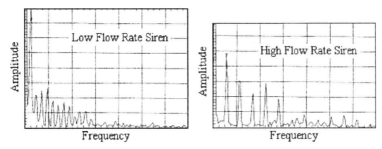

Figure 9.10a. Signal strength amplitude spectrum measured in wind tunnel.

10. We expand on the results in Figure 9.10a. When these figures represent Δp's, only the "primary harmonic" represents the useful signal, e.g., the 12 Hz component of a siren rotating at 12 Hz. The Δp's associated with 24, 36, 48, 60 Hz and so on, are wasted, in the sense that they are destroyed by the drillpipe telemetry channel (signals associated with very, very high frequencies are eliminated by destructive wave interference since the pipe acts as a low-pass filter, while those for lower high frequencies are simply removed by thermodynamic attenuation). Intermediate harmonic signals that arrive at the surface can be problematic since they appear as echoes. The "true signal" is the one with the lowest frequency. It can be extracted from the measured transient waveform using standard methods in FFT (Fast Fourier Transform) analysis. Figure 9.10b provides an illustrative (non-MWD) calculation. The transient time domain waveform at the left appears somewhat complicated, but upon Fourier analysis, actually consists of discrete periodic components as shown at the right. Detailed discussion can be found in standard signal processing references, e.g., the practical textbook by Stearns and David (1993) with source code in Fortran and C.

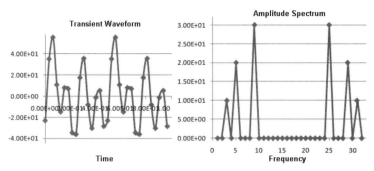

Figure 9.10b. Time (left) and frequency domain (right) displays.

Test procedures for source strength are state-of-the-art. Determining signal strength experimentally, that is, finding $p_s = p_s(Q,\omega)$ for a given siren, is very challenging. Some researchers have assumed, following Bernoulli's equation, that p_s must vary like Q^2, however, this equation describes constant density and not compressible flows. On the other hand, from water hammer considerations, p_s might vary linearly with Q and ω. Montaron, Hache and Voisin (1993) and Martin *et al* (1994) speculate that pressure pulse amplitudes created are roughly independent of frequency, however, recent CNPC experiments indicate a monotonic decrease with frequency for the same flow rate, e.g., as is evident from experimental data in Figure 9.10c taken to about 60 Hz. This result is expected physically: at higher rotation rates, rapid rotor movement does not provide enough time for the fluid to come to a complete stop and recover.

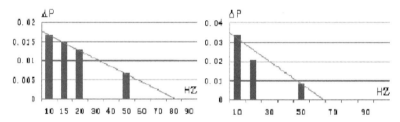

Figure 9.10c. Experimental CNPC results, Δp versus frequency.

Additional discussion is available for positive pulsers. In one-dimensional acoustics, a semi-infinite pipe with a piston at one end that is quickly struck will create a propagating wave with strength $p_s = \rho Vc$. The piston model, an exact solution attributed to the nineteenth century physicist Joukowsky, applies more to positive pulsers than to sirens. Here, ρ is fluid density, c is sound speed and V is the instantaneous piston speed. If we consider water, with $\rho = 1.935$ lbf sec^2/ft^4 and c = 5,000 ft/sec, and assume a flow rate of 500 gpm in a 6 inch diameter pipe, we find that $p_s = 381$ psi. This is a realistic downhole value for positive pulsers since surface values range up to 100 psi. If this acoustic model is valid, then p_s is linearly proportional to Q (and also to ρ and c) as opposed to Q^2. But because siren signal generation is both acoustic and hydraulic (due to leakage through the gap), the truth, most likely, lies somewhere in between, with U depending on a combination of Q and rotation rate ω.

9.3.5   Predicting mud flow Δp's from wind tunnel data.

How to analyze and extend wind tunnel pressure results to actual mud flowrates and densities raises questions best answered through discussion. We motivate the mathematical ideas using our wind tunnel methods for turbine stall torque analysis (see Chapter 8 for a complete discussion). In any physical problem, it is important to identify the key parameters. Since torque is primarily determined by the inviscid character of the fluid, viscosity and rheology are unimportant to leading order (these parameters are, of course, important in assessing viscous losses). This being the case, the primary variables needed to characterize the torque T are the oncoming speed U, the density ρ, the surface area A and the mean moment arm R. This observation, while simple, is key.

The "dimensional analysis" methods used in engineering argue that natural phenomena can and should be expressible in terms of fundamental dimensionless relationships. The quantity $\rho U^2 AR$ is the only product that can be formed with dimensions of torque. Thus, the ratio $T/(\rho U^2 AR)$ can only depend on the remaining dimensionless parameter, namely, the *shape* of the turbine. This idea is expressed by the equation $T/(\rho U^2 AR) = C_T$ where C_T is a dimensionless torque coefficient associated with the geometry being considered. Now suppose that measurements are made in both air and mud. Since C_T is invariant, one must have $T_{air}/(\rho_{air} U_{air}^2 A_{air} R_{air}) = T_{mud}/(\rho_{mud} U_{mud}^2 A_{mud} R_{mud})$. For full-scale testing, we have $A_{air} = A_{mud}$ and $R_{air} = R_{mud}$. Hence, we find that the equation $T_{mud} = (\rho_{mud}/\rho_{air})(U_{mud}/U_{air})^2 T_{air}$ can be used to determine mud torque at any flow rate and density from any set of wind tunnel data.

Now, we turn to MWD signal generation and attempt to understand the underlying physics using similar techniques. Consider the semi-infinite pipe at the top in Figure 9.11 with a piston at the left end. If this piston is struck, for instance, by a hammer, an acoustic "water hammer" pressure wave is created which propagates to the right. Fortunately, in this example, an exact, analytical solution to the governing one-dimensional wave equation can be developed. Again, we consider the fluid displacement u(x,t), which satisfies $u_{tt} - c^2 u_{xx} = 0$ where c is the sound speed. The general solution takes the form $u = f(t - x/c)$ which represents a right-going wave. At the piston face where x = 0, this reduces to the boundary condition $u(0,t) = f(t)$ for piston position. Recall that the acoustic pressure P(x,t) satisfies $P(x,t) = -B u_x(x,t)$ where B is the bulk modulus. Then, $P(x,t) = (B/c) f'(t - x/c)$ or, since $c^2 = B/\rho$ where ρ is fluid density, a simple $P(0,t) = \rho c f'(t) = \rho c V(t)$ holds, where V(t) is piston speed.

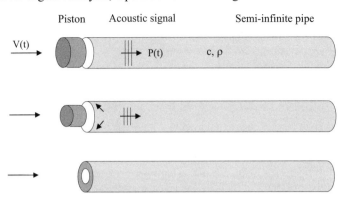

Figure 9.11. Three pressure disturbance modes.

In summary, the acoustic pressure developed by a piston that fully occupies the cross-section of the pipe is $P(0,t) = \rho c V(t)$. The middle drawing in Figure 9.11 shows a piston with nonzero wall clearance, which allows flow leakage as the piston moves. One therefore expects that the pressure predicted by the foregoing formula cannot be realized fully, and it is reasonable to modify that relationship in the form $P(0,t) = \rho c V(t) G$ where $G < 1$ is a dimensionless efficiency factor associated with piston geometry. Thus, we write $P/(\rho c V) = G$, where G can be determined from wind tunnel or mud flow loop analysis. Because the dimensionless G represents a unique descriptor for the piston, then, as in our torque analysis, we can infer that $P_{air}/(\rho_{air}c_{air}V_{air}) = P_{mud}/(\rho_{mud}c_{mud}V_{mud})$ or $P_{mud}/P_{air} = (\rho_{mud}/\rho_{air})(c_{mud}/c_{air})(V_{mud}/V_{air})$. For example, a heavy mud might have $\rho_{mud}/\rho_{air} = 1,500$ and $c_{mud}/c_{air} = 3$. If the wind tunnel test is carried out at 300 gpm and the desired downhole volume flow rate is 1,200 gpm, then P_{mud}/P_{air} $= 1,500 \times 3 \times 4 = 18,000$. Thus, a Δp of 0.01 psi measured in the wind tunnel translates to 180 psi under the downhole conditions assumed.

Once G is determined from wind tunnel analysis, one can use the model $P_{air} = \rho_{air}c_{air}V_{air}G$ for air flow and $P_{mud} = \rho_{mud}c_{mud}V_{mud}G$ for mud. We caution that our discussion so far assumes that the only physically important phenomenon is acoustical. In the bottom sketch of Figure 9.11, we do not have a piston; instead, any analogous piston action is deemed to be so slow that the left end acts as an orifice. When this is the case, the sound speed c is no longer important, and ρU^2 is the only quantity with dimensions of pressure. Thus, the ratio $P/(\rho U^2)$ must depend on the geometrical details of the orifice only, characterized by a dimensionless constant H, also measured in the wind tunnel. In fact, we can write $P_{air}/(\rho_{air}U^2) = P_{mud}/(\rho_{mud}U_{mud}^2) = H$, from which it follows that $P_{air} = \rho_{air}U^2 H$ and $P_{mud} = \rho_{mud}U_{mud}^2 H$. Note that in the purely acoustic model, the pressure depends linearly on V(t), while in the constant density hydraulic model, it depends quadratically on V(t).

Of course, the mud siren, with its piston-like action and orifice similarities, will have a pressure dependence that is linear with density and varies with "$\alpha V + \beta V^2$" in a way determined experimentally. From a modeling perspective, one hopes that a simpler relation can be found, so that "αV" or "βV^2" alone holds. The dependence on the sound speed c is not obvious, and recourse to mud loop testing may be necessary after wind tunnel test data have been evaluated. Care should be exercised with the extrapolation procedures defined above.

9.4   Long Wind Tunnel for Telemetry Modeling

We recapitulate the ideas developed so far. Short wind tunnels provide inexpensive, fast-turnaround tests for properties like torque, erosion, stable-open versus stable-closed, and turbine no-load rotation rate, torque and power; they are used to evaluate hydraulic effects associated with constant density flows. Intermediate wind tunnels recognize the wave propagation, reflective and reverberant aspects of siren-induced wave motions, which arise from compressible fluid effects. They too, are inexpensive, providing simple means to determine signal strength. When integrated with dynamic torque and azimuthal position measurements, valuable data can be obtained that will be useful in subsequent feedback and control loop design. We had discussed the effects wave interference. These are both good and bad. For instance, it can be used to enhance MWD signals without incurring the usual power and erosion penalties. Alternatively, less-than-optimal placement of the siren in the collar, or the use of inappropriate frequencies, can reduce the signals that ultimately travel up the drillpipe by virtue of destructive interference.

9.4.1   Early construction approach – basic ideas.

To evaluate wave interactions, it is not possible to use our short and intermediate wind tunnels. Instead, a "long wind tunnel" must be used, which provides enough "leg room" for concept evaluation. But how long is long? The sound speed in air is approximately 1,000 ft/sec. Suppose we consider a maximum signaling frequency of 100 Hz. Then, the wavelength associated with this motion is (1,000 ft/sec)/(100 Hz) or 10 ft. Thus, a total length of 2,000 ft should suffice for most wave evaluation purposes. Because 10 ft still greatly exceeds a typical diameter, the waves are long in an acoustic sense. Therefore they will not reflect at bends, even ninety-degree bends (of course, the hydraulic pressure drops required to move air at prescribed volume flow rates may be large). A long wind tunnel built with steel tubing wound several times around a building is shown in Figure 9.2b. But since reflections are almost non-existent, as can be validated experimentally, the long wind tunnel can also be constructed from flexible plastic tubing wound on 3-4 ft diameter reels, as in Figure 9.12a, for an early "concept" long wind tunnel that is no longer in use. Preliminary studies in this facility and an understanding of its limitations led to the CNPC design described extensively in this book.

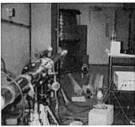

Figure 9.12a. Early long wind tunnel wound on small diameter reels.

Consider the long wind tunnel and apparatus in Figure 9.12a, whose development was funded in part by the Gas Research Institute (Gardner, 2002). That public domain report utilized fixtures summarized in the present write-up. At the left is a test fixture allowing simultaneous siren rotation and rotor-stator gap axial reciprocation for amplitude modulation. The upstream "drillpipe" in the middle drawing, in reality a thousand feet of plastic tubing wound on reels, is connected to an air compressor to provide the oncoming flow; the downstream "annulus" in the far right drawing is likewise wound on reels. Because pressures are large up to and including the solid plastic test section, dangerous cracking with flying debris is likely. For this reason, a pressure relief valve is installed, which is no more than a "cork" located in a suitably drilled orifice; ring clamps are also used to reinforce the plastic test section.

We note that signal distortion due to desurger bladder elasticity can also be evaluated. The wind tunnel "desurger" is simply a balloon or rubber "hot water bottle" installed along the flow path. Desurger charge levels are modeled using different types of balloon skin stiffnesses. Expansion and contraction of the plastic tubing may also distort pressure signals. This occurs with strong pulser movements such as those associated with poppet valves, but the effects are much less so with siren pulsers. For Figure 9.12a, the test section is operated hydraulically from a management and data acquisition station in the same laboratory outside of the flow loop. The mechanical layout for the research test siren, which is driven hydraulically, is shown in Figure 9.12b, together with its control system. The control system, used for research purposes, rotates the siren while, and at the same time, axially reciprocates rotor-stator gap if desired. This allows testing of phase-shift-keying (PSK), frequency-shift-keying (FSK) and multi-level amplitude modulation (AM) of constant frequency carrier waves. Details of the design are offered in Gardner (2002), developed under the lead author's guidance. We emphasize that care must be undertaken to eliminate area changes within the tubing system, which can lead to spurious reflections. Also, flexible tubing may not be suitable for positive pulser testing at small closing gaps, since flow rates cannot be held steadily under these circumstances. We emphasize that bit and annular reflections are not always accurately studied. Very often, math models like those in Chapter 2 are preferable.

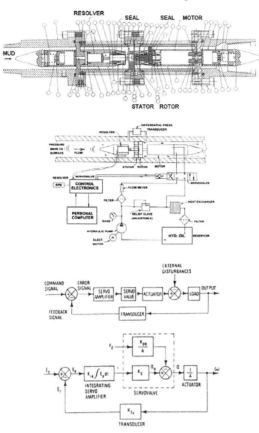

Figure 9.12b. Mechanical siren layout, hydraulic and control system, electro-hydraulic and angular velocity control servomechanisms from Gardner (2002).

Control systems in actual tools would drive compact hydraulic or brushless DC motors. They need to be robust, since mud flow rates are rarely constant and torque characteristics change with tool erosion and random debris impingement. Torques will also change if edges of brittle wear plates bonded to rotors and stators are lost to the flow or if solid particles lodge within rotor-stator gaps. Recent improvements in data rate are due mainly to incremental improvements, e.g., advances in control, state-of-the-art microprocessors, phase shift changes that can be carried out in two (as opposed to, say, three) wave cycles because low-torque downstream rotors are used with powerful motors. Our long wind tunnels are intended to evaluate game-changing telemetry concepts. As noted, these include novel uses of signal augmentation by constructive wave interference, the use of multiple sirens placed in series, operating at identical or different frequencies, and so on.

Figure 9.13 shows standard laboratory instrumentation, e.g., spectrum analyzers, digital oscilloscopes, strip chart recorders and computers used to control siren rotor actions, and to record multi-channel pressures at piezoelectric transducers installed along the tubing. The tape recorder is used to play back noise sources obtained during actual drilling jobs. Our controls also allow us to evaluate, in real time, the performance of new software filters such as those developed in earlier chapters, e.g., echo and noise cancellation algorithms using multiple transducer methods like those in Chapter 4.

Figure 9.13. Siren control and signal processing station.

Installed in our test section may be different prototype sirens, e.g., the two, three and six lobe sirens shown in Figure 9.14. These tests aimed at answering the following questions. For a given signal frequency, is it best to rotate a two-lobe siren quickly or a six-lobe siren slowly? What are the signal strength consequences? What are the harmonic contents of the signals? What are the relative acoustic efficiencies? Figure 9.15 clearly demonstrates the existence of higher order harmonics when the rotor turns at constant frequency.

Figure 9.14. Siren test concepts, test section for both air and mud.

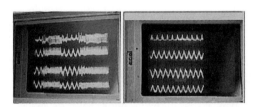

Figure 9.15. Measured frequency-shift-keying and periodic signals (not quite the perfect sinusoids shown in common illustrations!).

Sirens selected for intermediate and long wind tunnel testing should be pre-screened using short wind tunnel tests for torque and erosion since the latter are fast and inexpensive. For example, the two and three-lobe metal sirens in Figure 9.14 were first studied as wood models in the short tunnel as indicated in Figure 9.16a. Similar comments apply to the miniaturized, two-part, self-spinning siren shown in Figure 9.4d. The turbosiren in Figure 9.16b provides still another example; here, plastic and metal models are shown side-by-side. Sirens-in-series, which may be used to increase signal when both operate at the same frequency, or to increase data rate when each operates at distinctly different frequencies, and sirens with "swept" lobes, which alter the harmonic content of the main signal, are shown in Figure 9.16c. It is interesting to note that siren signals sound like Harley-Davidson motorcycle engines in the room during testing. This is especially interesting because, in long wave acoustics, very little of the acoustic energy actually escapes the open end. Thus we are led to conclude that the signal within the tunnel is much greater. The signal in the corresponding mud flow would be approximately one thousand times more, since the ratio of mud to air density takes that value.

Figure 9.16a. Wood prototypes used for torque and erosion testing.

Figure 9.16b. Wind tunnel forerunner to turbosiren (U.S. Patent 5,831,177).

Figure 9.16c. Sirens in series (left) and "swept" rotors (right, middle).

9.4.2 Evaluating new telemetry concepts.

In existing tools, the turbine is located beneath the pulser because it is (incorrectly) believed that it would otherwise block the MWD signal from traveling uphole. That little or no reflection at the turbine occurs is easily verified in the long wind tunnel and should be apparent from its large "see through" area. This opens up numerous possibilities for telemetry design. Earlier we discussed the use of constructive wave interference for signal enhancement without incurring the usual erosion and power penalties. A siren placed nearer to the drillbit would also allow faster wave reinforcement and, by implication, higher data rate. As we have seen in Chapter 2, the drilling telemetry channel should support frequencies much higher than the 12 – 24 Hz range currently used; in Chapter 10, we cite independent confirmation that much higher frequencies can be transmitted in the drilling channel. A higher frequency, aside from increased data rate, would be beneficial because MWD signals can be more easily distinguished from slower acting mudpump noise. Standard frequency filtering can be used for preliminary signal processing.

9.5 Water and Mud Flow Loop Testing

The use of a ruggedized steel test fixture for the wind tunnel system in Figure 9.12a should be noted. This was designed so that tests performed in wind can be effortlessly repeated in mud loop and field tests, for instance, as shown in Figure 9.17. The siren control and signal processing station in Figure 9.13 was developed with the same philosophy in mind. Signal processing algorithms and telemetry tests planned and programmed in software for the wind tunnel are transferred to the field with minimal modification. Since actual test times are small compared to "down times" typically encountered in setting up equipment, our approach makes optimal use of expensive test resources and labor. From this author's experience with wind tunnel and mud loop testing, wind tunnel problems can be diagnosed and solved much more quickly than those arising in the field, because plumbing issues like leaks, blown hydraulic lines, opening and closing flanges, and simply "getting wet" do not arise.

Figure 9.17. Wind test fixture used in mud and field tests (Gardner, 2002).

9.5.1   Real-world flow loops.

A number of long mud flow loops have been built by various organizations for MWD telemetry testing. Many are proprietary, but perhaps the best known is the 9,460 feet loop at Louisiana State University. This facility is described in "MWD Mud Pulse Telemetry System," by W.R. Gardner, Gas Research Institute Report GRI-02/0019, April 2002, and also, "MWD Transmission Data Rates Can be Optimized," by Desbrandes, R., Bourgoyne, A.T., and Carter, J.A., Petroleum Engineer International, June 1987, pp. 46-50. The information quoted here is publicly available.

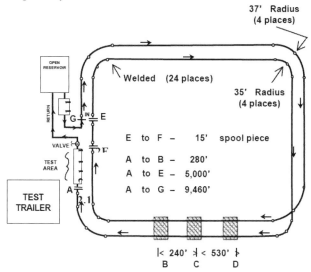

Figure 9.18. Lousiana State University 9,460 feet flow loop.

As shown in Figure 9.18, the buried loop overlaps twice, with each side of the square layout measuring about 1,400 ft. Despite the apparent simplicity, time domain results needed to understand downhole and surface reflection physics can be difficult to interpret. For example, the PEI paper states that drillpipe used for the main loop has an ID of 3.64 inch. Discharge from the pulser, however, is arranged through a return line consisting of 300 feet of 2 inch pipe, so that a large area mismatch factor of 3.3 is found – the flowloop, in this sense, resembles a telescoping waveguide that may amplify or reduce incident signals depending on wavelength. A downstream valve is also used to control pressure. In addition, fluid is discharged into a open reservoir, as opposed to a semi-infinite annulus that supports wave propagation. Thus, the reflections and time domain waveforms obtained bear no resemblance to those encountered downhole and cannot be used to study echo cancellation and wave interference methods. The six-segment acoustic waveguide model derived and exactly solved in Chapter 2, with true radiation conditions in both pipe and annulus, therefore, provides a more rigorous and accurate means for evaluating advanced telemetry concepts. Of course, final hardware testing and evaluation must be performed under high pressure mud conditions, but given the problems associated with a number of long flow loops, and the fact that they offer no telemetry advantages, the author recommends the use of simpler test vessels.

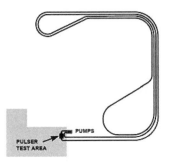

Figure 9.19. Halliburton 11,000 feet flow loop.

The details associated with one industry flow loop, an 11,000 feet facility shown in Figure 9.19 and described in "High Data Rate MWD Mud Pulse Telemetry," W.R. Gardner, United States Department of Energy Natural Gas Conference, Houston, Tx, March 25, 1997, a public domain document, are equally vague. As far as the lead author is aware, little attention had been paid to wave propagation issues, and unless details related to impedance mismatches, and inlet and outlet boundary conditions, and so on, are understood, care must be undertaken in designing and interpreting experiments. Long flow loops designed without attention to wave acoustics can lead to incorrect field predictions. Obviously, it is simpler to design long flow loops correctly at the outset, rather than having to correct measurements later using software.

In assessing the test-worthiness of long flow loops and long wind tunnels, great care should be taken in eliminating spurious reflections and reverberations. For example, even small area mismatches between the flexible and solid plastic tubing in Figure 9.12a can lead to reverberations within the test section that lead to errors in evaluating signal strength. On the other hand, area mismatches between drillpipe and MWD collar cross-sectional areas which lead to acoustical reverberations, modeled in Chapters 2 and 5, should and can be validated experimentally. The presence of impedance discontinuities is easily detected. If the pump piston quickly excites the mud column (or a balloon "pops" at the inlet or outlet of a long wind tunnel), the existence of early reflections indicates problems. For instance, if the sound speed is 5,000 ft/sec in a 10,000 ft flow loop, the round-trip transit time should be 4 seconds – earlier arrivals time would indicate spurious reflections that degrade data quality. Attenuation measurements should be made "following the wave." If a pulser is operated by sweeping frequencies from low to high, standing waves are created in the loop whose nodes and antinodes move spatially. A transducer located at a fixed location would "see" continuous node movements. Residence at a node would measure zero energy when in fact there may be little attenuation. These considerations have not been considered in attenuation studies and published decay rates are often questionable. This author has in fact performed experiments where signals have "mysteriously" reappeared beyond 50 Hz, indicating that channel attenuation is not as severe as is commonly accepted.

9.5.2   Solid reflectors.

The "100 ft hose" patent, arguably the simplest MWD signal processing invention ever, is based on acoustic principles. Recall that the Lagrangian fluid displacement $u(x,t) = f(x - ct)$ represents a traveling wave as it moves past the standpipe transducer. Its pressure is proportional to $u_x(x,t) = f'(x - ct)$, whose transducer value is $u_x(0,t) = f'(- ct)$. Next consider a situation where a solid termination exists at $x = 0$. Now the general solution is given by a superposition of left and right-going waves satisfying $u = 0$ at $x = 0$. The revised wave solution $u(x,t) = f(x - ct) - f(-x - ct)$ yields $u(0,t) = f(- ct) - f(- ct) = 0$ with the result that $u_x(x,t) = f'(x - ct) + f'(-x - ct)$, thus yielding $u_x(0,t) = 2f'(- ct)$. Therefore, the pressure at a solid reflector is twice that in the incident wave.

At one field test in the 1990s, the lead author was witness to a logging situation with weak standpipe signals. Knowing the foregoing result, he removed the standpipe transducer, installed a 100 feet hose into the fitting, and placed the transducer at the opposite termination of the hose. The result was an expected doubling of the MWD signal! Pump noise also amplified, but since this was higher in frequency, its attenuation was greater. This demonstration resulted in two inventions, "MWD Surface Signal Detector Having Enhanced Acoustic Detection Means," U.S. Patent No. 5,459,697, by Chin, W.C. and Hamlin, K.H., Oct. 17, 1995, and "MWD Surface Signal Detector Having Enhanced Acoustic Detection Means," U.S. Patent No. 5,535,177 by Chin, W.C. and Hamlin, K.H., July 9, 1996. The method is illustrated in Figure 9.20. We note that 50 feet originally did not work, but 100 feet finally did – apparently, the pressure disturbance is not wavelike unless the hose is sufficiently long.

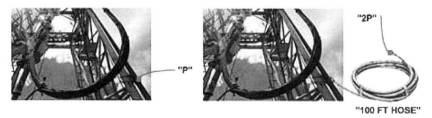

Figure 9.20. Simple "one hundred feet hose" analogue signal amplifier.

For the same reason, it is better to install MWD pressure transducers near the mudpump pistons rather than faraway, in order to guarantee pressure doubling. The lead author has in fact demonstrated this counter-intuitive method in recent field work. At higher data rates, wavelengths decrease; faraway standpipe pressures do not necessarily see this doubling due to phase cancellations, and in fact, the presence of the reflected wave may introduce complications that require multiple transducer processing. This is clearly avoided by placing the transducer near the piston face, a somewhat counter-intuitive situation given the "bad reputation" associated with noisy mudpumps.

While mud pump pistons, being solid reflectors, will double an incident signal, one can prove that low data rate signals in the presence of centrifugal mud pumps will go almost undetected because the boundary condition $u_x = 0$ holds. This signal loss has often confused drilling engineers running MWD tools that are in apparently perfect mechanical condition. At higher data rates, the cancellation between left and right-going waves is less perfect and some signal may be detectable and recoverable using multiple transducer methods. Often, confusing issues are easily explained using simple acoustic arguments.

9.5.3 Drillbit nozzles.

The effects of solid and open reflectors on surface signals have been emphasized. Similar considerations apply downhole at the bit. Figure 9.21 shows two bits, one with greater nozzle area than the other. In either case, we caution against "solid versus open" based on appearances alone. While areas are important from a mud transport perspective, they enter only indirectly in the case of wave reflections. As shown in Chapter 2, the pertinent parameters are relative axial and cross dimensions, wavelengths and sound speeds – whether a drillbit is solid or open as a reflector should be ascertained using the rigorous six-segment waveguide model previously developed. In the mid-1990s, the lead author obtained his constructive interference patent assuming the bit as a solid reflector, not knowing at the time that this "obvious" conclusion was rarely correct. Again, Chapter 2 explains the subtleties in detail.

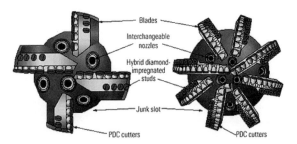

Figure 9.21. Drillbits with different nozzle sizes.

Sirens in series. In the same way that multiple conversations are carried along modern phone lines, the drilling telemetry channel supports multiple transmissions provided each is relegated to its own narrow frequency band. "Measurement-While-Drilling System and Method," U.S. Patent No. 5,583,827, awarded on December 10, 1996 to the lead author, describes this mode of operation, noting that "the pulse generation system preferably includes a plurality of mud sirens in tandem to further enhance signal level and to provide multiple amplitude levels to increase data transmission rate."

Multiple sirens can also be operated at the same frequency to augment amplitude, e.g., two sirens each with larger rotor-stator gaps may be better than a single siren with a tight gap from an erosion perspective. The two need not operate all the time – one might "kick in" only when needed, say, under attenuative circumstances in deep wells or cold environments. Figure 9.22, duplicated from U.S. Patent No. 5,583,827, shows one possible configuration. Figure 9.23 shows two "turbosirens" with different lobe numbers in tandem; these self-spinning sirens rotate without motor drive, drawing upon the energy of the mud flow, and would be modulated by controlled mechanical braking. Their telemetry characteristics, first tested in the long wind tunnel, were later confirmed by mud loop testing using the metal model shown.

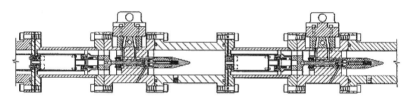

Figure 9.22. Multiple mud sirens in series (U.S. Patent No. 5,583,827).

Figure 9.23. Turbosirens in series (U.S. Patent 5,831,177).

9.5.4   Erosion testing.

Here we address the important subject of erosion evaluation. Although we have described how wind tunnel analysis can provide useful clues related to wear, e.g., streamline convergence indicates locally high sand convection speeds, it does not provide complete information. Short of detailed testing in which actual muds and metal specimens are used to obtain wear patterns and life span information – *and we emphasize that there are no substitutes for such "real world" testing* – it is possible, however, to develop test methods that offer considerably greater insight than wind tunnel analysis. A conceptual "erosion flow loop" is shown in Figure 9.24. Here, a lightweight mud (or water flow with 1-3% sand) recirculates in a system requiring minimal space or pump power. Test models, cleaned and coated with epoxy paint, are run for several minutes, before they are removed for visual examination. Results are almost always identical to those obtained from damaged field parts.

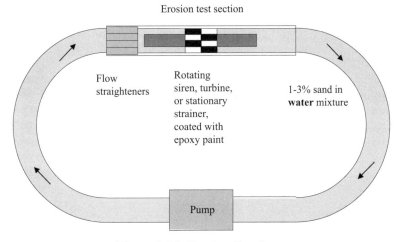

Erosion test section

Flow straighteners

Rotating siren, turbine, or stationary strainer, coated with epoxy paint

1-3% sand in **water** mixture

Pump

Figure 9.24. Erosion flow loop.

If the above erosion flow loop is operated in the horizontal position, it is possible for sand to accumulate at the bottom due to gravity segregation. This can be avoided by running the system vertically, so that sand is uniformly dispersed throughout the cross-section of flow. We emphasize that wear patterns obtained from wind tunnel visualization apply only if the metal never erodes. In practice, it is the interaction between fluid erosion and actual metal removal that results in observed erosion. Wind tunnel testing will help in understanding erosion, but will never completely solve the problem.

9.5.5   Attenuation testing.

Elsewhere in this book, we have pointed out the dangers inherent in drilling mud signal attenuation evaluation using finite-length flow loops excited by periodic mudpump pistons. Essentially, the effects of amplitude variations due to standing wave node patterns must be removed in order to provide true estimates for thermodynamic loss. This process is difficult computationally. Even when it can be accurately performed, there is always lingering doubt that the buried flowloop used may contain undocumented area discontinuities which may render the acoustic data useless.

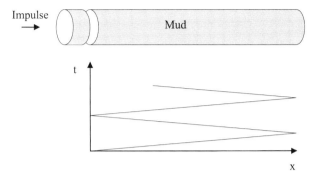

Figure 9.25. Conceptual attenuation test fixture.

A conceptual "attenuation test fixture" is sketched in Figure 9.25 that avoids the problems associated with acoustic standing waves. Essentially, one has a very long, thick-walled metal tube that is rigid at the right end while a movable piston is installed at the left end. Pressurized mud free of bubbles fills the test chamber within the pipe. A sudden impulse excitation, e.g., a bullet fired from an oblique angle, is used to displace the piston momentarily. The pipe is instrumented with piezoelectric transducers that are connected to a digital oscilloscope. The speed of sound in the mud can be readily measured from the first few wave traverses by dividing total travel distance by total time. Then, attenuation can be inferred by counting the number of reflection cycles until the acoustic signal completely disappears. The effects of fluid shear are not accounted for in this procedure, so the results are optimistic. Still, the data so obtained is less problematic that that found in standing wave environments.

9.5.6   The way forward.

We have addressed important issues related, first, to mud siren hardware design, and second, to telemetry optimization and evaluation. These issues draw, respectively, on the constant density and compressible flow properties of fluids, which are in turn studied using short and much longer wind tunnels. Throughout this book, we have justified new methods using physical arguments and first principles, and then, extended our capabilities using mathematical and numerical models where possible. In the next and final chapter, we integrate many of the ideas introduced earlier to create a "technology roadmap" to 10 bits/sec for very deep wells. Comprehensive mud siren signal strength testing at the China National Petroleum Corporation, e.g., see Figures 9.2a to 9.2h, point to an unfortunate reality: pressure signals produced by conventional, single-siren MWD tools have reached their technological limits because Δp's cannot be significantly increased. Thus, every means must be employed to increase the signal traveling up the drillpipe: constructive wave interference based on reflection from the bit and also by employing sirens-in-series, use of lower attenuation muds and larger diameter pipe if possible, introduction of highly accurate multiple transducer echo cancellation and pump noise removal schemes, novel transducer placement methods at the surface, application of piezoelectric transducers, and so on. The mud pulse telemetry tool of the future must be designed as a system where all components work together, reinforcing strengths and eliminating weaknesses. The discussion presented next, of course, represents a first (and not final) attempt at this challenging endeavor.

10

Advanced System Summary and Modern MWD Developments

In this final chapter on Measurement-While-Drilling analysis and design, we discuss the basic elements of a hypothetical high-data-rate MWD system which embodies the ideas in this book and the authors' field and modeling experiences. Our exposition applies key telemetry and acoustics concepts which are also useful in other possible prototypes and highlights the pitfalls and fallacies behind commonly accepted engineering design rules-of-thumb.

At the present time, several MWD service companies provide mud pulse logging worldwide. Three largely operate in the 1 bit/sec (or less) range, one operates siren tools at 3-6 bits/sec (bps), possibly faster, and the last, in recent product literature related to siren-like pulsers, claims extraordinarily data rates. The exact specifications in all cases are not well documented, that is, technical details related to mud type, borehole depth and environment are not available, but the high data rate at least points to mechanical capabilities which we believe, importantly, are nowhere near present technology limits.

Can we do better? Yes. And just what *is* required to send and receive signals over large distances successfully at high rates? Three key factors apply: strong amplitudes, high frequencies and low torque. First, the acoustic signal should be large and ideally created "intelligently" with minimal impact on erosion and power demand. Second, the telemetry sequence should be simple to operate and decode, with very small possibilities for error. Third, surface signal processing should be robust and based on well-defined acoustics principles, allowing sensitive piezoelectric transducers to "see" beneath high levels of noise. Fourth, the pulser should require minimal mechanical power for operation. And finally, as we will explain, drilling mud should be selected for minimal attenuation. We will discuss each of these ideas in detail.

273

10.1 Overall Telemetry Summary

For convenience, we discuss the foregoing points separately to emphasize key ideas in depth. Doing so allows us to develop supporting arguments and suggestions more clearly.

10.1.1 Optimal pulser placement for wave interference.

Conventional misunderstandings are common to hardware design. Perhaps the most misunderstood is the requirement for downhole turbines to remain at the bottom of the MWD collar, for fear that uphole turbines would block the upgoing signal. But this is impossible because rotor-stator configurations are always open and will pass both flow and signals. Thus, there are no acoustic restrictions on turbine location. This conclusion is important for one crucial reason: by placing say, a siren or positive pulser closer to the bottom and by applying the correct phasing, the use of constructive wave interference for signal enhancement is rendered faster and more practical.

The time required for downgoing waves to reflect upward and add to later upgoing waves is significantly decreased and fewer wave cycles are required to establish stronger signals that can be clearly seen. This implies higher data rates. At the same time, the use of constructive interference for signal enhancement means reduced erosion penalties at the pulser and decreased power demands on the turbine or any batteries – all of which imply increased life span and reliability. The increased signal strengths mean that attenuation, while problematic at higher frequencies, will be less of a problem, thus permitting greater travel than is otherwise possible. Of course, sometimes there are other considerations, for instance, connections to rotary steerable systems may require top-mounted sirens, that preclude this arrangement.

But how close to the drillbit can we position our siren? Certainly, from a wave or signal strength perspective alone, the closer the better. This would, however, defeat the purpose of Measurement-While-Drilling, which attempts to use near-bit logging information quickly and effectively for geosteering. To address this concern, and to determine practical numbers, we allow placement of a resistivity-at-bit (or, "RAB") tool between the mud siren collar and the drillbit (the use of any other tool, e.g., a turbodrill, is acceptable for our analysis). From the vendor advertisement for a RAB tool in Figure 10.1, a typical near-bit tool length might be on the order of 10 ft. For our analysis, we assume that the MWD collar is 20 ft long, which is consistent with several commercial designs.

We now apply the six-seqment acoustic waveguide model in Chapter 2, which we emphasize includes the correct outgoing wave radiation conditions for both drillpipe and borehole annulus, and hence, is representative of real drilling scenarios. Different perspectives of our computed color results are shown in Figures 10.2a,b,c. The parameters in Figure 10.3 are assumed for the calculations (RAB geometry appears in "mud motor drill collar" input text boxes).

Logging-While-Drilling Resistivity-at-Bit Tool

Description

The Schlumberger Resistivity-at-Bit tool (LWD-RAB) makes azimuthal resistivity and gamma ray measurements while drilling. Using an azimuthal positioning system, measurements are acquired around the borehole to create a high resolution, 360-degree resistivity image of the drilled hole. The RAB tool may be connected directly behind the bit or further back in the bottom hole assembly. It may be run with other LWD tools such as the ADN, ISONIC or NMR. Typically, the RAB data are stored in memory within the tool and retrieved at the end of the bit run. As an option, the data may be transmitted to a surface acquisition system if a Measurement-While-Drilling (MWD) power pulse tool is run in conjunction with the RAB.

Applications

♦ Detection of resistivity heterogeneity via azimuthal resistivity images
♦ Lithology estimation
♦ Instantaneous detection of casing and coring points
♦ Accurate resistivity when mud is salty or formation resistivity is high
♦ Detection of early invasion of borehole fluids into the formation

Assembly of the RAB tool aboard
the *JOIDES Resolution*

OCEAN
DRILLING
PROGRAM   ODP Logging Services, Lamont-Doherty Earth Observatory, Rt. 9W, Palisades, NY 10964

Schematic illustration of the Resistivity-at-Bit tool (LWD-RAB).

Specifications

Tool weight:	1200 lbm
Tool length:	10.1 ft (3.08m)
Temperature range:	-13° to 300° F (-25° to 150° C)
Drill collar nominal outside diameter:	6.75 in.
Drill collar max. outside diameter (azimuthal):	8.125 in.
Power:	Lithium battery pack (200+ hrs)
Recommended maximum torque:	16,000 ft-lbf

Resistivity measurement accuracy:

Range	Bit & focused electrode	Azimuthal
0.2 - 1	+/- 20%	+/- 20%
2 - 1000	+/- 5%	+/- 5%
1000 - 2000	+/- 10%	+/- 20%
2000 - 20,000	+/- 20%	---

Gamma ray specifications:

Range:	0 - 250 API units
One sigma statistical repeatability: (10-API formation, 3-level averaging)	< 3% at 100 ft/hr < 2% at 50 ft/hr
Maximum flow rate:	800 gal/min
Maximum operating pressure:	18,000 psi
Maximum weight on bit:	F = 74,000,000/L² lbf (where L is distance between stabilizers in feet)
Maximum jarring load:	330,000 lbf

Figure 10.1. A typical resistivity-at-bit (or, "RAB") tool.

An unweighted mud is taken with a typical sound speed of 3,500 ft/sec and we considered frequencies up to 100 Hz. The results indicate that a 60 Hz carrier with the pulser located 0–5 feet on top of the RAB tool provides the greatest constructive interference, i.e., an amplification factor of approximately 1.7 (sensitivity analyses, performed with minor adjustments to the assumed parameters, lead to similar conclusions). The "1.7" factor is relevant, of course, assuming that a good underlying signal can be created and that turning torques are not high – these, in turn, require detailed short and intermediate wind tunnel siren design and analysis.

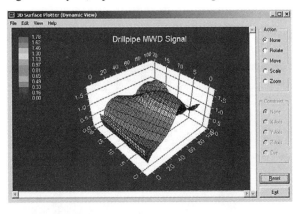

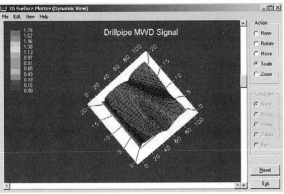

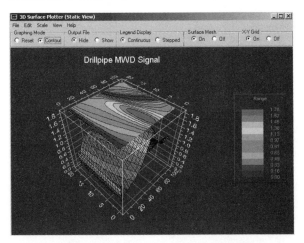

Figure 10.2a,b,c. Six-segment acoustic waveguide results.

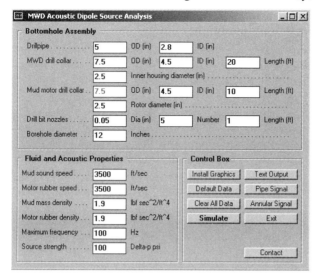

Figure 10.3. Six-segment acoustic waveguide assumptions.

10.1.2 Telemetry design using FSK.

We have emphasized that signals are created at the siren or positive pulser that travel both upward and downward as the valve opens and closes. Those traveling downward reflect at the drill bit and add to later upgoing signals and will, in general, interfere constructively or destructively in a more or less random manner (transmissions also enter the annulus, as explained previously). When a phase-shift-keying (PSK) scheme is employed, ghost reflections are created which also travel uphole, confusing and degrading surface signal processing. While we have developed schemes such as those in Chapter 5 to recover true, fully transient $\Delta p(t)$'s from net downhole signals, it is best, whenever possible, to avoid PSK methods to begin with. Methods based on constant frequencies, which we term "optimized FSK," are therefore ideal. There are no distracting phase shifts to deconvolve – waves of the assumed frequency are always found everywhere, although their amplitudes will vary – surface reflection removal is also necessary to decipher transmissions properly.

To provide concrete results, we select a baseline frequency of 60 Hz and locate the mud siren 5 ft from the bottom of the MWD collar, that is, 5 ft + 10 ft or 15 feet from the drillbit. The time required for the downgoing pulser signal to reflect upward and interact with later upgoing signals – a requirement for constructive interference – is 2 (15 ft)/ 3,500 ft/sec or 0.00857 sec. Now, since our frequency is 60 Hz, as suggested by the results of Figure 10.2a,b,c, each wave cycle is 0.0167 sec long. Suppose we wish to achieve 10 bps in a FSK scheme. One possible way to transmit the binary sequence " $1 - 0 - 1 - 0 - 1 - 0 - 1 - 0 - 1 - 0$" in one second would be our alternating of carrier frequencies between 60 Hz and 0 Hz, that is, bringing the rotor to a complete stop with the rotor fully open, using the frequency sequence "60 – 0 – 60 – 0 – 60 – 0 – 60 – 0 – 60 – 0." Each "60 – 0" interval would take 6 cycles or 0.1 sec, so that 60 cycles are used per second. From the above, 0.00857 sec (or roughly, 0.01 sec) is required to establish constructive interference and the 0.1 sec interval would waste only 2 × 0.01 sec in noise tails at the beginning and end of each interval. This leaves 0.08 sec of pure harmonic signal (or four wave cycles) for signal identification. Thus, 10 bps is achievable as described; a higher rate is possible if each frequency interval requires less than six wave cycles. The amplitude pattern would be a wavelike with alternating bands with and without signal. The 60 Hz target carrier is doable, in practice, because 24 Hz is already realizable from several service companies using the lead author's low-torque "rotor downstream" designs.

What is 60 Hz in terms of siren rotation speed? If a rotor with N lobes rotates at M rpm, then it will create MN cycles in 60 sec, that is, MN/60 cycles per second. If N = 4 and 60 Hz is required, then M = 900 rpm is needed. This may be demanding in terms of inertia and torque, since we have argued that the rotor is brought to a complete stop between 60's. But we need not do this. From Figures 10.2a,b,c, 40 Hz provides enough signal contrast to that at 60 Hz, and we can consider alternatively the frequency sequence "60 – 40 – 60 – 40 – 60 – 40 – 60 – 40 – 60 – 40." The 60 Hz would be associated with high amplitude, owing to constructive interference, while the 40 Hz would be associated with easily distinguishable waves of much smaller amplitude, owing to destructive interference. Because the rotor is not brought to a full stop, a very low torque siren may not be necessary, and mechanical inertia demands on the drive motor would be reduced. Also, because time is saved by not completely stopping, data rate can increase since more frequency cycles can be performed. A siren with azimuthally wider rotors, which in the author's experience produces larger Δp's, is associated with higher torques; because one does not completely stop the rotor, the torque issue is now less of an issue.

It is important to emphasize the roles played by wave reflections. In a PSK scheme where information is conveyed by introducing phase shifts to a carrier wave, the downward wave from the pulser, upon reflection at the bit, adds to later waves traveling uphole and introduces phase shift uncertainties associated with ghost signals. The result is a type of randomness traveling up the drillpipe that is not easily deconvolved. This issue is not discussed in the literature, perhaps intentionally, but processing the downhole signal in any event adds to signal processing demands. Even if multiple transducer surface signal processing methods perfectly remove mudpump, desurger and uphole reflections, the uncertainty created near the downhole pulser remains.

In theory, the processed signal must be further deconvolved to account for wave interactions in the MWD collar. This is possible in principle, however, the models available so far are crude and take the drillbit either as an open or solid reflector.   The use of constant frequency methods again eliminates phase randomness.   A frequency f at the source will lead to an identical surface frequency f for whatever wave interactions are present (nonlinearities will, of course, add harmonics, but these are damped by the telemetry channel).   We note that, in practice, it is not necessary to use the model of Chapter 2 during field operations.   A simpler "self-optimizing" procedure would have the pulser sweep frequencies from, say 0 to 100 Hz, with amplitudes monitored from the surface.   Suppose high and low amplitudes are associated with the frequencies f_h and f_l.   Then, the pulser can be instructed to operate with the frequency sequence " $f_h - f_l - f_h - f_l - f_h - f_l - $ " and so on using a downlink procedure employing, say, mudpump flow variations or a surface-based siren device. Multiple frequencies, say "40-50-60-70 Hz," may provide more than simple 0's and 1's by adding 2's and 3's.   Of course, additional bandwidth is possi ble by clever encoding, a subject not addressed in this book.

10.1.3   Sirens in tandem.

Our use of FSK *plus* constructive interference serves twofold purposes: simplified signal processing and additional signal strength without incurring erosion and power penalties.   A second type of constructive wave interaction is possible for signal enhancement which can be used in addition to our telemetry scheme.   Signal reinforcement is accomplished by placing two (or more) sirens or positive pulsers in series as suggested in Figure 10.4.   The mechanical system would be designed so that the volume between the two pulsers never closes completely, so as to isolate contained fluid from that in the drilling channel.   In fact, all sirens would open and close "in step," that is, if one is 10% open, so are the others.   This allows acoustic signals to superpose and thus reinforce each other.   This implementation is important to transmissions from deep wells. Other uses for "sirens in series" are possible.   For instance, high data rates can be achieved if individual sirens operate at different frequencies.   Since the drilling channel is linear, these transmissions act independently and are processed without difficulty.   This usage, of course, involves mechanical complexities beyond the scope of our present discussion.

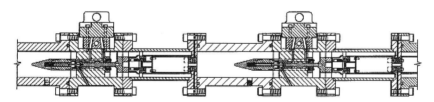

Figure 10.4. Section view, a pair of ganged or tandem mud sirens
(e.g., see U.S. Patent No. 5,583,827 for details).

Following Chapter 7, each siren would have downstream rotors, which are known for restoring torques that diminish significantly from those of upstream rotors for the same flowrate. Special tapers can be added to rotor sides which additionally reduce torque, as described in the lead author's work in U. S. Patent Nos. 4,785,300 and 5,787,052. In fact, torque requirements might be further reduced if one rotor, or all rotors, were self-spinning in the sense of the turbosirens discussed previously, although the systems would now require non-conventional mechanical design. Rather than complete dependence on a brushless DC motor, which may lack sufficient power to turn all sirens, the self-spinning system could be modulated by a mechanical braking system.

Magneto-rheological fluids-based braking provides one possibility. A mud pulser is controlled by an electric field which may be applied to an electro-active fluid. The electro-active fluid is employed to act as a rapid-response brake to slow or interrupt the rotation of a mud motor or mud siren, thus creating pressure pulses in a circulating fluid. In short, the applied electric field alters the molecular orientation of constituent fluid molecules and very rapidly changes its viscosity or resistance. In certain embodiments, the electro-active fluid is used as a direct brake acting on a shaft rotating in a volume of electro-active fluid where the shaft is coupled to the mud motor or siren. The application of a field to the electro-active fluid impedes the rotation of the shaft, thus slowing the mud motor and creating a pressure pulse in the circulating fluid. In another embodiment, a Moineau pump circulating an electro-active fluid is coupled to the mud motor. The application of a field to the electro-active fluid slows the rotation of the pump, thus slowing the mud motor and creating a pressure pulse in the circulating fluid. Further details are offered in the lead author's U.S. Patent No. 7,082,078 entitled "Magnetorheological Fluid Controlled Mud Pulser."

Again, the principles underlying "sirens in series" designs are developed by the lead author in U.S. Patent No. 5,583,827, "Measurement-While-Drilling System and Method." Essentially, the work done by the rotor on the flowing mud should result in increased signal. Since the distance between sirens is small, say one foot, the phase difference between the created signals can be neglected when compared to that associated with reflections from the drillbit. Two sirens would create twice the signal of a single siren. Together with, say, the "1.7" gain arising from the constructive interference due to drillbit reflections, a pulser system with 3.4 times the signal of a single unoptimized siren is possible, resulting in significant increases in transmission distance.

10.1.4 Attenuation misinterpretation.

Serious misconceptions in MWD design are found in conventional perceptions underlying attenuation. The paper of Desbrandes, Bourgoyne and Carter (1987) describes tests of a fluidic pulser in the flow loop of Figure 9.18 and concludes that signals beyond 25 Hz suffer from great attenuation. In the early 1990s, this author used the same flow loop, however, with siren and positive MWD pulsers, and reached identical conclusions. All three pulser types are dipole sources so that the consistent results obtained were at first reassuring.

Frequencies were subsequently increased up to 50 Hz, to the point where the hydraulically driven system inefficiently created *smaller* Δp's – surprisingly, measured pressures unexpectedly *increased* noticeably from those at 25 Hz. Thus, the author was led to conclude that the experiments largely measured amplitude changes associated with standing wave node and antinode movement. This important conclusion, drawn by the lead author, was summarized in Gardner (2002), which reported our experimental results – "Very high data rate signals can be transmitted through drilling mud with a relatively small amount of signal attenuation. We found that what has generally been attributed to non-recoverable attenuation is really the effect of wave interference."

This conclusion is also independently confirmed in Figure 10.5, supporting our assertions that attenuation, while not negligible, is not as overwhelming as previously thought. In fact, with signal amplification via constructive interference as described above, plus suggested changes in the telemetry scheme and optimized encoding, real data rates in excess of 10 bits/sec are achievable.

The precise effects of attenuation cannot be determined without development of still another model for the flowloop used, but the published conclusions of Desbrandes *et al* are suspect. The author, supported by attenuation models similar to those in Chapter 6, has separately determined from detailed measurements at a separate proprietary long flow loop facility that wave attenuation may not be as severe as the industry presently believes.

Interestingly, the signal processing website in Figure 10.5, current as of April 2011, independently supports the author's contention that attenuation is not the primary culprit for low data rate transmissions. Quoting directly, "the first practical problem we were asked to resolve was the high incidence of bad signal quality for a series of shallow (5,000 feet) wells in the North Sea. This was blamed on any number of factors such as bad mud valves, bad software, electrical problems, and so forth. After looking through the data, we concluded that the problem was due to lack of attenuation in their signaling band. That is, the shallow wells suffered from multi-path phenomena similar to those which cause ghosting in TV images. Also, our examination of the data indicated the presence of higher frequency bands (up to 100 Hz) which had low attenuation rates and were thus suitable for communication. This insight was confirmed just the next year by an independent university laboratory (Ph.D. dissertation) and it has since provided the basis for greatly increased MWD throughput."

This explanation suggests that our use of a higher 60 Hz is very reasonable. In addition, the "ghosting" which the website alludes to could be due to multiple reflections realized at shallow depths. However, it could also be explained by the up and down-going signals created at the source within the MWD drill collar. More than likely, both explanations apply, illustrating the severity of problems encountered when short wavelength transmissions are predominant.

Measurement While Drilling

Modern oil and gas wells are far more sophisticated than a simple vertical bore. In order to maximize recovery from the oil bearing strata, the drill head is actively steered to follow the geologic formation lines, often resulting in horizontal drilling once the appropriate depth has been reached. In order to locate and evaluate the correct geologic formations, a sensor package is installed just behind the drill head. Over the years, quite a variety of sensors have been used to detect gamma rays, temperature, soil resistivity, pressure, drill angle, and so forth. The technical challenge is to obtain these measurements in real time and transmit them to the surface for analysis. This is called measurement while drilling (MWD).

Mud Pulse Telemetry

The image below depicts the basic components of the MWD system. The heart of the rig is the drill stem - a steel pipe which is driven mechanically at the top end and carries the drill head (cutter) at the bottom. Although we commonly think of a 6 or 10 inch steel pipe as being mechanically stiff, consider a 25,000 foot deep well. The drill stem for such a well has the same aspect ratio as a piece of #30 wire wrap wire that is 40 feet long. Imagine trying to transfer torque and vertical load from the end of a wire wrap wire to a tiny cutter located forty feet below!

It's not practical to run electrical or optical cable down to the measurement package near the cutter. The drill stem is made up of 40 to 60 foot segments of pipe that get screwed together as the well progresses. As each new pipe segment is added, the communication path needs to get extended too. About 40 years ago, MWD innovators developed the concept of mud pulse telemetry. Their communication "channel" is based on the mud slurry (often bentonite clay suspended in water) which is pumped down the center of the drill stem to the cutter head. This slurry cools the cutter head and clears drilling debris away, carrying it to the surface through the outer annulus of the bore hole. Maintaining positive pressure in the bore hole also helps prevent collapse of the walls. The drilling slurry is typically supplied by a triplex pump which operates at a few hertz, developing a pressure of several thousand PSI.

The pressurized mud slurry provides a low frequency acoustic channel which can be used to send signals from the down-hole measurement package back to the surface. Just behind the cutter head, a mud turbine steals a bit of energy from the slurry stream to power the measurement and communications package. Data transmission is by means of a valve which periodically constricts the mud flow, sending a pressure pulse back up the mud column to the top. A pressure sensor acts as the signal receiver for the top side data logging equipment. Early mud telemetry systems operated below the two Hz fundamental frequency of the slurry pumps, typically providing a communication rate of 0.1 to 0.5 baud.

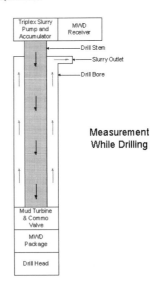

Figure 10.5. From Presco Inc. website, on siren pulsers (http://www.prescoinc.com/science/drilling.htm in April 2011 – – this page is no longer available).

An Improved MWD Receiver

Presco's client was the world's largest supplier of MWD systems to the oil industry. Their goal was to increase the data transmission rate so that more instruments and higher sampling rates could be used in the MWD sensor package. They also needed to improve the reliability of their signaling mechanism so that it would work in wells of greatly varying depths and topologies.

A first step at improved signal quality was to upgrade the pressure sensor used to receive the mud pulses. The triplex pumps deliver hundreds of horsepower to the mud slurry and operate at thousands of PSI. Pulses from the down hole package are typically less than 1 PSI so it is difficult to discern them in the presence of the pump noise. Also, the oil rigs are well known for their bad electrical grounding and huge ground loop currents. All of this makes it difficult to recover the signals of interest. Our client used a 16 bit A/D converter installed in their computer chassis to monitor mud pressures at several points in the system. However, their data recovery algorithm showed poor SNR and a series of tests showed that the digitized data exhibited only 8 bits of true content.

Presco designed a new analog front end that was specialized for low frequency operation, high signal to noise ratio, and a bad operating environment. Each sensor input was received by an Analog Devices 295 isolation amplifier. This part provides at least 1000 volt common mode capability, as well as containing its own isolated power converter for the input side circuitry. . Each converter was supplied from a separate "dirty" supply to avoid contamination of the clean +-15 volt supplies in the quiet section of circuitry. The individual "dirty" supplies and power filters were chosen to suppress injection of low frequency components due to beating of the internal oscillators in the converters. Supplies and layout were also chosen to minimize stray capacitance to suppress noise coupling.

Each of the primary channels was passed through a resistor programmable anti-aliasing filter before being sent to the A/D card inside the DSP chassis. Filters were of the Bessel (constant time delay) type to preserve waveform shape for the benefit of computer based correlation detection methods. The low corner frequency (4 Hz) and four pole configuration provided the required attenuation of unwanted high frequency components, including any residual feed-through of the modulator frequency from the isolation amps. Connection to the A/D card was made by flat cable with a full coverage shield and metal connector shells for EMI resistance.

Digital control signals from the computer were received by RS-422 receivers and latched inside a special digital section of the card. Control signals were then filtered upon entering the analog section of the circuit to further reduce the chance of EMI contamination from the computer. While these design techniques might appear extremely conservative, our precautions were rewarded during final acceptance testing by achieving a SNR of 105 dB in an end-to-end test.

Increasing MWD Bandwidth

Our client had dominated the MWD business for years without being forced to increase their channel bandwidth, but changes in the industry forced a reassessment of their MWD system. On being introduced to the problem, Presco's initial response was to ask for information concerning the bandwidth and attenuation characteristics of the acoustic channel. To our great surprise, there was no hard data about the mud channel, just a lot of folklore about how the mud was impossibly lossy and how the frequency response rolled off "forever" starting below one hertz. It was also "common knowledge" that Manchester coding was the only secure signaling method for the mud channel and that data compression would produce unacceptably high error rates.

The first practical problem we were asked to resolve was the high incidence of bad signal quality for a series of shallow (5,000 feet) wells in the North Sea. This was blamed on any number of factors such as bad mud valves, bad software, electrical problems, and so forth. After looking through the data, we concluded that the problem was due to lack of attenuation in their signaling band. That is, the shallow wells suffered from multi-path phenomena similar to those which cause ghosting in TV images. Also, our examination of the data indicated the presence of higher frequency bands (up to 100 Hz) which had low attenuation rates and were thus suitable for communication. This insight was confirmed just the next year by an independent university laboratory (PhD dissertation) and it has since provided the basis for greatly increased MWD throughput.

Since our client's mud valve didn't have the frequency response to access the higher communication bands in the slurry channel, we concentrated our attention on using the lower bands more effectively. The first point of attack was to double the effective data rate by abandoning the practice of Manchester coding. This coding scheme is commonly used for magnetic tapes because it insures at least one signal transition in each coding cell. Because of the guaranteed transitions, it's easy to phase lock to a Manchester data stream and retrieve the bits, but this coding method uses twice the minimum bandwidth. There are alternative coding methods such as the 4B/5B scheme used in FDDI fiber optic links and the various run-length-limited codes used for disk recording which provide good clock recovery without wasting so much bandwidth. Also, we demonstrated that good data integrity could be maintained while using data compression to remove redundancy from the data stream. The trick was to start with maximally compressed data and then use an overall coding method (data packets with CRC) to inject intentional redundancy to improve data link integrity. In total, these changes permitted an increase of 8:1 in data rate without changing the telemetry hardware. Hence, an enormous investment in down hole equipment was given a major end-of-life extension before becoming obsolete.

Figure 10.5. From Presco Inc. website, on siren pulsers, continued (http://www.prescoinc.com/science/drilling.htm in April 2011 – – this page is no longer available).

10.1.5 Surface signal processing.

The depth over which MWD transmissions can operate successfully depends not only on signal strength created at the source, but importantly, on the "signal to noise ratio" (S/N) found at the surface. To emphasize this point, we suggest that a small 0.1 psi signal in itself might not be entirely detrimental if noise did not exist. But it does and, very often, overwhelms the upcoming signal. Different types of noise are found at the surface, e.g., mud flow noise of a random nature, noise associated with drillstring vibrations and rig operations, and so on, many of which can be removed using conventional frequency filtering methods.

However, a major source of problems is propagating noise traveling in a direction opposite to the upcoming signal, e.g., MWD signals reflecting from mudpump pistons, shape-distorted signals reflected from the desurger and rotary hose, and very large noise amplitudes created by moving duplex and triplex pump pistons themselves. In principle, these are filtered by multiple transducer signal processing methods, but several service-company schemes used to this author's knowledge are derived under dubious assumptions. For instance, some unrealistically assume sinusoidal time variations, while others casually invoke "common sense" subtraction methods. None apply the degree of rigor found in seismic processing, which is based on exacting geophysical models.

The multiple transducer methods in Chapter 4, however, are based on formal wave equation manipulations and results. For instance, in Method 4-4, the one-dimensional equation separating left from right-going waves can be finite-differenced in space and time – pressures at spatial nodes are interpreted as those at specific transducer locations – values available at different time levels are interpreted as values stored in different computer locations. Multilevel and multi-node schemes are easily developed which can be as complicated as the need warrants. One such implementation is given by the author in his U.S. Patent No. 5,969,638, "Multiple Transducer MWD Surface Signal Processing," awarded Oct. 19, 1999. However, the method is incomplete in that the formulas terminate with the time derivative of the signal when it is really its integral that is important. The model in Chapter 4 remedies this by augmenting the basic approach with a highly robust integrator that successfully handles the sudden starts and stops associated with short high-data-rate pulses.

When this surface signal processing method is used, all downgoing noise regardless of shape and amplitude is virtually eliminated, allowing standpipe mounted piezoelectric transducers to detect minute MWD signals accurately, knowing only the local sound speed, which is separately measured or estimated. We give examples of our success with the scheme. Recall that we had previously considered the upcoming test signal

```
Internal MWD upgoing (psi) signal available as

P(x,t) = +      5.000 {H(x-  150.000-ct)  - H(x-  400.000-ct)}
         +     10.000 {H(x-  600.000-ct)  - H(x- 1000.000-ct)}
         +     15.000 {H(x- 1400.000-ct)  - H(x- 1700.000-ct)}
```

consisting of three closely spaced and short rectangular pulses (H is the Heaviside step function). At time t = 0, the pressure P(x,0) contains three rectangular pulses with amplitudes (a) 5 for 150 < x < 400, (b) 10 for 600 < x < 1000, and (c) 15 for 1400 < x < 1700. Thus, the pulse widths and separations, going from left to right, are

```
•    400  -  150  =  250 ft
•    600  -  400  =  200 ft
•   1000  -  600  =  400 ft
•   1400  - 1000  =  400 ft
•   1700  - 1400  =  300 ft
```

The average spatial width is about 300 ft. If the sound speed is 5,000 ft/sec (as assumed below) then the time required for this pulse to displace is 300/5,000 or 0.06 sec. Since sixteen of these are found in a single second, this represents 16 bps, approximately. Below we define the noise function, which propagates in a direction opposite to the upgoing signal. For our upgoing signal we have 16 bps. In our noise model below, we assume a 15 Hz sinusoidal wave (for convenience, though not a requirement) with an amplitude of 20 (which exceeds the 5, 10, 15 above). These equal frequencies provide a good test of effective filtering based on directions only – conventional frequency methods will not work since both signal and noise frequencies are similar.

For the MWD pulse, the far right position is 1,700 ft. We want to be able to "watch" all the pulses move by in our graphics, so we enter "1710" (>1700 below). We also assume a transducer separation of 30 ft. This is about 10% of the typical pulse width above, and importantly, is the length of the standpipe; thus, we can place two transducers at the top and bottom of the standpipe. Recall that Method 4-4 is based on derivatives. The meaning of a derivative from calculus is "a small distance." Just how small is small? The results seem to suggest that 10% of a wavelength is small enough.

```
Downward propagating noise (psi) assumed as
N(x,t) = Amplitude * cos {2π f (t + x/c)} ...
o  Enter noise freq "f"  (hz):   15
o  Type noise amplitude (psi):   20
o  Enter sound speed c (ft/s):   5000
o  Mean transducer x-val (ft):   1710
o  Transducer separation (ft):   30
```

Note that the noise amplitude is not small, but is chosen to be comparable to the MWD amplitudes, although only large enough so that all the line drawings fit on the same graphical display. The method actually applies to much larger amplitudes as we will shortly demonstrate. After SAS14D.exe executes, it creates two output files, SAS14.DAT and MYFILE.DAT. The first is a text file with a "plain English" summary. The second is a data file used for plotting. To plot results, run the program FLOAT32, which will give the results in Figure 10.6a where an index related to time appears on the horizontal.

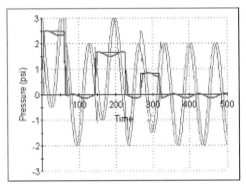

Figure 10.6a. Recovery of three step pulses from noisy environment.

In Figure 10.6a, black represents the clean upgoing MWD three-pulse original signal. Red is the recovered pulse – this result is so good that it partially hides the black signal. The green and blue lines are pressure signals measured at the two pressure transducers, again separated by thirty feet. From these two traces individually, one would not surmise that the red line can be recovered; the green and blue signal curves differ only through minor "bumps." The algorithm handles very small signal-to-noise ratios extremely well. Below, we take the foregoing three-pulse signal as input again; a noise amplitude of 200 is assumed, so that the S/N ratio ranges from 0.025 to 0.075, all of which are small. Calculated results in Figure 10.6b again show excellent signal recovery.

```
Downward propagating noise (psi) assumed as
N(x,t) = Amplitude * cos {2π f (t + x/c)} ...
o  Enter noise freq "f"  (hz):   5
o  Type noise amplitude (psi):  200
o  Enter sound speed c (ft/s):  5000
o  Mean transducer x-val (ft):  1700
o  Transducer separation (ft):   30
```

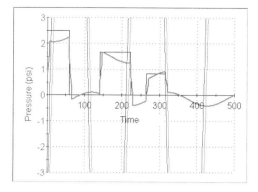

Figure 10.6b. Recovery of three step pulses from *very* noisy environment.

10.1.6 Attenuation, distance and frequency.

The most important design questions in MWD relate to attenuation as it depends on the combined and competing effects of fluid properties, drillpipe geometry, signal generator characteristics and surface transducer specifications. Let ρ represent fluid density, μ the viscosity, c the sound speed, D the drillpipe inner diameter and L the transmission distance along the pipe. If P_0 is the source signal strength and P_{xdcr} is the surface transducer sensitivity (that is, the smallest detectable pressure after successful noise removal), then it can be shown that a "critical carrier frequency" (expressed in Hertz) satisfies $f_{crit} = \rho c^2 D^2 (\log_e P_0/P_{xdcr})^2 /(4\pi\mu L^2)$ where the attenuation of plane waves propagating in a laminar flow is modeled.

This critical frequency is the value necessary to support PSK or FSK transmission over the assumed distance L. It provides the needed operational bound; for frequencies $f > f_{crit}$, signal transmission is not possible. The formula highlights several interesting physical features behind MWD wave propagation, in particular, its dependence on the kinematic viscosity μ/ρ and not on μ alone, not to mention the role played by pipe cross-sectional area ($\sim D^2$).

Interestingly, a dimensionless pressure P_0/P_{xdcr} controls successful reception. For instance, a 300 psi downhole source coupled with a surface transducer unable to "see" with 1 psi resolution is less effective than a 100 psi source working with a transducer having 0.25 psi capabilities. The strategy required to optimize this ratio is obvious. First, reduce P_{xdcr} as much as possible by (a) using the most sensitive piezoelectric gauges available, (b) employing effective *surface* noise cancellation methods such as those for Figures 10.6a,b, (c) applying downhole reverberation algorithms as in Chapter 5, or perhaps, all of the foregoing. And second, increase P_0 to the maximum extent permissible by (1) using FSK schemes that take advantage of constructive wave interference, (2) developing ganged sirens arranged in series, (3) optimizing siren geometries, e.g., small rotor-housing clearances, reduced rotor-stator gap distances, good rotor-stator azimuthal overlap, or possibly, all of the foregoing.

Just how optimally can a new high-data-rate mud pulse MWD system perform? And how would this compare with the best systems currently available? To answer these questions, we first describe calculated results used to validate our formula for critical frequency. Figure 10.7 shows the software interface developed to host the calculation. Here, the input values P_0 (145 psi), P_{xdcr} (0.4 psi), specific gravity (1.7, for 14 lb/gal mud) and viscosity (50 cp) are taken from Hutin, Tennent and Kashikar (2001) in "New Mud Pulse Telemetry Techniques for Deepwater Applications and Improved Real-Time Data Capabilities." The pipe diameter (4.0 in), depth (25,000 ft) and sound speed (3,000 ft/s) are our estimates. These assumptions, for our Run A, give a critical frequency of 13.8 Hz, which would be consistent with the 12 Hz siren carrier frequency used by Schlumberger at such depths to achieve 3 bps. In this respect, the model for critical MWD carrier frequency provides reasonable results, which have been further supported by sensitivity analyses.

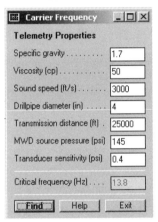

Figure 10.7. Critical frequency calculation.

In Run B, we repeat Run A except that the transmission depth is increased to 35,000 ft. The result demonstrates that the critical frequency is reduced to f_{crit} = 7.0 Hz so that the 3 bps obtainable at 12 Hz is no longer possible – it is known that the company instead transmits at approximately 1 bps. Again, our model conclusions are reasonable.

We use this model to evaluate other published data. Only limited information on BakerHughes's system is available, e.g., "15 bps in the North Sea to a depth of 24,000 ft," "20 bps at depths less than 6,000 m, and just over 3 bps from depths of more than 10,000 m," and "rates of 30 bps have been achieved from 3,000 m onshore and 40 bps from 900 m in a test well." Corresponding mud properties and pipe diameter data are not published, so that it is difficult to ascertain true system performance. However, we can estimate the conditions under which these claims are realistic. If we assume that its rotary shear valve possesses characteristics similar to that of the mud siren, not an unreasonable assumption, a simple reduction of the viscosity from 50 cp to 20 cp would allow transmission depths to 35,000 ft, as the results for Run C show – the calculated value of 17.6 Hz would easily support 3 bps but no more. In Run D, we increase our viscosity to 50 cp but limit travel distance to 3,600 ft. The critical frequency increases to 665 Hz, demonstrating at 40 bps is not unexpected. Cumulative results are given in Figure 10.8.

	SG	CP	C	DIA	L	P0	P	Fcrit
A	1.7	50	3000	4	25000	145	0.4	13.8
B	1.7	50	3000	4	35000	145	0.4	7.0
C	1.7	20	3000	4	35000	145	0.4	17.6
D	1.7	50	3000	4	3600	145	0.4	664.9

Figure 10.8. Cumulative results, critical carrier frequencies.

Finally, we ask, "What data rates are possible using all the technology elements developed in this book for mud pulse telemetry?" Again, we emphasize that maximizing the critical carrier frequency requires us to optimize the ratio P_0/P_{xdcr} following the strategies indicated earlier. Some experimental evidence suggests that delta-p source strengths are independent of siren frequency at higher frequencies exceeding 10 Hz. Now let us assume the same baseline numbers used in Run A for Schlumberger's mud siren. For Run E, instead of PSK, we operate FSK with constructive interference with the siren optimally positioned and assume a 1.7 factor increase in signal output as might be suggested by Figure 10.2. Thus, the $P_0 = 145$ psi used previously is replaced by $P_0 = 1.7 \times 145$ or 246.5 psi, increasing the critical frequency to 16.4 Hz – not enough to increase data rates substantially.

In Run F, we additionally apply the "sirens in series" design suggested in Figure 10.4, which would double the 246.5 psi to 493 psi. This only increases the critical frequency to 20.1 Hz. However, if we increase the drillpipe diameter to 5 in as in Run G, f_{crit} increases to a remarkable 31.4 Hz. In Run H, we decrease the mud viscosity to 20 cp, showing in increase in critical frequency to 78.5 Hz for the assumed 25,000 ft transmission – or, per a prior analysis, at least 10 bps. In Run I, depth is increased to 35,000 ft, and our 78.5 Hz decreases to 40.1 Hz, which should allow 6-7 bps.

The software model of Figure 10.7 can be used to select drilling muds that facilitate high-data-rate transmissions too. In Run J, let us formulate a mud with a specific gravity of 2, a plastic viscosity of 40 and a sound speed of 4,000 ft/s. Then, employing a single siren, but with the use of FSK and constructive interference, we have a high value of 34.1 Hz at 35,000 ft. In our final Run K, we reduce P_{xdcr} to 0.2 psi and demonstrate that the critical frequency increases to 41.9, for efficient 6-7 bps operation (piezoelectric transducers with such sensitivities are readily available and need to be used together with the directional cancellation schemes developed here). Our results, summarized in Figure 10.9, demonstrate that 10 bps is possible, and how they might be accomplished by using constructive wave interference, increasing transducer sensitivity, and by altering drilling system properties.

	SG	CP	C	DIA	L	PO	P	Fcrit
A	1.7	50	3000	4	25000	145.0	0.4	13.8
E	1.7	50	3000	4	25000	246.5	0.4	16.4
F	1.7	50	3000	4	25000	493.0	0.4	20.1
G	1.7	50	3000	5	25000	493.0	0.4	31.4
H	1.7	20	3000	5	25000	493.0	0.4	78.5
I	1.7	20	3000	5	35000	493.0	0.4	40.1
J	2.0	40	4000	5	35000	246.5	0.4	34.1
K	2.0	40	4000	5	35000	246.5	0.2	41.9

Figure 10.9. Critical frequencies, hypothetical MWD tools.

Software reference, MWDFreq.vbp and datarate*.for are used for critical frequency analysis. Results are based on attenuative acoustic wave model allowing fluid flow in pipe.

10.1.7 Ghost signals and echoes.

While the dangers of surface reflections are well known, we have repeatedly emphasized the existence of downhole reflections which, if not properly addressed, can destructively interfere with upgoing signals, or introduce drill collar reverberations, or both. Interestingly, this problem was also identified in "An Overview of Acoustic Telemetry" by Drumheller (1992) in the context of drillpipe telemetry research conducted at Sandia National Laboratories. Many authors, until then, had assumed simply that materials waves created downhole simply traveled upwards. Drumheller remarks, "Unfortunately, this over-simplified picture is extremely misleading. As the early results of this project illustrate, real hammer blows in real drillstrings do not result in this kind of response at all. The first complication which arises is that unless the hammer blow occurs exactly at the top or bottom end of the drillstring, two pulses are generated. One pulse travels up the drillstring while the other travels down. If the hammer is placed near the bottom of the drillstring, the downward traveling pulse will quickly reflect off the drill bit and follow directly behind or possibly overlap the leading upward-traveling pulse. This results in an unwanted echo of the original pulse. In a similar fashion, if the receiver is near the top of the drillstring, the two pulses will pass the receiver, reflect off the top of the drillstring, and pass the receiver again. This process will continue creating more echoes until attenuation weakens the echoes to an undetectable level." In the time-domain view, echoes create false pulses which are indistinguishable from and confused with the true data pulse. This has been our assertion throughout this book, however, we have put these echoes to good use by having them reinforce (as opposed to canceling) upgoing waves.

10.2 MWD Signal Processing Research in China

MWD research in China is active, openly pursued and published. Because the program is key to the country's national agenda, it is well funded. These factors provide an atmosphere conducive to and fostering innovation and, importantly, supporting the training of engineers and researchers capable of improving and extending old and new ideas. In the lead author's opinion, the enthusiasm with which these efforts is pursued is unmatched anywhere else. The record is clear: a significant body of original research, pursued by both government and national oil companies, exists, focusing on modern MWD issues in signal processing, hardware design and environment characterization. And it is growing rapidly. In the "screen shots" presented below, we have captured representative research, all available publicly, reflecting a diversity of work related to topics in this book, e.g., wind tunnel simulation, phase-shift-keying, adaptive filtering, reflection deconvolution, echo removal, turbine design, torque and so on. Copies of these papers are available from the authors.

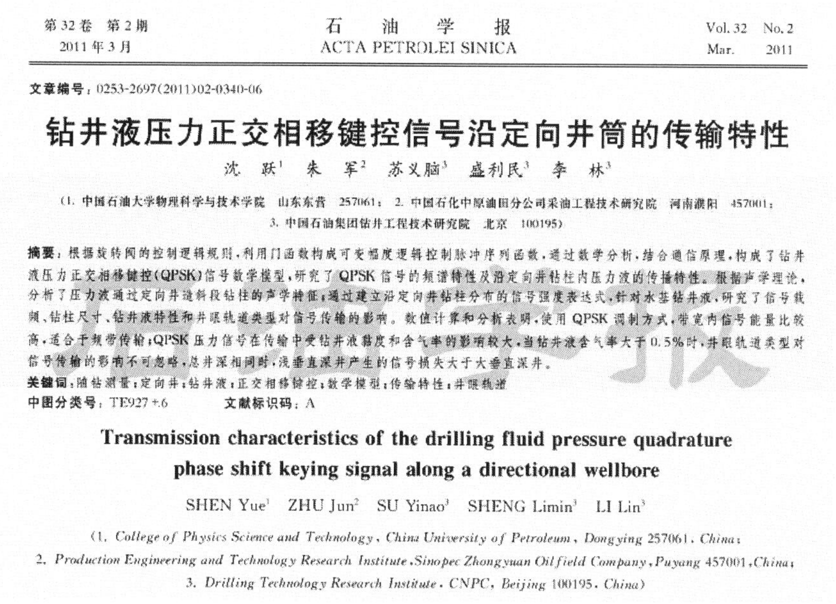

第 32 卷 第 2 期　　　　　　石　油　学　报　　　　　Vol. 32　No.2
2011 年 3 月　　　　　　ACTA PETROLEI SINICA　　　　Mar.　2011

文章编号：0253-2697(2011)02-0340-06

钻井液压力正交相移键控信号沿定向井筒的传输特性

沈 跃[1]　朱 军[2]　苏义脑[3]　盛利民[3]　李 林[3]

(1. 中国石油大学物理科学与技术学院 山东东营 257061；2. 中国石化中原油田分公司采油工程技术研究院 河南濮阳 457001；
3. 中国石油集团钻井工程技术研究院 北京 100195)

摘要：根据蝶转阀的控制逻辑规则，利用门函数构成可变幅度逻辑控制脉冲序列函数，通过数学分析，结合通信原理，构成了钻井液压力正交相移键控(QPSK)信号数学模型，研究了 QPSK 信号的频谱特性及沿定向井钻柱内压力波的传播特性。根据声学理论，分析了压力波通过定向井造斜段钻柱的声学特征；通过建立沿定向井钻柱分布的信号强度表达式，针对水基钻井液，研究了信号载频、钻柱尺寸、钻井液特性和井眼轨道类型对信号传输的影响。数值计算和分析表明，使用 QPSK 调制方式，带宽内信号能量比较高，适合于宽带传输；QPSK 压力信号在传输中受钻井液黏度和含气率的影响较大。当钻井液含气率大于 0.5%时，井眼轨道类型对信号传输的影响不可忽略；且井深相同时，浅垂直深井产生的信号损失大于大垂直深井。

关键词：随钻测量；定向井；钻井液；正交相移键控；数学模型；传输特性；井眼轨道
中图分类号：TE927+.6　　　文献标识码：A

Transmission characteristics of the drilling fluid pressure quadrature phase shift keying signal along a directional wellbore

SHEN Yue[1]　ZHU Jun[2]　SU Yinao[3]　SHENG Limin[3]　LI Lin[3]

(1. College of Physics Science and Technology, China University of Petroleum, Dongying 257061, China;
2. Production Engineering and Technology Research Institute, Sinopec Zhongyuan Oilfield Company, Puyang 457001, China;
3. Drilling Technology Research Institute, CNPC, Beijing 100195, China)

Figure 10.2.1. Signal processing, PSK methods (CNPC).

Research Article

Propagation of Measurement-While-Drilling Mud Pulse during High Temperature Deep Well Drilling Operations

Hongtao Li,[1] Yingfeng Meng,[1] Gao Li,[1] Na Wei,[1] Jiajie Liu,[1] Xiao Ma,[1] Mubai Duan,[1] Siman Gu,[1] Kuanliang Zhu,[2] and Xiaofeng Xu[2]

[1] *State Key Laboratory of Oil and Gas Reservoir Geology and Exploration, School of Petroleum Engineering, Southwest Petroleum University, Chengdu 610500, China*
[2] *Drilling and Production Technology Institute, PetroChina Jidong Oilfield Company, Tangshan 063000, China*

Figure 10.2.2. High temperature environmental effects.

Available online at www.sciencedirect.com

SciVerse ScienceDirect

Procedia Engineering

ELSEVIER

Procedia Engineering 24 (2011) 319 – 323

www.elsevier.com/locate/procedia

2011 International Conference on Advances in Engineering

Development of Downlink Communication System for Steerable Drilling Application

Dang Rui-rong, Yin Guang, Gao Guo-wang, Liang Lu[*]

Key Laboratory of photoelectric logging and detecting of oil and gas, Ministry of Education Xi'an Shiyou University, Xi'an 710065, China

Figure 10.2.3. Downlink strategies for rotary steerable systems.

266

Pet.Sci.(2009)6:266-270

DOI 10.1007/s12182-009-0042-8

Numerical modeling of DPSK pressure signals and their transmission characteristics in mud channels

Shen Yue[1], Su Yinao[2*], Li Gensheng[3], Li Lin[2] and Tian Shouceng[3]

[1] College of Physics Science and Technology, China University of Petroleum, Dongying, Shandong 257061, China
[2] CNPC Drilling Research Institute, Beijing 100083, China
[3] State Key Laboratory of Petroleum Resources and Prospecting, China University of Petroleum, Beijing 102249, China

Abstract: A numerical model and transmission characteristic analysis of DPSK (differential phase shift keying) pressure signals in mud channels is introduced. With the control logic analysis of the rotary valve mud telemetry, a logical control signal is built from a Gate function sequence according to the binary symbols of transmitted data and a phase-shift function is obtained by integrating the logical control signal. A mathematical model of the DPSK pressure signal is built based on principles of communications by modulating carrier phase with the phase-shift function and a numerical simulation of the pressure wave is implemented with the mathematical model by MATLAB programming. Considering drillpipe pressure and drilling fluid temperature profile along drillpipes, the drillpipe of a vertical well is divided into a number of sections. With water-based drilling fluids, the impacts of travel distance, carrier frequency, drillpipe size, and drilling fluids on the signal transmission were studied by signal transmission characteristic analysis for all the sections. Numerical calculation results indicate that the influences of the viscosity of drilling fluids and volume fraction of gas in drilling fluids on the DPSK signal transmission are more notable than the others and the signal will distort in waveform with differential attenuations of the signal frequent component.

Figure 10.2.4. Signal processing (CNPC).

Delay pressure detection method to eliminate pump pressure interference on the downhole mud pressure signals

Yue Shen,[1] Ling-Tan Zhang,[1] Shi-Li Cui,[1] Li-Min Sheng,[2] Lin Li[2] and Yi-Nao Su[2]

[1] *School of Science, China University of Petroleum, Qingdao 266580, China*
[2] *Drilling Technology Research Institute, CNPC, Beijing 100195, China*

Correspondence should be addressed to Yue Shen; sheny1961@aliyun.com

The feasibility of applying delay pressure detection method to eliminate mud pump pressure interference on the downhole mud pressure signals is studied. Two pressure sensors mounted on the mud pipe in some distance apart are provided to detect the downhole mud continuous pressure wave signals on the surface according to the delayed time produced by mud pressure wave transmitting between the two sensors. A mathematical model of delay pressure detection is built by analysis of transmission path between mud pump pressure interference and downhole mud pressure signals. Considering pressure signal transmission characteristics of the mud pipe, a mathematical model of ideal low-pass filter for limited frequency band signal is introduced to study the pole frequency impact on the signal reconstruction and the constraints of pressure sensor distance are obtained with pole

Figure 10.2.5. Delay line methods for pump noise removal (CNPC).

Research on Mechanism of Continuous Wave Signal Generator Controlled by DSP and the Wind Tunnel Simulation Test
Posted on April 9, 2010 by China Papers

Abstract: MWD is a new logging technology which is developed in recent years, it can improve large displacement wells, horizontal wells in difficult engineering control and formation evaluation capabilities, improve the rate of oil drilling encountered. Data signal transmission plays a pivotal role in the MWD system design, and is the core in the system design. For low MWD data transmission rates and the difficult test questions in design process, the paper researched on mechanism of continuous wave signal generator controlled by DSP and the wind tunnel simulation test, made some progress and useful conclusions.In this paper, we explore the use of theoretical analysis, system design, computer simulation and wind tunnel simulation method of combining. At the basis of research on telemetry signal transmission mechanism, continuous wave signal generator working mechanism, the principle of wind tunnel tests and other analytical studies, completed a signal generator, DSP control system, wind tunnel test model of the structure and control system design. On this basis, wind tunnel tests and computer simulation experiments were done, and built a variable frequency pressure wave transmission predictive control model and the optimization algorithm model of impeller design. The main goal of the paper is to achieve higher data rates, improve the reliability of data transmissions, as well as enhance the system environmental adaptability, and to lay a foundation at theoretical and experimental methods for our own independent intellectual property rights of continuous mud wave MWD system.Analysis of the bottom telemetry signal transmission mechanism showed that the mud continuous wave represents the development direction of MWD wireless data transmission technology. By analyzing the data encoding, M-ary frequency modulation mode was selected to modulate the down-hole information. Compared with M-ary phase modulation transfer mode, M-ary frequency modulation mode has relative simple sending and receiving equipment and control system, and relative low bit error rate is conducive to improve data transfer rate. Ground to down-hole and down-hole to ground data

Figure 10.2.6. Wind tunnel simulation for signal processing (CNPC).

Available online at www.sciencedirect.com

SciVerse ScienceDirect

Procedia Engineering 15 (2011) 2364 – 2368

Procedia Engineering

www.elsevier.com/locate/procedia

Advanced in Control Engineering and Information Science

Continuous-wave mud telemetry digital communication system design and the simulation test

Xinping Liu[a],Xiwen Xue[b],a*

[a]Xinping Liu,Computer and communication engineering college, China University of Petroleum, Dongying Shandong 257061
[b]Xiwen Xue,Computer and communication engineering college, China University of Petroleum, Dongying Shandong 257061

Abstract

This paper researched on the continuous wave mud telemetry MWD system based on the frequency modulation (FM) transmission mode. The digital communication system based on the continuous wave mud telemetry was designed. The system architecture design includes the ground signal transceiver devices, the bottom signal transceiver devices, as well as the third part of data transmission channel. In the initial stage of the system design, the wind tunnel simulation tests could be employed. The structure of the wind tunnel test model was designed according to the similarity principle, and a series of wind tunnel simulation tests were carried out for data transmission. Test results showed that the continuous wave mud telemetry MWD system based on the FM transmission mode could achieve higher data transfer rate, improve job reliability, and enhance the adaptability to the environment.

Figure 10.2.7. Wind tunnel simulation methods for signal processing.

2010 International Conference on Computing, Control and Industrial Engineering

The Variable Frequency Data Transmission Technology Based on Artificial Neural Network Applying in Measurement while Drilling System

Wuhan, China
June 05-June 06
ISBN: 978-0-7695-4026-9
Xinping Liu
Youhai Jin
Jun Fang

DOI Bookmark: http://doi.ieeecomputersociety.org/10.1109/CCIE.2010.185

The current continuous wave mud telemetry measurement while drilling (MWD) systems commonly apply binary phase shift keying (BPSK) modulation data transmission. Its communication system is complex, the data transfer rate is low, and the data transmission is not flexible enough to adapt to the environment. Accordingly, the paper explored to use the M-ary frequency shift keying (MFSK) modulation data transmission mode, to research the training algorithm of the neural network and the variable frequency data transmission model. The data reception bit error rate of different transmission modes would be predicted with well trained network, and then chose the data transmission method which had the highest data transfer rate to send and receive the down-hole data in the allowed range of bit error rate. The data reception bit error rate in different drilling conditions and different transmission modes would be recorded. The neural network would be retrained periodically. Thereby the adaptability of the network would be increased. Test results show that the variable frequency data transmission technology based on neural network can achieve higher data transfer rate, improve job reliability, and enhance the adaptability to the environment.

Figure 10.2.8. Neural network approaches (CNPC).

《Journal of China University of Petroleum(Edition of Natural Science)》 2011-04

Continuous-wave drilling fluid telemetry measurement while drilling system design and wind tunnel simulation test

LIU Xin-ping,XUE Xi-wen (College of Computer and Communication Engineering in China University of Petroleum,Dongying 257061,China)

The main problems of the current measurement while drilling(MWD) systems are low data transmission rate and the difficult tests in design process.The continuous wave drilling fluid telemetry MWD system based on the frequency modulation(FM) transmission mode was designed and the wind tunnel simulation tests were carried out.The digital communication system based on the continuous wave drilling fluid telemetry was designed.Multiple frequency shift keying mode was selected to modulate the down-hole information and send to the ground.The system architecture design includes the ground signal transceiver devices,the bottom signal transceiver devices,and data transmission channel.Two types of fan-shaped and round-shaped valve port structures were designed.The structure of the wind tunnel test model was designed according to the similarity principle,and a series of wind tunnel simulation tests were carried out for data transmission.The test results show that the continuous wave drilling fluid telemetry MWD system based on the FM transmission mode can achieve high data transmission rate,improve job reliability,and enhance the adaptability to the environment.

【Key Words】： measurement while drilling continuous-wave drilling fluid telemetry frequency modulation data transmission data transmission rate wind tunnel simulation test

Figure 10.2.9. Wind tunnel analysis.

第 29 卷 第 4 期	石 油 学 报	Vol. 29 No. 4
2008 年 7 月	ACTA PETROLEI SINICA	July 2008

文章编号：0253-2697(2008)04-0596-05

去除随钻测量信号中噪声及干扰的新方法

赵建辉[1] 王丽艳[1] 盛利民[2] 王家进[2]

(1. 北京航空航天大学仪器科学与光电工程学院 北京 100083; 2. 中国石油勘探开发研究院 北京 100083)

摘要：由于受到钻井特殊条件的限制,井底和地面之间的随钻测量信号难以通过有线方式进行传输,目前工程中主要采用基于钻井液脉冲位置编码调制的无线传输方式。但是,由于泥浆压力、井深、泵压等信道传输特性会对信号造成影响,随钻测量中在井口采集到的钻井液脉冲信号往往包含了各种噪声和干扰,直接影响地面信号解码结果的正确性。对钻井液脉冲脉位编码无线通信原理、噪声及干扰进行了分析,根据干扰与噪声的特点,采用线性滤波方法还原脉冲信号,又利用一种非线性"平顶消除"的方法对现场采集的信号进行了处理。与线性滤波方法处理结果的综合对比表明,该方法能更有效地去除钻井泵噪声及其他干扰。现场应用结果表明,该方法对脉冲信号位置复原误差影响小,解码过程简单实用,具有误码率低、可靠性高等特点。

关键词：随钻测量;钻井液脉冲信号;非线性滤波;信号处理;解码
中图分类号：TE24   文献标识码：A

A nonlinear method for filtering noise and interference of pulse signal in measurement while drilling

ZHAO Jianhui[1] WANG Liyan[1] SHENG Limin[2] WANG Jiajin[2]

Figure 10.2.10. Nonlinear signal processing approach (CNPC).

第 34 卷 第 1 期　　　　　石　油　学　报　　　　　Vol. 34　No. 1
2013 年 1 月　　　　　ACTA PETROLEI SINICA　　　　　Jan.　2013

文章编号: 0253-2697(2013)01-0178-06　DOI: 10. 7623/syxb201301023

钻井液压力脉宽及脉位多进制相移键控信号分析

李　翠[1]　高德利[1]　沈　跃[2]

(1. 中国石油大学石油工程教育部重点实验室　北京　02249; 2. 中国石油大学理学院　山东青岛　266580)

摘要: 在构建出钻井液压力脉宽及脉位多进制相移键控(MPSK)调制信号数学模型的基础上, 分析了 MPSK 信号的频谱特性, 从频域研究了将脉宽及脉位调制方式应用于钻井液压力 MPSK 信号传输的可行性。通过建立 MPSK 信号沿定向井钻柱分布的传递函数, 研究了钻井液信道参数和调制方式对信号传输的影响。数值分析表明, 钻井液压力脉宽 MPSK 调制信号的频谱主瓣带宽内信号的相对能量过低, 从频谱角度看不适于频带传输; 钻井液脉位 MPSK 调制可用于井下数据的频带方式传输, 但信号的传输损失相对较大, 因此对信号的检测、发射及旋转阀的转速控制要求较高。
关键词: 随钻测量; 载波调制; 多进制相移键控; 脉宽调制; 脉位调制; 传递函数
中国分类号: TE 249　　文献标识码: A

Analysis of drilling fluid pressure MPSK signals for PWM and PPM

LI Cui[1]　GAO Deli[1]　SHEN Yue[2]

Figure 10.2.11. Signal processing.

TELKOMNIKA, Vol. 11, No. 6, June 2013, pp. 3028 ~ 3035
e-ISSN: 2087-278X

■　3028

Eliminating Noise of Mud Pressure Phase Shift Keying Signals with A Self-Adaptive Filter

Yue Shen[*1], Lingtan Zhang[2], Heng Zhang[3], Yinao Su[4], Limin Sheng[5], Lin Li[6]
[1,2,3]School of Science, China University of Petroleum, Qingdao, 266580, P. R. China
[4,5,6]Drilling Technology Research Institute, CNPC, Beijing, 100195, P. R. China
*Corresponding author, e-mail: sheny1961@yahoo.com.cn[*1], zhanglt@upc.edu.cn[2],
zhang66h@163.com[3], suyinao@petrochina.com.cn[4], slmdri@cnpc.com.cn[5], lilin550703@yahoo.com.cn[6]

Abstract

The feasibility of applying a self-adaptive filter to eliminate noise in the downhole mud pressure phase shift keying (PSK) signals is studied. The self-adaptive filter with carrier wave as the filter input signal and mud pressure PSK signal including noise as the filter expected input signal in structure was adopted to process the mud pressure PSK signals with the broadband signal characteristic in communication. Mathematical model of the filter was built to reconstruct the mud pressure PSK signals based on the evaluation criterion of least mean square error (LMS) and the mathematical model of mud pressure PSK signals. According to the filter mathematical model, a special self-adaptive control algorithm was adopted to realize the filter by adjusting the filter weight coefficients self-adaptively and the impacts of

Figure 10.2.12. Signal processing, adaptive filtering methods (CNPC).

第29卷 第6期　　石 油 学 报　　Vol.29 No.6
2008年11月　　ACTA PETROLEI SINICA　　Nov. 2008

文章编号: 0253-2697(2008)06-0907-06

井下随钻测量涡轮发电机的设计与工作特性分析

沈 跃[1,2]　苏义脑[1,3]　李 林[3]　李根生[1]

(1. 中国石油大学石油工程教育部重点实验室 北京 102249; 2. 中国石油大学物理科学与技术学院 山东东营 257061;
3. 中国石油集团钻井工程技术研究院 北京 100097)

摘要: 依据经典的二维涡轮设计理论, 研究了井下随钻测量涡轮发电机涡轮设计过程中应遵循的原则及约束条件。根据发电机电枢的反应原理, 采用相量法建立了发电机研究模型和内阻数学模型, 通过数值分析手段研究了发电机外特性和电压调整率与电枢反应电抗之间的关系。研究结果表明, 发电机内阻具有的电流负反馈和功率因数角负反馈特性, 在一定程度上改善了电压调整率, 并确定出了获得最佳电压调整率时电枢反应电抗的变化区域。根据涡轮发电机实验测量数据, 结合发电机电磁特性的理论分析以及涡轮设计理论, 计算分析了涡轮的实际工作特性参数。
关键词: 涡轮发电机; 外特性; 转矩; 电枢反应电抗; 电压调整率; 有功功率; 数值模拟; 随钻测量技术
中图分类号: TE242.4　　文献标识码: A

Design of downhole turbine alternator for measurement while drilling and its performance analysis

SHEN Yue[1,2]　SU Yinao[1,3]　LI Lin[3]　LI Gensheng[1]

Figure 10.2.13. Downhole turbine design (CNPC).

第32卷 第5期　　石 油 学 报　　Vol.32 No.5
2011年9月　　ACTA PETROLEI SINICA　　Sept. 2011

文章编号: 0253-2697(2011)05-0887-06

无线随钻测量系统的数据报协议

李忠伟[1]　王瑞和[1]　房 军[2]

(1. 中国石油大学石油工程学院 山东东营 257061; 2. 中国石油大学石油工程教育部重点实验室 北京 102249)

摘要: 为了满足无线随钻测量对高传输速率的迫切需要, 避免采用现有固定结构的数据报, 通过分析井下参数的特点, 提出了一种适用于无线随钻测量系统的通用、高效、灵活、可扩展的数据报协议。针对系统传输效率提出的增量传输技术避免了冗余数据, 缩短了数据报长度。设计的数字排列表减小定界符在对传统数据报效果的改进, 能够根据参数的实际取值来动态调整相应的数据域大小, 有效地利用了数据报空间。数据报协议等方法能够根据需要自行定制数据报格式, 提高了系统的灵活性, 并使系统具有可扩展性, 能够满足随钻测量系统测量更多参数的要求。
关键词: 随钻测量; 连续波; 数据报协议; 增量数据报; 完整数据报; 数据域定量
中图分类号: TE243　　文献标识码: A

Datagram protocol of wireless MWD system

LI Zhongwei[1]　WANG Ruihe[1]　FANG Jun[2]

Figure 10.2.14. Theoretical considerations (CNPC).

Available online at www.sciencedirect.com

SciVerse ScienceDirect

Procedia Engineering 15 (2011) 2072 – 2076

Procedia Engineering

www.elsevier.com/locate/procedia

Advanced in Control Engineering and Information Science

CST: Compressive Sensing Transmission for Real Time M/LWD Communication

Yu Zhang[a,b], Ke Xiong[a,c]*, Dandan Li[a,b], Zhengding Qiu[a,b], Shenghui Wang[a,b]

[a]Institute of Information Science, Beijing Jiaotong University, Beijing 100044, China
[b]Beijing Key Laboratory of Advanced Information Science and Network Technology, 100044, China
[c]Department of Electronic Engineering, Tsinghua University, Beijing 100084, China

Abstract

The Measurement/Logging while Drilling (M/LWD) system is used to collect and transmit the logging data while drilling oil/gas wells, in which logging data is transmitted using Mud Pulse Telemetry. Due to the one-way non-feedback erasure channel property, it is hard to correct transmission errors caused by Mud channel noises in real time. This paper proposes a Compressive Sensing (CS) based Transmission (CST) scheme for logging data delivering in M/LWD systems. The CTS scheme is tolerant to transmission errors without Automatic Request for Repetition mechanism and error correcting code, and also provides an interface for field technicians to adjust the effective

Figure 10.2.15. Signal processing, modern "compressive sensing."

2010 年 第 34 卷
第 1 期

中国石油大学学报（自然科学版）
Journal of China University of Petroleum

Vol. 34 No. 1
Feb. 2010

文章编号：1673-5005（2010）01-0099-06

钻井液连续波信号发生器转阀水力转矩分析

贾　朋[1]，房　军[1]，苏义脑[2]，李　林[2]

（1.中国石油大学 机电工程学院，山东 东营 257061；2.中国石油集团 钻井工程技术研究院，北京 100097）

摘要：转子运动过程中在转子背面产生的回流以及随转子的转动而不断变化的入口速度角导致了转阀的水力转矩变化较大，影响了转阀的动态性能。利用动量矩定理建立转阀水力转矩的理论模型，然后利用 CFD 方法，针对所设计的实验样机进行三维流场仿真，对转阀水力转矩的几种影响因素进行仿真分析。研究结果表明：在转阀关闭的起始阶段，水力转矩使转阀趋于打开，其余阶段，水力转矩使转阀趋于关闭；采用曲线阀口可以使水力转矩变化平缓；适当增大定转子轴向间隙，增加转阀阀瓣个数，减小转子阀瓣厚度均可以减小水力转矩，改善电机的控制性能。
关键词：随钻测量；钻井液连续波；射流角；水力转矩；CFD 方法
中图分类号：TE 927　　文献标志码：A

Analysis on rotary valve hydraulic torque of drilling fluid continuous wave signal generator

JIA Peng[1], FANG Jun[1], SU Yi-nao[2], LI Lin[2]

(1. College of Mechanical and Electronic Engineering in China University of Petroleum, Dongying 257061, China;
2. Drilling Research Institute, CNPC, Beijing 100097, China)

Figure 10.2.16. Computational siren torque analysis (CNPC).

Applied Mechanics and Materials Vols. 278-280 (2013) pp 1107-1113
Online available since 2013/Jan/11 at www.scientific.net
© 2013 Trans Tech Publications, Switzerland
doi:10.4028/www.scientific.net/AMM.278-280.1107

Coherent demodulation of the mud pressure DPSK signal and analysis of noise impact on the signal demodulation

Yue Shen[1,a], Lingtan Zhang[1,b], Shili Cui[1,c], Yinao Su[2,d], Limin Sheng[2,e], Lin Li[2,f]

[1]College of Science, China University of Petroleum, Qingdao Shandong 266580, China

[2]Drilling Technology Research Institute, CNPC, Beijing 100195, China

[a]sheny1961@yahoo.com.cn, [b]zhanglt@upc.edu.cn, [c]borndate@163.com,
[d]suyinao@petrochina.com.cn, [e]slmdri@cnpc.com.cn, [f]lilin550703@yahoo.com.cn

Keywords: mud pressure differential phase shift keying (DPSK) signal; coherent demodulation; noise; signal to noise ratio (SNR); bit error rate (BER); frequency band

Abstract. Based on the mathematical analysis of band transmission signal coherent detection theory and mud pressure DPSK signal modulation process, by establishing the coherent detection mathematical model of mud pressure DPSK signal, this paper studies the practicability of broadband mud pressure DPSK signal coherent detection. By analyzing the noise impact on input signal parameter of demodulation system, based on probabilistic statistics theory and coherent demodulation mathematical model, the bit error rate and antinoise ability of mud pressure DPSK signal is studied. Theoretical analysis and numerical simulation indicate that, by adding arc cosine and derivation segment to basic demodulation process, the coherent demodulation effect of mud pressure DPSK signal is satisfied, but the demodulation process is more complex than conventional communication system. Theoretical analysis shows that mud pressure DPSK signal has the same theatrical bit error rate as conventional communication binary phase shift keying (2PSK) signal, but due to the difference of signal demodulation means, the reconstruction of rotary valve control pulse will be affected seriously by the derivate of vertical component of the noise within frequency band, causing that the antinoise ability of mud pressure DPSK signal demodulation system is far lower than 2PSK signal system, and numerical calculation shows that the actual bit error rate is far higher than latter; so under the condition of large carrier frequency, to obtain relative small actual bit error rate and relative larger information transmission rate, the mud pressure DPSK signal input signal-to-noise ratio should be raised as far as possible.

Figure 10.2.17. Signal processing (CNPC).

10.3 MWD Sensor Developments in China

Well logging involves multiple sensors, among them acoustic, resistivity, NMR, formation testing (for permeability and pore pressure), annular pressure, direction and inclination, and so on. A significant amount of information is available from oil service company websites, e.g., those of Schlumberger, Halliburton, BakerHughes and others. In this section, we describe recent CNPC engineering and research activities in logging and "geosteering" (that is, directional drilling guided using integrated real-time geological information) – data generating functions that drive the demand for high-data-rate telemetry.

These efforts mirror those of the West, and are becoming commercially important in many parts of the world – and especially so, given Chinese advantages in labor and manufacturing costs. Little information is available in the West about Chinese activities and here we provide insight into the company's MWD activities. Photographs of tools, laboratory and field work are presented, courtesy of CNPC, but detailed hardware and software specifications are omitted because they are rapidly evolving, given the continual drive toward deeper HPHT offshore wells. The greater part of this book focused on theory and laboratory telemetry work and it is worthwhile in this closing chapter to consider related sensor development and field testing.

10.3.1 DRGDS Near-bit Geosteering Drilling System.

10.3.1.1 Overview.

CNPC's Near-bit Geosteering Drilling System or "DRGDS" is jointly developed by CNPC Drilling Research Institute, Beijing Petroleum Machinery Factory and CNPC Well Logging Company Limited. This system integrates advanced drilling, well logging and reservoir engineering technologies and utilizes near-bit geologic and engineering parameter measurement with drilling control to optimize wellbore placement within petroleum-bearing layers. Based on information acquired while drilling, DRGDS adjusts and controls in real-time the trajectory of the well so that the drill bit follows payzones as closely as possible. In other words, it recognizes hydrocarbon zones while drilling and provides complementary geosteering functions.

10.3.1.2 DRGDS tool architecture.

DRGDS is composed of several elements, namely, Near-bit Measurement and Transmission Steering Motor (CAIMS), Wireless Receiving System (WLRS), Positive Pulser MWD (DRMWD) and Ground Information Processing and Steering Decision-making Software System (CFDS), as shown in Figure 10.3.1-1. CAIMS architecture is given in Figure 10.3.1-2 and it is composed of Motor Assembly (bypass valve, screw motor and shaft assembly), Near-Bit Measurements and Transmission Sub (NBMTS), Ground Adjustable Bent

Housing Assembly and Driving Shaft Assembly with Near-bit Stabilizer. NBMTS is composed of resistivity, natural gamma ray and deviation sensors, electromagnetic wave transmitting antenna, control circuits and battery packs. This measures bit resistivity, azimuthal resistivity, azimuthal natural gamma, hole angle, gravity tool face (GTF) angle, temperature and other parameters. Measured near-bit parameters are wirelessly transmitted to the WLRS.

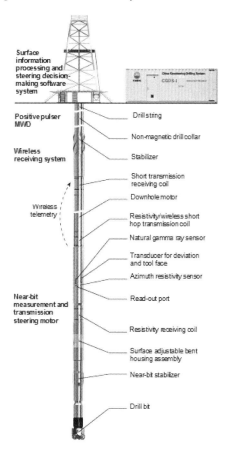

Figure 10.3.1-1. DRGDS tool architecture.

WLRS consists of an uploading data connection assembly, stabilizers, batteries and control circuits, short-distance transmission receiving coil and lower connector, as shown in Figure 10.3.1-3. It is connected upwardly to DRMWD and downwardly to CAIMS. It receives electromagnetic wave signals wirelessly from the transmitter coil below the motor. Signals are then passed to the DRMWD unit via the uploading data connection assembly.

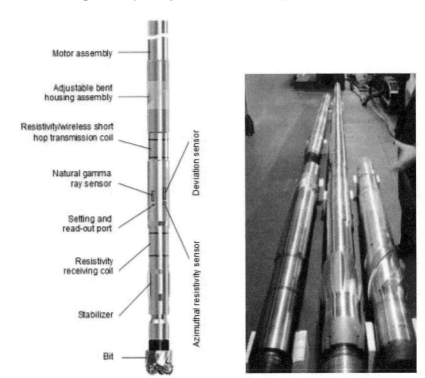

Figure 10.3.1-2. CAIMS tool architecture and hardware.

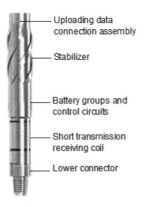

Figure 10.3.1-3. WLRS tool architecture.

The DRMWD includes DRMWD-MD downhole instruments and DRMWD-MS surface devices, as shown in Figure 10.3.1-4, which communicate with each other through pressure pulse signals in the drilling mud channel; they assist each other in real-time monitoring of downhole tool status, operating conditions and related measurement parameters (including inclination, azimuth, tool surface and other orientation parameters, gamma and resistivity and similar geologic parameters as well as pressure and engineering parameters).

Surface Devices are composed of ground sensors (e.g., pressure, depth and pump stroke sensors), instrument shed, front-end receiver and ground signal processing devices, host computer and peripherals, and associated software, with sophisticated signal processing capabilities permitting operations up to and beyond 4500m. Downhole Instruments is composed of non-magnetic drilling collar, positive pulse generator installed in the non-magnetic drilling collar, driver sub, power supply unit, directional sub and a downloading data connection assembly. It is upwardly connected to a common (or non-magnetic) collar and downwardly to WLRS. Since it adopts an open bus design, DRMWD-MD is compatible with different types of pulsers. In addition to its application with DRGDS, DRMWD can be used in other operations.

CFDS is mainly composed of data processing and analysis software, and drilling trajectory design and steering decision-making software. Additionally, it has analysis, data management and chart output modules, and so on. This software system helps process and analyze near-bit resistivity, natural gamma and other geologic parameters uploaded during drilling in real-time, and explain and assess the economic potential of the newly-drilled strata. Simulations for the geological formation to be drilled (ahead of the drill bit) are made, and then, necessary adjustments to well trajectory are calculated. This geosteering increases the probability for exploration success, as downhole recommendations are implemented by direct drilling control. Surface and downhole system architecture are shown immediately below.

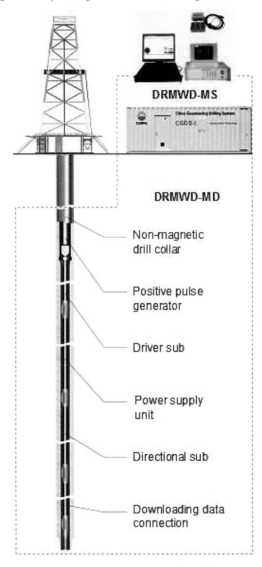

Figure 10.3.1-4. DRMWD tool architecture.

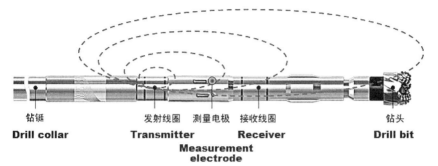

钻铤　　　　　发射线圈　测量电极　接收线圈　　　　　　钻头

Drill collar　　　　**Transmitter**　　　**Receiver**　　　　**Drill bit**

Measurement electrode

Figure 10.3.1-5. Near-bit resistivity measurement.

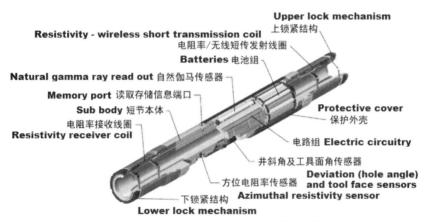

Upper lock mechanism
上锁紧结构

Resistivity - wireless short transmission coil
电阻率/无线短传发射线圈

Batteries 电池组

Natural gamma ray read out 自然伽马传感器

Memory port 读取存储信息端口

Sub body 短节本体

电阻率接收线圈
Resistivity receiver coil

Protective cover
保护外壳

电路组 **Electric circuitry**

井斜角及工具面角传感器
Deviation (hole angle) and tool face sensors

方位电阻率传感器
Azimuthal resistivity sensor

下锁紧结构
Lower lock mechanism

Figure 10.3.1-6. Near-bit measurement sub configuration.

Figure 10.3.1-7. Near-bit measurement sub.

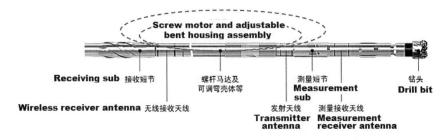

Figure 10.3.1-8. Downhole information wireless electromagnetic short transmission overview.

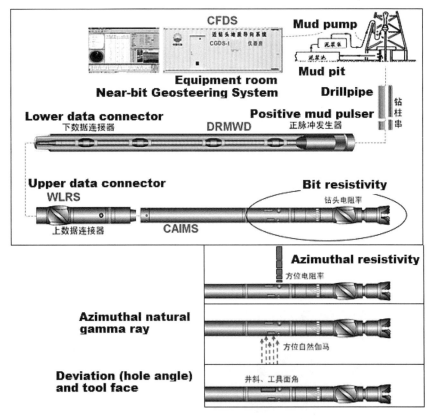

Figure 10.3.1-9. Near-bit parameter measurement.

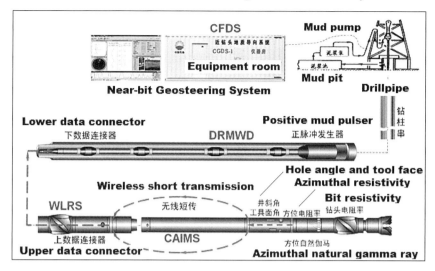

Figure 10.3.1-10. Near-bit information wireless short transmission.

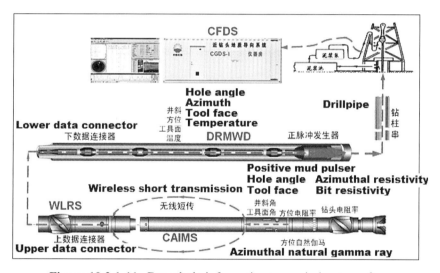

Figure 10.3.1-11. Downhole information transmission to surface.

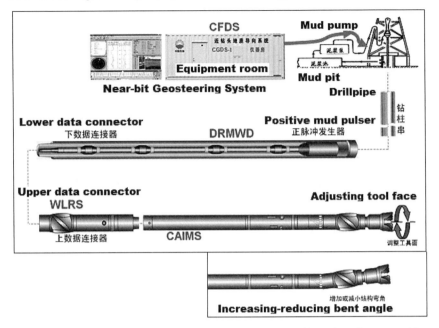

Figure 10.3.1-12. Adjusting tool face and bent angle of bent housing assembly.

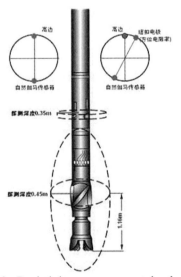

Figure 10.3.1-13. Resistivity measurement depth of investigation and transmitter position.

10.3.1.3   Functions of DRGDS.

DRGDS has three major functions: measurement, transmission and steering. These are briefly summarized next.

Measurement. Resistivity, natural gamma ray and well inclination sensors are installed in NBMTS, and a receiving coil is installed in WLRS. The NBMTS measures bit resistivity, azimuthal resistivity, azimuthal natural gamma, near-bit well inclination angle and GTF angle; these parameters can be converted into electromagnetic wave signals and then sent by the transmission coil in NBMTS to the receiving coil in WRLS over the screw motor in a time-sharing manner.

Transmission. After the wireless receiving coil has received information from below the motor, it is incorporated in the DRMWD by the Uploading Data Connection Assembly; then DRMWD will activate the positive pulse generator to produce the pressure pulse signals in the drilling column and transmit the measured near-bit information to the ground processing system while uploading the information measured by DRMWD, which includes well inclination, azimuth, tool face and downhole temperature and other parameters.

Steering. After receiving and collecting the mud pressure pulse signals uploaded by the downhole instrument (DRMWD-MD), the surface processing system filters the information, reduces noise, checks, identifies, decodes, displays and stores the information and then, transmits decoded data to the driller display for the engineer to read. At the same time, the CFDS steering and decision-making system will judge it and make a decision regarding use of the downhole motor as the steering tool – it instructs the steering tool to drill into the oil and gas layer or to continue to drilling as before.

Figures 10.3.1-14a to 10.3.1-14d show scenes captured at product launch. Figures 10.3.1-15a and 10.3.1-15b importantly display wellbore trajectory differences between "conventional geosteering" versus "near-bit geosteering." The former is not unlike "driving from the rear seat" – without immediate, close-up information, the driver is likely to steer away from the target. Course corrections can be expensive and time-consuming. The simple need to follow the payzone as closely as possible drives the design of modern near-bit sensors and high-data-rate MWD telemetry systems.

Figure 10.3.1-14a. CNPC DRGDS geosteering system.

Figure 10.3.1-14b. Dr. Yinao Su, co-author, introducing
CNPC's DRGDS geosteering and EILog surface systems.

Figure 10.3.1-14c. Mr. Ning Sun, President, CNPC Drilling Engineering
Technology Research Institute, addressing conference attendees.

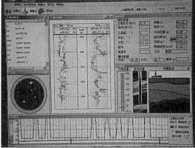

Figure 10.3.1-14d. EILog surface system and sample CFDE screen.

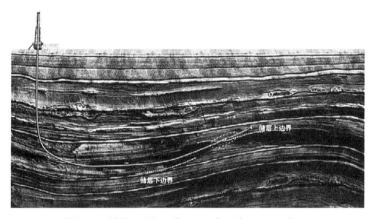

Figure 10.3.1-15a. Conventional geosteering.

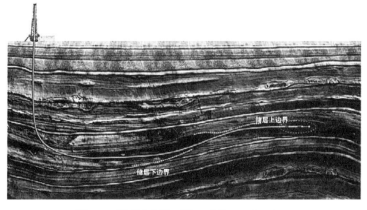

Figure 10.3.1-15b. Near-bit geosteering.

Figure 10.3.1-16a. Field test activities.

Figure 10.3.1-16b. Field test activities.

10.3.2 DRGRT Natural Azi-Gamma Ray Measurement.

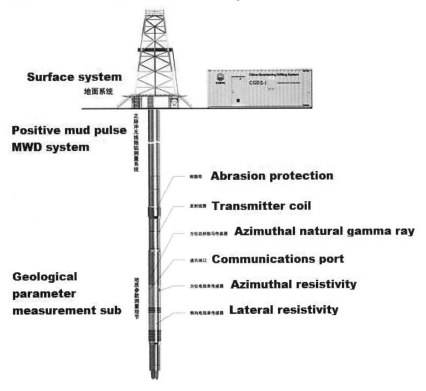

Figure 10.3.2-1a. DRGRT configuration (English).

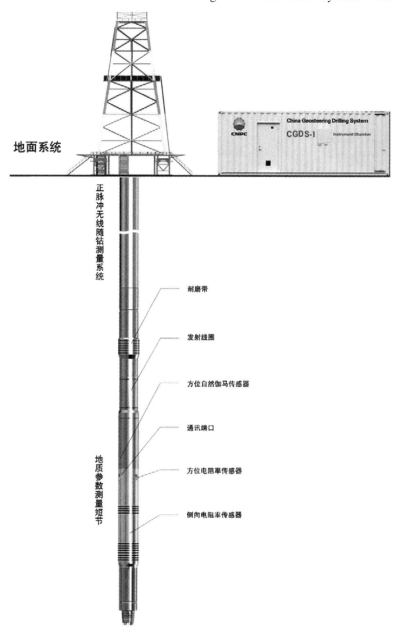

Figure 10.3.2-1b. DRGRT configuration (Chinese).

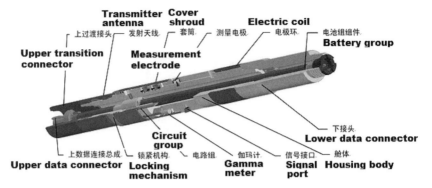

Figure 10.3.2-2a. Geological parameter measurement sub (English).

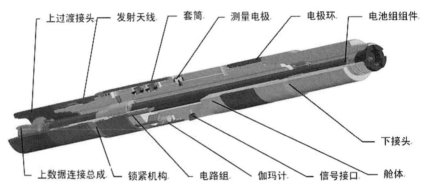

Figure 10.3.2-2b. Geological parameter measurement sub (Chinese).

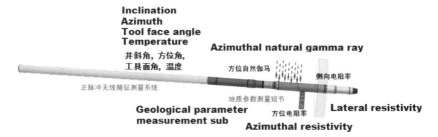

Figure 10.3.2-3a. Data measurement overview (English).

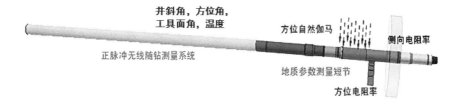

Figure 10.3.2-3b. Data measurement overview (Chinese).

Figure 10.3.2-4. Field test and tool preparation for downhole operation.

10.3.3 DRNBLog Geological Log.

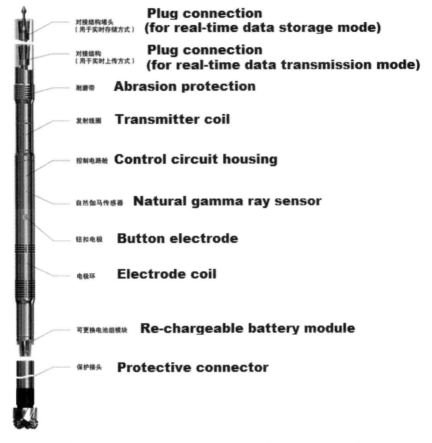

对接结构堵头
（用于实时存储方式） **Plug connection (for real-time data storage mode)**

对接结构
（用于实时上传方式） **Plug connection (for real-time data transmission mode)**

耐磨带 **Abrasion protection**

发射线圈 **Transmitter coil**

控制电路舱 **Control circuit housing**

自然伽马传感器 **Natural gamma ray sensor**

钮扣电极 **Button electrode**

电极环 **Electrode coil**

可更换电池组模块 **Re-chargeable battery module**

保护接头 **Protective connector**

Figure 10.3.2-5a. DRNBLog tool configuration (English).

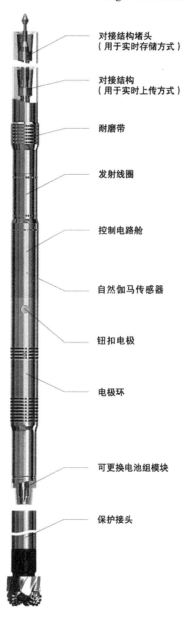

对接结构堵头
（用于实时存储方式）

对接结构
（用于实时上传方式）

耐磨带

发射线圈

控制电路舱

自然伽马传感器

钮扣电极

电极环

可更换电池组模块

保护接头

Figure 10.3.2-5b. DRNBLog tool configuration (Chinese).

10.3.4 DRMPR Electromagnetic Wave Resistivity.

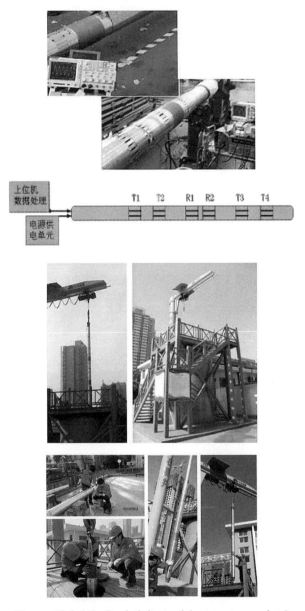

Figure 10.3.4-1. Resistivity tool (note antenna slots).

10.3.5 DRNP Neutron Porosity.

Figure 10.3.5-1. Well calibration.

Figure 10.3.5-2. Well calibration -Indoor adjustments and tank test.

Figure 10.3.5-3. Neutron generator performance test.

Figure 10.3.5-4. First, second and third tests, top to bottom.

Figure 10.3.5-5. Data read-out.

10.3.6 DRMWD Positive Mud Pulser.

Figure 10.3.6-1. High pressure test facility and test, top; full-scale MWD system simulation facility and test site, center and bottom.

10.3.7 DREMWD Electromagnetic MWD.

Figure 10.3.7-1. DREMWD system transmission path.

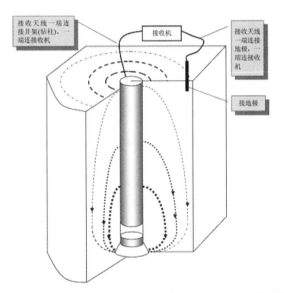

Figure 10.3.7-2. Drill collar and surface electric field.

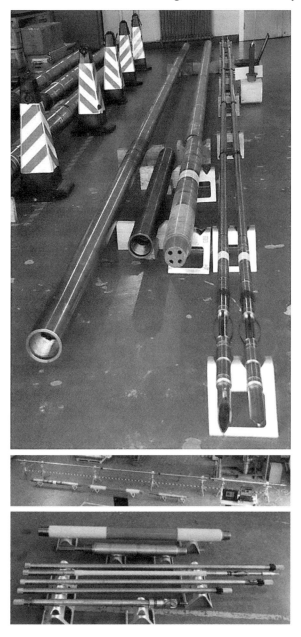

Figure 10.3.7-3a. Hardware display.

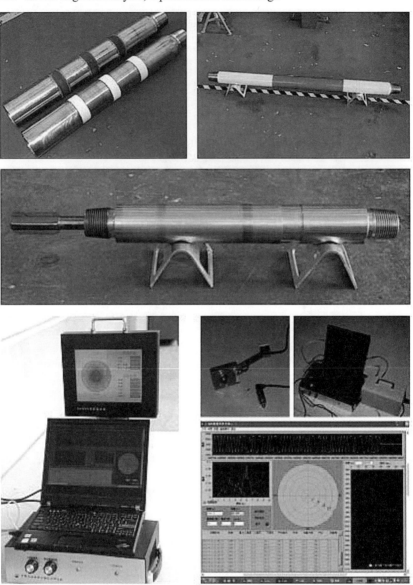

Figure 10.3.7-3b. Hardware display.

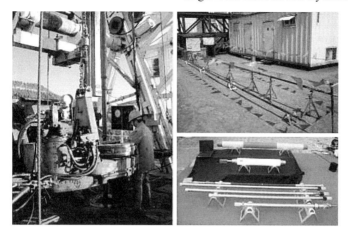

Figure 10.3.7-3c. Hardware display.

10.3.8 DRPWD Pressure While Drilling.

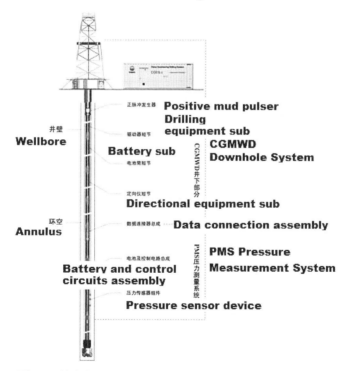

Figure 10.3.8-1a. DRPWD system configuration (English).

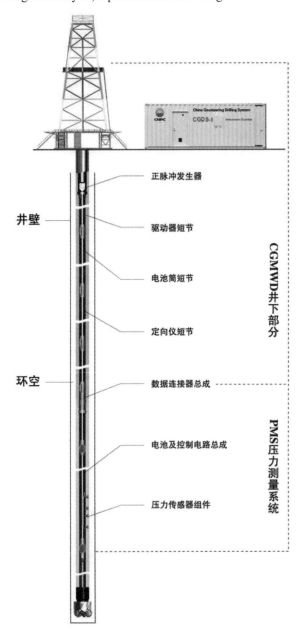

正脉冲发生器

驱动器短节

电池筒短节

定向仪短节

数据连接器总成

电池及控制电路总成

压力传感器组件

井壁

环空

CGMWD井下部分

PMS压力测量系统

Figure 10.3.8-1b. DRPWD system configuration (Chinese).

Figure 10.3.8-2. Downhole equipment assembly.

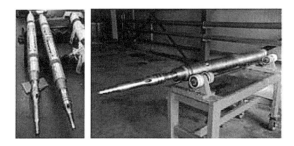

Figure 10.3.8-3. 6.75 in (left) and 4.75 in (right) downhole
annular pressure measurement sub.

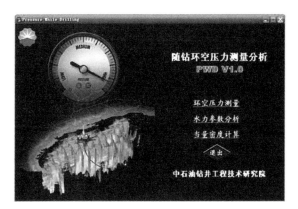

Figure 10.3.8-4. Product announcement.

10.3.9 Automatic Vertical Drilling System – DRVDS-1.

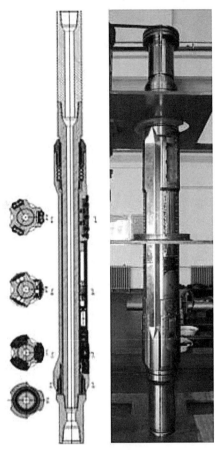

Figure 10.3.9-1. DRVDS-1 tool architecture and hardware.

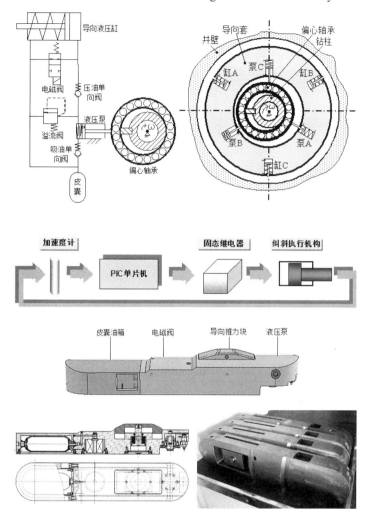

Figure 10.3.9-2. DRVDS-1 tool architecture and hardware.

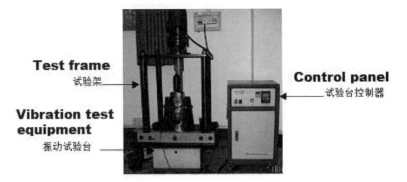

Figure 10.3.9-3a. Test fixtures.

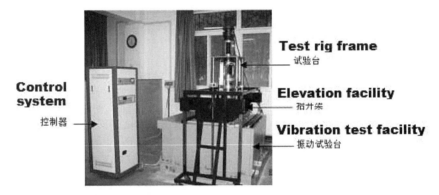

Figure 10.3.9-3b. Test fixtures.

Figure 10.3.9-3c. Test fixtures.

Figure 10.3.9-4. Field testing.

10.3.10 Automatic Vertical Drilling System – DRVDS-2.

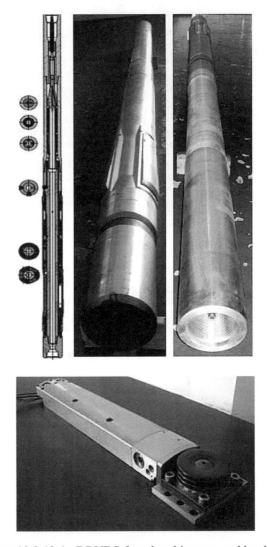

Figure 10.3.10-1. DRVDS-2 tool architecture and hardware.

10.4 Turbines, Batteries and Closing Remarks

In this closing section, we offer comments on hands-on tool design, again, for siren-type MWD pulser systems capable of 10–20 "real" bits/sec (as opposed to "compressed" bits).   Our advice is practical, but not meant to be comprehensive; nor do we claim to have performed exhaustive trade-off studies – our observations are simply offered "as is."

10.4.1 Siren drive.

Assumed in our discussion is a turbine alternator/generator for supplying the required power to operate the siren.   Very likely, the siren should be powered by a brushless-DC stepper-type motor.   This type of motor has the desirable mechanical characteristics of very high starting torque and extremely low inertia – ideally suited for rapid angular accelerations.   This same type of motor is used in computer data drives, robotics, etc.   Included with this motor is a high resolution resolver for precise angular positioning and control.   Although we have not performed calculations on power requirements, our belief is that this motor will require 300–500 W of instantaneous power to provide the desired data rate results. These motors require a rigid supply of electrical power, say 50 VDC or perhaps as high as 200 VDC. Sophisticated switching electronics, probably microprocessor controlled, will be required to run the stepper motor.

10.4.2 Turbine–alternator system.

The turbine power supply is somewhat more direct.   As is well known, the mud flow provides an abundance of available power.   The mechanical difficulties include providing a blade design suitable for a wide range of flow rates.   Some experimentation using the wind tunnel methods of Chapter 8 will be required or, perhaps, novel airfoil concepts involving twisted blades.   In addition, problems with rotary seals are well known.   The major difficulty is providing consistent sealing in the presence of pressure fluctuations contributed, not only by the drilling environment, but more severely by the pulser itself which will create variations in hundreds of psi downhole.

The alternator/generator is more straightforward. There are sophisticated alternator designs available, e.g., homo-polar, but a basic rotating field design is preferred. A major difficulty in the alternator design is the conditioning electronics. The peak-to-peak AC voltage out of the alternator is proportional to the rotational speed, and thus, clever electrical engineering will be required to convert this varying input into a rigid DC output for powering the siren pulser.

If multiple turbine stages cannot be used because of size constraints and high power is required to turn "sirens in series," self-spinning sirens which draw on the energy of the flowing mud are imperative.   In this case, modulation can performed by using mechanical or magneto-rheological braking as discussed previously – an electric motor is still required to regulate rotation rates precisely or to provide "assists" when additional torques are required momentarily.

10.4.3 Batteries.

Discussions related to power are not complete without some mention of batteries. We need not dwell on well known limitations: toxicity, explosiveness and impractical handling. These negatives must be balanced with needs for turbine maintenance and repair, which introduce inefficiencies of their own. New to the market are rechargeable batteries, whose implications are discussed in a recent article of Pitt (2010). Quoting from the article –

"MWD is a highly demanding application that creates a challenging environment for the batteries. They must be able to operate over a wide temperature range – from well below 0°C at the surface in Arctic oil and gas exploration projects to well over 100°C during drilling – while enduring very high vibrations (20 g rms). Another requirement is complete reliability and long life time, typically from a few hours to more than 20 hours. Each time the bottom-hole assembly has to return to the surface it costs tens of thousands of dollars in downtime, so premature withdrawal to replace a failed MWD battery adds significant costs to the operation.

During drilling operations, there is continuous mud flow and the battery delivers a low idle current to the MWD tool. Drilling is often stopped and then restarted, resulting in frequent battery replacement. If, for any reason, the drilling operation has to be stopped early to replace a drill bit, the MWD tool must return to the surface and the battery must be replaced to ensure there will be a sufficient safety margin of battery power to maintain operation of the MWD tool through to completion. Often, a primary battery might have to be discarded with much of its capacity unused.

Until recently, specialized primary lithium batteries were the only product capable of providing reliable, cost effective operation in harsh MWD conditions. Saft offers primary lithium batteries for the oil & gas market and also recently launched the world's first rechargeable lithium-ion (Li-ion) cell capable of operating at temperatures up to 125°C under drilling conditions. This is a significant increase in the previous Li-ion operating temperature – 65°C – opening new horizons for MWD tool manufacturers. For the first time, MWD tool developers can incorporate a high performance rechargeable battery into their designs. This development eliminates the need to withdraw a MWD tool for replacement of a spent battery, with the benefit of improved continuity for the drilling operation.

Saft's VL 25500-125 C-size and new VL 32600-125 D-size cells are intended for use in constructing batteries to be integrated into sophisticated MWD tools that incorporate onboard alternator technology, driven by the mud flow, to power their electronic systems. When the mud is flowing, the battery will be charged. When the flow stops, such as when drilling is halted, the battery will be discharged to provide power for the MWD electronics. When the mud flow restarts, the battery is recharged.

The fast-charging, deep discharge and high-cycling capability of the Li-ion electrochemistry will enable the MWD tool to remain in continuous downhole operation. The C-size cell functions as an energy buffer and is commonly used in oil & gas drilling applications, while the higher power D-size cell was designed for oil exploration operations. The Li-ion cells are integrated into customized, rugged, cylindrical MWD battery staves. A key part of the stave design is to provide complete mechanical integrity, even under extreme temperatures, vibration and pressure. This involves careful selection of the construction materials and specialized manufacturing techniques, such as the cross-ply, tape-wrapping process. The staves also incorporate electronic controls, such as diodes to protect primary cells or balancing circuits to manage rechargeable cells."

10.4.4 Tool requirements.

We have addressed the issues encountered in designing high-data-rate MWD systems with 10 bps capability or more. Several actions are required. P_0 must be increased by constructive interference using "smart FSK" telemetry without complete rotor stoppage or multiple sirens, or both, additionally employing "turbine on top of siren" designs, and optimizing sirens for high Δp and low torque. P_{xdcr} should be minimized by using sensitive piezoelectric transducers and advanced multiple transducer noise removal methods. Low rotor torques (which allow rapid changes in frequency) with low erosion should be designed in the wind tunnel which are also consistent with high signals. Larger diameter drillpipe should be used if possible, and optimum low-attenuation muds should be employed in field operations with proper values of ρ, μ and c selected with the model in Figure 10.7. A well-designed, integrated system should embody all of the design principles in this book, which we emphasize are based on rigorous acoustics and fluid-dynamics principles.

Signal processing efforts deserve special mention. While excellent conventional treatises exist on digital signal processing methods, e.g., the now classic books by Oppenheim and Schafer (1975, 1989), the MWD environment poses extremely difficult challenges. The telemetry channel is essentially one-dimensional and supports strong plane wave noise emanating from both downhole and uphole ends, i.e., mud motor "thumping" and mudpump "banging." In between, shape-distorting reflections are found at desurgers while reverberations are induced at collar-pipe impedance mismatches. These effects are not addressed by existing methods and provide a fertile area for continuing work for all researchers. Because "high data rates" for mud pulse telemetry are still slow by modern standards, e.g., cell phone or Internet connections, real-time processing for relatively complicated algorithms, e.g., not unlike those disclosed in this book, is possible, since time steps for digitization finer than 0.001 sec are likely not required.

10.4.5 Design trade-offs.

Many engineers and managers have asked, "Is there one 'standard' design we can recommend?" This innocuous question is not unreasonable. After all, in the low-data-rate positive pulser world populated by 60-70 manufacturers internationally, all designs (with minor exceptions) are generic "me too" products. High-data-rate mud pulse telemetry, however, is different, because the design options are numerous. And given that each set of options implies significant engineering development, testing and manufacturing, and millions of dollars in subsequent inventory, maintenance and repair costs, the choice is not easily made. Here, we will summarize the options discussed in this book – the mechanical and electrical implications will be obvious. These are not given in any particular order:

- Single siren, or multiple "sirens-in-series"?

- If multiple sirens, will all operate at one frequency from a single shaft, or at different frequencies built on a more complicated design?

- Single-stage turbine, or multiple stages?

- Turbine versus batteries?

- Turbine, alternator, rechargeable battery option?

- Brushless DC or hydraulic motor drive for siren(s)?

- Self-spinning plus mechanical brake for modulation?

- Self-spinning plus magneto-rheological brake for modulation and motor assist?

- FSK versus PSK?

- Top versus bottom mounted siren(s)?

- Collar versus probe-based design?

- Sensor types used, power demands, data density and logging speed?

- Surface signal process requirements?

- Downhole telemetry scheme, feedback and control requirements?

The above design options must be selected with care. Different companies have different market needs, e.g., onshore versus offshore, number of sensors, hole depth and attenuation constraints, power requirements and so on. Each design represents a unique set of specifications. Development costs easily exceed millions of dollars, and final designs are likely to remain in field use for years. It is these practical but important objectives that our equations, wind tunnels, test methods and advice address, and in the final analysis, the authors hope that this book has contributed meaningfully to solving these problems.

10.5 References

Desbrandes, R., Bourgoyne, A.T. and Carter, J.A., "MWD Transmission Data Rates Can be Optimized," *Petroleum Engineer International*, June 1987, pp. 46-50.

Drumheller, D.S., "An Overview of Acoustic Telemetry," in *Geothermal Energy and the Utility Market – Opportunities and Challenges for Expanding Geothermal Energy in a Competitive Supply Market, Meeting Proceedings, U.S. Department of Energy, Geothermal Program Review 10*, March 24-26, 1992.

Fripp, M., Skinner, N.G. and Chin, W.C., "Magneto-rheological Fluid Controlled Mud Pulser," U.S. Patent No. 7,082,078, July 25, 2006.

Gardner, W.R., "High Data Rate MWD Mud Pulse Telemetry," United States Department of Energy Natural Gas Conference, Houston, Texas, Mar. 25, 1997.

Gardner, W.R., "MWD Mud Pulse Telemetry System," Final Report GRI-02/0019, Gas Research Institute, April 2002.

Pitt, W., "Rechargeables to Expand MWD Horizons," *OEDigital* (*Offshore Engineer*), April 2010.

Cumulative References

Ashley, H. and Landahl, M.T., *Aerodynamics of Wings and Bodies*, Addison-Wesley, Reading, Massachusetts, 1965.

Batchelor, G.K., *An Introduction to Fluid Dynamics*, Cambridge University Press, 1970.

Chin, W.C., "Algorithm for Inviscid Flow Using the Viscous Transonic Equation," *AIAA Journal*, Aug. 1978.

Chin, W.C., "Why Drill Strings Fail at the Neutral Point," *Petroleum Engineer International*, May 1988.

Chin, W.C., *Borehole Flow Modeling in Horizontal, Deviated and Vertical Wells*, Gulf Publishing, Houston, 1992.

Chin, W.C., *Modern Reservoir Flow and Well Transient Analysis*, Gulf Publishing, Houston, 1993.

Chin, W.C., *Wave Propagation in Petroleum Engineering, with Applications to Drillstring Vibrations, Measurement-While-Drilling, Swab-Surge and Geophysics*, Gulf Publishing, Houston, 1994.

Chin, W.C., *Formation Invasion, with Applications to Measurement-While-Drilling, Time Lapse Analysis and Formation Damage*, Gulf Publishing, Houston, 1995.

Chin, W.C., "Measurement-While-Drilling System and Method," U.S. Patent No. 5,583,827, Dec. 10, 1996.

Chin, W.C., "Multiple Transducer MWD Surface Signal Processing," U.S. Patent No. 5,969,638, Oct. 19, 1999.

Chin, W.C., *Computational Rheology for Pipeline and Annular Flow*, Butterworth-Heinemann, Reed Elsevier, London, 2001.

Chin, W.C., *Quantitative Methods in Reservoir Engineering*, Elsevier Science, London, 2002.

Chin, W.C., "MWD Siren Pulser Fluid Mechanics," *Petrophysics*, Journal of the Society of Petrophysicists and Well Log Analysts (SPWLA), Vol. 45, No. 4, July – Aug. 2004, pp. 363-379.

Chin, W.C., *Formation Testing Pressure Transient and Contamination Analysis*, E&P Press, Houston, 2008.

Chin, W.C., *MWD Signal Analysis, Optimization and Design*, E&P Press, Houston, 2011.

Chin, W.C., *Managed Pressure Drilling: Modeling, Strategy and Planning*, Elsevier Scientific Publishing, London, 2012.

Chin, W.C., *Electromagnetic Well Logging: Models for MWD/LWD Interpretation and Tool Design*, John Wiley & Sons, New Jersey, 2014.

Chin, W.C., Y. Zhou, Y. Feng, Q. Yu and L. Zhao, *Formation Testing: Pressure Transient and Contamination Analysis*, John Wiley & Sons, New Jersey, 2014.

Chin, W.C., Gardner, W.R., and Waid, M., "MWD Surface Signal Detector Having Bypass Loop Acoustic Detection Means," U. S. Patent No. 5,515,336, May 7, 1996.

Chin, W.C., Golden, D. and Barber, T., "An Axisymmetric Nacelle and Turboprop Inlet Analysis with Flow-Through and Power Simulation Capabilities," *AIAA Paper No. 82-0256, AIAA 20th Aerospace Sciences Meeting*, Orlando, FL, Jan. 1982.

Chin, W.C. and Hamlin, K., "MWD Surface Signal Detector Having Enhanced Acoustic Detection Means," U. S. Patent No. 5,459,697, Oct. 17, 1995.

Chin, W.C. and Hamlin, K., "MWD Surface Signal Detector Having Enhanced Acoustic Detection Means," U.S. Patent No. 5,535,177, July 9, 1996.

Chin, W.C., Presz, W., Ives, D, Paris, D. and Golden, D., "Transonic Nacelle Inlet Analyses," *NASA Lewis Workshop on Application of Advanced Computational Methods*, Nov. 1980.

Chin, W.C. and Ritter, T., "Turbosiren Signal Generator for Measurement While Drilling Systems, U.S. Patent No. 5,586,083, Dec. 17, 1996.

Chin, W.C. and Ritter, T., "Turbosiren Signal Generator for Measurement While Drilling Systems," U.S. Patent No. 5,740,126, April 14, 1998.

Chin, W.C. and Trevino, J.A., "Pressure Pulse Generator," U.S. Patent No. 4,785,300, Nov. 15, 1988.

Chin, W.C., Y. Zhou, Y. Feng, Q. Yu and L. Zhao, *Formation Testing: Pressure Transient and Contamination Analysis*, John Wiley & Sons, New Jersey, 2014.

Desbrandes, R., Bourgoyne, A.T. and Carter, J.A., "MWD Transmission Data Rates Can be Optimized," *Petroleum Engineer International*, June 1987, pp. 46-50.

Drumheller, D.S., "An Overview of Acoustic Telemetry," in *Geothermal Energy and the Utility Market – Opportunities and Challenges for Expanding Geothermal Energy in a Competitive Supply Market, Meeting Proceedings, U.S. Deparment of Energy, Geothermal Program Review 10*, March 24-26, 1992.

Fripp, M., Skinner, Skinner, N.G. and Chin, W.C., "Magnetorheological Fluid Controlled Mud Pulser," U.S. Patent No. 7,082,078, July 25, 2006.

Gardner, W.R., "High Data Rate MWD Mud Pulse Telemetry," United States Department of Energy Natural Gas Conference, Houston, Texas, Mar. 25, 1997.

Gardner, W.R., "MWD Mud Pulse Telemetry System," Gas Research Institute Final Report GRI-02/0019, April 2002.

Gardner, W.R. and Chin, W.C., "Snap Action Rotary Pulser," U.S. Patent No. 5,787,052, July 28, 1998.

Gavignet, A.A., Bradbury, L.J. and Quetier, F.P., "Flow Distribution in a Tricone Jet Bit Determined from Hot-Wire Anemometry Measurements," SPE Paper No. 14216, *60th Annual Technical Conference and Exhibition of the Societ of Petroleum Engineers*, Las Vegas, Nevada, Sept. 22-25, 1985.

Gavignet, A.A., Bradbury, L.J. and Quetier, F.P., "Flow Distribution in a Roller Jet Bit Determined from Hot-Wire Anemometry Measurements," *SPE Drilling Engineering*, Mar. 1987, pp. 19-26.

Gilbert, G.N. and Tomek, M.L., "Screen and Bypass Arrangement for LWD Tool Turbine," U.S. Patent No. 5,626,200, May 6, 1997.

Hawthorne, W.R., *Aerodynamics of Turbines and Compressors*, Volume X, High Speed Aerodynamics and Jet Propulsion, Princeton University Press, 1964.

Hutin, R., Tennent, R.W. and Kashikar, S.V., "New Mud Pulse Telemetry Techniques for Deepwater Applications and Improved Real-Time Data Capabilities," SPE/IADC Paper No. 67762, *SPE/IADC Drilling Conference*, Amsterdam, The Netherlands, Feb. 27 – Mar. 1, 2001.

Kinsler, L.E., Frey, A.R., Coppens, A.B. and Sanders, J.V., *Fundamentals of Acoustics*, Fourth Edition, John Wiley and Sons, New York, 2000.

Lippert, W.K.R., "The Measurement of Sound Reflection and Transmission at Right-Angled Bends in Rectangular Tubes," *Acustica*, Vol. 4, 1954, pp. 313-319.

Lippert, W.K.R., "Wave Transmission around Bends of Different Angles in Rectangular Ducts," *Acustica*, Vol. 5, 1955, pp. 274–278.

Llosa, R., "Advances of MWD Technology for HPHT Wells," ONS 1994.

Martin, C.A., Philo, R.M., Decker, D.P. and Burgess, T.M., "Innovative Advances in MWD," IADC/SPE Paper No. 27516, *1994 IADC/SPE Drilling Conference*, Dallas, Texas, Feb. 15-18, 1994.

Montaron, B.A., Hache, J.M.D. and Voisin, B., "Improvements in MWD Telemetry: 'The Right Data at the Right Time," SPE Paper No. 25356, *SPE Asia Pacific Oil & Gas Conference & Exhibition*, Singapore, Feb. 8-10, 1993.

Morse, P.M. and Ingard, K.U., *Theoretical Acoustics*, McGraw-Hill, New York, 1968.

Norton, D.J., Heideman, J.C. and Mallard, W.W., "Wind Tunnel Tests of Inclined Circular Cylinders," *Society of Petroleum Engineers Journal*, Feb. 1983, pp. 191-196.

Oates, G.C., *The Aerothermodynamics of Aircraft Gas Turbine Engines*, Air Force Aero Propulsion Laboratory Report No. AFAPL-TR-78-52, July 1978.

Oates, G.C., Private communication, 1982.

Oates, G.C., *Aerothermodynamics of Aircraft Engine Components, AIAA Education Series,* American Institute of Aeronautics and Astronautics, New York, 1985.

Oppenheim, A.V. and Schafer, R.W., *Digital Signal Processing*, Prentice-Hall, New Jersey, 1975.

Oppenheim, A.V. and Schafer, R.W., *Discrete-Time Signal Processing*, Prentice-Hall, New Jersey, 1989.

Patton, B.J., "Torque Assist for Logging-While-Drilling Tool," U.S. Patent No. 3,867,714, Feb. 18, 1975.

Patton, B.J., Prior, M.J., Sexton, J.H. and Slover, V.R., "Logging-While-Drilling Tool," U.S. Patent No. 3,792,429, Feb. 12, 1974.

Patton, B.J., Gravley, W., Godbey, J.K., Sexton, J.H., Hawk, D.E., Slover, V.R. and Harrell, J.W., "Development and Successful Testing of a Continuous-Wave, Logging-While-Drilling Telemetry System," *Journal of Petroleum Technology*, Oct. 1977, pp. 1215 – 1221 (originally, SPE Paper No. 6157, *SPE-AIME 51*[st]

Annual Fall Technical Conference and Exhibition, New Orleans, Oct. 3-6, 1976).

Pitt, W, "Rechargeables to expand MWD horizons," *OEDigital* (*Offshore Engineer*), April 2010.

Press, W.H., Teukolsky, S.A., Vetterling, W.T. and Flannery, B.P., *Numerical Recipes: The Art of Scientific Computing*, 3rd Edition, Cambridge University Press, Cambridge, 2007.

Sexton, J.H., Slover, V.R., Patton, B.J. and Gravley, W., "Logging-While-Drilling Tool," U.S. Patent No. 3,770,006, Nov. 6, 1973.

Staff, Schlumberger, "The Anadrill IDEAL," *Euroil*, Feb. 1994.

Stearns, S.D. and David, R.A., *Signal Processing Algorithms in Fortran and C*, Prentice Hall, Englewood Cliffs, New Jersey, 1993.

Su, Y., Sheng, L., Li, L., Bian, H., Shi, R., and Chin, W.C., "High-Data-Rate Measurement-While-Drilling System for Very Deep Wells," Paper No. AADE-11-NTCE-74, American Association of Drilling Engineers' *2011 AADE National Technical Conference and Exhibition*, Houston, Texas, April 12-14, 2011.

Waid, M., Chin, W.C., Anders, J. and Proett, M., "Fluid Driven Siren Flowmeter," U.S. Patent No. 5,831,177, Nov. 3, 1998.

Index

A

AADE, 9-10, 229
AADE-11-NTCE-74, 9-10, 229
Accordion-like, 163, 171
Aerodynamics, 186-187, 193, 199, 205-206, 211, 213-214, 223
Aerospace, 1, 8, 177, 183, 188, 191-192, 195, 198, 205, 207, 220-221, 225, 228, 238
Airfoil, 170, 183-187, 189-190, 193-194, 196, 211, 222-223, 337
Alternator, 33, 98, 206, 213, 337-338, 340
Amplitude spectrum, 255
Antenna, 301, 320
Anti-node, 4
Antinode, 281
Architecture (tool), 9, 300-304, 332-333, 336
Attenuation, 4, 7, 9-10, 15, 24, 27-28, 54, 56-57, 59-61, 67, 69, 76, 81, 84, 87, 101-102, 105, 129, 142-146, 153, 158, 164-169, 210, 240, 254-255, 266-267, 270-274, 280-281, 287, 290, 340
Automatic vertical drilling, 332, 336

B

Ball in cage, 91, 246-247
Batteries, 274, 301, 337-338, 340
Battery, 179, 231, 301, 338-340
BHA, 76-77
Bottomhole assembly, 3, 6, 10, 14, 47-48, 52, 57-58, 61-62, 67, 77-78, 80, 86, 90, 92, 145, 153, 164, 172, 174
Brushless DC motor, 261, 280, 337, 340
Butterworth filter, 153, 174
Bypass loop, 125

C

CAIMS, 300-302
Carrier frequency, 2, 80-81, 104-105, 111, 131-132, 159, 210, 287-289
Carrier wave, 9, 11, 53, 75, 77, 105-108, 137-138, 147-151, 168, 278
Cascade (turbine), 190, 193-195, 197, 199
CFDS, 300, 303, 309
China National Petroleum Corporation, 10, 45, 229, 272
CNPC, 9-10, 22, 27, 41, 45, 182, 229, 237, 239, 245, 256, 259, 291-300, 310
Constructive interference, 14, 27-28, 36, 44, 52, 54, 75-76, 78, 81, 89-90, 140, 225, 240, 268, 274-275, 277-281, 289, 339
Continuous wave, 45, 47-50, 52, 58, 68, 108, 115, 185
Control system, 34, 163, 238, 260-261
Crests, 3, 106, 132
Critical carrier frequency, 287, 289
Critical frequency, 167, 287-290

D

Data acquisition, 18, 34-35, 260
Deconvolution, 7, 60, 98-100, 108-109, 114, 126, 133, 291
Delta-p, 5, 50-51, 65-66, 90, 136, 146-148, 170-171, 184, 250, 254, 289
Deltap, 126, 133, 147-149
Density, 8-10, 14, 22, 24, 33, 55, 64-66, 68, 70, 87, 128, 145, 155, 159, 164, 177, 181, 185, 193, 209-210, 212, 214-216, 224, 226-227, 232,

235, 242-243, 245-246, 254, 256-259, 263, 272, 287, 340
Destructive interference, 36, 47, 53-54, 61, 80-81, 89-90, 92, 151, 164, 169, 210, 232, 259, 278
Desurger, 10, 14, 48, 54, 60-61, 97, 108, 112-114, 119-121, 125-127, 144, 153-154, 156-161, 168, 260, 278, 284
Difference equation, 98, 112
Differential detection, 6
Differential equation, 12, 18, 67, 69-70, 98, 112, 120, 157-158, 188, 190
Differential pressure, 15, 22, 24, 27, 34, 55, 57, 91, 125, 231-232, 234, 240, 248-249, 252-255
Dimensionless, 5, 8, 54, 60, 63, 76, 94, 159, 209, 211, 215, 225-226, 245-246, 257-258, 287
Dipole, 4, 15, 50-51, 55, 58, 64-67, 76, 87-89, 91-92, 98, 126-129, 133-134, 142, 146, 157, 170-171, 249-250, 281
Directional filter, 124
Distortion (signal), 60, 101, 108, 112-114, 120, 139, 153-154, 156-159, 168, 260
Drag, 183, 185, 189, 227
DREMWD, 326
DRGDS, 300-301, 303, 309-310
Drill bit, 62, 64, 136, 277, 290, 300, 303, 338
Drill collar, 6-7, 10, 13-15, 19, 47, 51, 54-55, 61-62, 64, 66-69, 77, 80, 86, 88, 92, 94, 98, 140, 142-147, 163, 167, 172, 178, 180, 182-183, 189, 192, 206-207, 236, 239, 241, 244, 274, 281, 290, 326
Drill pipe, 10, 13, 139
Drillbit, 2-3, 5-6, 12, 14-15, 26, 44, 47-48, 52-57, 59, 61-62, 64, 75, 84, 86, 88-90, 92-94, 96, 98, 113, 126-128, 130, 132-133, 135, 137-139,

141-142, 144, 146-147, 150-151, 161-163, 167, 170-171, 174, 225, 238, 249, 264, 268, 274, 277-278, 280
DRMPR, 320
DRMWD, 300-301, 303-304, 309, 325
DRNP, 321
DRPWD, 329-330
DRVDS, 332-333, 336
Dynamic similarity, 8, 209, 225-227, 246

E

Echo cancellation, 24, 27, 36, 48, 97, 109, 164, 265, 272
Echo removal, 291
Elastic wave, 153, 170-172
Erosion, 3, 8-10, 16, 19-21, 24, 28, 39, 41, 47, 49, 52, 54, 56, 76, 84, 140, 170, 177, 179, 181-183, 185, 187, 189, 191, 193, 195-199, 201-208, 210, 219-220, 224-225, 227-229, 231, 233-235, 239, 246, 248, 253, 259, 261, 263-264, 269-270, 273-274, 279, 339

F

Feedback, 24, 232, 259, 340
FFT, 97, 153, 174, 255
Flow loop, 2-4, 22, 27, 47, 51, 56-58, 84, 159, 182, 232, 238, 243, 245, 248, 258, 260, 264-266, 270, 280-281
Flow rate, 22, 27, 32-34, 42, 51, 55, 59, 65, 68-69, 76, 97, 128, 163, 165, 176, 179-182, 184, 199, 208, 210, 215-217, 227, 229, 235, 237-243, 245, 254-258
Flow straightener, 32, 240
Flowloop, 56-57, 265, 270, 281

Flowrate, 68, 181, 218, 246, 280
Frequency shift keying, 77, 86, 90, 97, 178
Frequency spectrum, 124, 176
FSK, 14, 24, 26, 48, 77, 83-84, 86, 90, 97, 260, 277, 279, 287, 289, 339-340

G

Gamma, 301, 303, 309, 314
Gap, 16, 19, 28, 84, 179-183, 199-200, 204, 229, 231, 248, 253-254, 257, 260, 269, 288
Gardner, 260-261, 264-266, 281
Gas Research Institute, 260, 265
Geosteering, 274, 300, 303, 309-311
Ghost signal, 12, 126, 141
GRI, 265
GTF, 301, 309

H

Harmonic, 28, 36, 39, 47, 49, 51, 53, 55, 57, 59, 61, 63, 65, 67-69, 71, 73, 75, 77, 79, 81, 83, 85-87, 89, 91-93, 95, 97, 140, 158, 170, 175, 224, 229, 249, 255, 262-263, 278
Harmonic distribution, 28, 39, 175
Harmonic generation, 28, 36, 175
Harmonics, 28, 42-43, 175, 255, 262, 279
High data rate, 5, 14, 28, 36, 51, 103-104, 107-108, 135, 147, 169, 231, 240, 266, 273, 281
High-data-rate, 2, 7, 9-10, 12-13, 27, 31, 44-45, 47-50, 52, 57-58, 64, 104, 108, 142, 174, 177, 182, 204, 206, 227, 229, 231-233, 240, 273, 275, 277, 279, 281, 283-285, 287-289, 291, 293, 295, 297, 299-301, 303, 305, 307, 309, 311, 313, 315, 317, 319, 321, 323, 325, 327, 329, 331, 333, 335, 337, 339-340
Humidity, 215, 243, 254
Hundred feet, 38, 77, 155, 227, 267
Hydraulic, 7, 23, 54, 62, 77, 125, 161, 179-181, 224, 234, 239-240, 244, 253, 257-259, 261, 264, 340

I

Impedance, 4, 7, 15, 52-55, 61-62, 66, 71, 76-77, 113, 140, 142-143, 149, 167, 172, 266, 339
Inertia, 14, 26, 83, 90, 198, 244, 278, 337
Instrumentation, 58, 251, 254, 262
Interference, 3-4, 6-7, 9-10, 14, 24, 26-28, 36, 40-41, 44, 47-48, 52-55, 61, 66-68, 75-78, 80-81, 84, 89-90, 92, 94, 97, 102-104, 106, 115-119, 130, 140, 151, 164, 168-169, 186, 210, 224-225, 228-229, 231-232, 240, 255, 259, 261, 264-265, 268, 272, 274-275, 277-281, 287, 289-290, 339
Intermediate wind tunnel, 234, 239, 248-249, 254, 275
Intersymbol interference, 104, 106
Inviscid, 170, 177, 183-185, 188-189, 194, 199, 204-205, 257

J

Jamming, 10, 20, 28, 179-180, 182, 208, 219, 221, 228, 240, 248
Jet engine, 198, 206-207

K

Kinematic viscosity, 8-9, 145, 164, 209-211, 226, 287

L

Lagrangian, 64, 66-67, 69, 71, 75, 87, 100, 128, 134, 143, 154-155, 171, 267
Laminar flow, 209, 287
Laplace equation, 19
Lift, 183-186, 188, 193, 196, 199, 211, 219, 225
Lift coefficient, 184, 211
Lippert, 62-63
Lithium, 338
Long wind tunnel, 10, 24, 31, 35, 37-39, 43-44, 210, 225, 227-230, 232, 237-238, 250, 254, 259-260, 263-264, 266, 269
Lost circulation material, 28, 180, 221
LSU flow loop, 4

M

Manometer, 159, 212-213, 237-238
Mechanical inertia, 14, 26, 90, 278
Monopole, 4, 51, 55, 66, 129, 171
Mud motor, 15, 26, 61, 64, 70, 72, 77-78, 175, 274, 280, 339
Mud pump, 13, 37-38, 44, 98-99, 108-109, 115-119, 156, 158, 228, 241, 268
Multiple transducer, 3, 6, 18, 24, 26, 36, 42, 45-46, 57, 99, 121, 125, 157, 160, 162, 169, 232, 262, 268, 272, 278, 284, 339

N

NBMTS, 300-301, 309
Near-bit, 178, 274, 300-301, 303, 305-307, 309, 311
Negative pulser, 4-5, 49-51, 57, 66, 68, 153, 170-171, 234
Newtonian, 144, 164-167, 233
Nodes, 4, 27, 57, 160, 164, 267, 284

Noise, 7, 10, 13, 18, 21-22, 24, 26-27, 34, 36-38, 42-43, 54, 60, 75, 91, 97-101, 108-120, 122-127, 140-141, 143-144, 153, 155, 157, 159-163, 165, 167-169, 171-173, 175-176, 232, 250-251, 253, 262, 264, 267, 272-273, 278, 284-287, 293, 309, 339
Non-Newtonian, 144, 164-167, 233
Notch filter, 174
Nozzle, 2-3, 5, 14, 53-54, 88, 98, 135, 150, 268

O

Open reflector, 6, 44, 53, 86, 90, 93-95
Organ, 7, 12, 65

P

Partial differential equation, 12, 67, 157-158, 188, 190
Periodic, 57, 60, 91, 112, 119-120, 157, 169, 178, 183, 192, 194-195, 197, 204, 255, 262, 270
Phase-shift-keying, 14, 48, 77, 84, 129, 131, 137-138, 147-151, 260, 277
Positive displacement, 7, 55-56, 58, 61, 99, 101, 112-113, 153, 158, 160-161, 164, 168, 206
Positive mud pulser, 325
Positive pulser, 4, 10, 12, 14, 49-51, 53, 57, 59, 64, 157, 171, 260, 274, 277, 300, 340
Potential flow, 208
Power, 3, 8, 10, 14, 16, 21-22, 24, 31, 33, 35, 47, 49-52, 54, 56, 62-63, 76, 83-84, 140, 161, 165, 179-180, 182, 184, 205-209, 213-221, 223-225, 227-228, 231, 233, 237, 243-

244, 246, 259, 264, 270, 273-274, 279-280, 303, 337-340

Presco website, 282-283

Pressure drop, 5, 32, 51, 68, 134, 179, 184, 193, 204, 233

Probe-based, 340

Propagating wave, 3, 38, 57-58, 71, 75, 88-89, 119, 143, 172-173, 256

PSK, 12, 14, 26, 48, 77, 84, 104-105, 107-108, 110-111, 129, 131, 146-147, 260, 277-278, 287, 289, 291, 340

R

Radiation, 15, 47, 56, 61-62, 66, 71, 88, 128, 143, 158, 266, 274

Rechargeable batteries, 338-340

Recirculating flow, 246

Reflection(s), 6-7, 13-14, 26, 38-39, 44, 47, 52-53, 57, 60, 62-63, 65, 75, 89, 93, 98-99, 101-102, 108-109, 111, 115-119, 126, 128, 133, 140, 144-145, 152-153, 156-159, 161, 167-168, 170, 176, 210, 251, 264-265, 271-272, 277-278, 291

Resistivity, 26, 55, 61, 130, 274-275, 300-301, 303, 305, 308-309, 320

Resistivity-at-bit, 26, 61, 274-275

Reverberation, 287

Reynolds number, 8, 209-210, 226, 233

Rheology, 145, 165, 184, 233, 257

Rotary hose, 48, 101, 113-114, 121, 144, 284

Rotating, 12, 24, 28, 32-33, 59, 162-163, 178, 183, 196, 204, 213-214, 235, 237, 239, 242, 244-245, 247-248, 250-251, 253-255, 270, 280, 337

Rotation, 8, 19, 27-28, 34-35, 41, 51, 59, 84, 94, 97, 127-128, 178, 210, 212-215, 217, 221, 223-224, 229,

232, 235, 237, 239-240, 244-246, 249, 254-257, 259-260, 278, 280, 337

Rotor, 14-16, 19-22, 26, 28, 31, 41, 50-51, 55, 64-65, 70, 77, 83-84, 91, 127, 161-162, 177-185, 188, 192, 194-196, 198-199, 204, 206-209, 219-221, 225, 229, 239-251, 253-254, 256, 260, 262, 269, 274, 278, 280, 288, 339

S

Savitzky-Golay filter, 153, 172-173

Seals, 4, 9, 179, 237, 337

Seismic, 6, 13, 284

Sensor development, 300

Shim stock, 223

Short wind tunnel, 10, 15, 19, 22-23, 28, 36, 39, 60, 206, 210, 221, 224, 228-234, 236-238, 240, 248, 253, 263

Shroud, 208, 219-220

Signal processing, 3, 6-7, 9-14, 18, 22, 26, 28, 36, 42, 44-46, 48, 52, 58, 75, 85, 96-97, 100, 112, 114, 119-121, 125, 129, 135, 139-141, 144, 152, 154, 157, 160, 162-163, 169, 172, 175-176, 182, 227-228, 232, 238, 248, 255, 262, 264, 267, 273, 277-279, 281, 284, 291-296, 298-299, 303, 339

Sirens in series, 39, 263, 269, 279-280, 289, 337

Sirens in tandem, 269, 279

Sirens-in-series, 10, 231, 240, 263, 272, 340

Six-segment waveguide, 44, 47, 55, 61, 87-88, 90, 162, 172, 268

Software reference, 85, 96, 99, 108, 112, 120, 126, 133, 142, 150, 164, 167, 172, 174, 204, 215, 290

Solid reflector, 3, 44, 53-54, 58, 61, 88-90, 93-94, 96, 98-102, 109, 111-112, 126, 128-129, 133-135, 141, 150-151, 154-155, 161, 168, 171, 249, 267-268

Sound speed, 3, 6-7, 9, 14, 24, 39, 47-48, 51-53, 55, 57, 59, 62-64, 68-69, 76-77, 80, 87, 94, 100, 112, 115, 121-125, 130, 140, 145-146, 164, 174, 210, 232, 249, 252, 256-259, 266, 275, 284-289

Source model, 66

Speaker, 65, 91, 250-253

Squirrel cage, 209, 228, 231, 237, 253

Stable closed, 19-20, 31, 33-34, 39, 196, 240-241

Stable open, 21, 33, 39, 196, 199, 224, 240, 244

Stable-closed, 178-179, 181-182, 184, 195-196, 204, 224, 240-241, 244, 259

Stable-open, 31, 181, 184, 196, 199, 204, 237, 240, 244, 259

Standing wave, 4, 56-57, 61, 84, 90, 160, 164, 250, 270-271, 281

Standpipe, 3, 6, 13, 18, 33, 48, 57, 60, 62, 77, 90, 99-102, 104, 108-109, 112-113, 115, 123, 125-127, 129, 135, 140, 144-145, 155, 158, 168-169, 267-268, 284-285

Stator, 15-16, 19, 28, 50, 55, 61, 65, 77, 84, 91, 161-162, 178-179, 181-185, 192, 194-196, 198-199, 206, 208-209, 219, 225, 229, 239, 242-244, 248, 250, 253-254, 260, 269, 274, 288

Streamfunction, 170, 195, 197

Streamline, 8, 20, 28, 80, 170, 177, 183, 193, 195-198, 201, 227-228, 234, 269

Streamlines, 197-198, 239

Strouhal, 8, 226

Surface reflection, 60, 140, 145, 265, 277

T

Tandem, 16, 27, 231, 269, 279

Telescoping waveguide, 77, 265

Temperature, 12, 65, 174, 209, 211, 215, 226, 243, 254, 292, 301, 309, 338

Thermodynamic loss, 59-60, 270

Tip speed ratio, 245

Tool face, 301, 308-309

Torque, 8-10, 14, 19, 21-22, 24, 26, 28, 31-33, 35, 39, 41-42, 84, 163, 170, 177, 179-185, 187-189, 191, 193, 195-199, 201, 203-225, 227-233, 235, 237, 239-246, 248, 253-254, 257-259, 261, 263, 273, 278, 280, 291, 298, 337, 339

Transducer, 3, 6, 10, 13, 18, 24, 26-27, 34, 36, 38, 42-43, 45-46, 48, 57, 60, 75, 90-91, 97-102, 104, 108-109, 112-114, 121-125, 127, 155, 157, 160, 162, 167-169, 175, 227, 232, 248, 250-254, 262, 267-268, 272, 278, 284-287, 290, 339

Transient, 6-7, 68-69, 92, 94, 97-99, 101, 103, 105, 107, 109, 111-113, 115, 117, 119, 121, 123, 125, 127-129, 131, 133, 135, 137, 139-145, 147, 149, 151, 158, 163-164, 244, 248, 254-255, 277

Transmission, 14-15, 19, 26, 44, 47-48, 50, 52-55, 60, 62-63, 68, 75-81, 83-84, 153, 171, 178, 184-185, 210, 225, 233, 265, 269, 280, 287-289, 300-301, 306-307, 309, 326

Transmission efficiency, 50, 53, 60, 76-81, 83-84

Troughs, 3

Turbine, 2, 8, 14-15, 22, 26, 31, 33, 50, 76, 98, 179-180, 183, 185, 192,

196, 198, 205-217, 219-225, 227-229, 231, 233-237, 239-240, 243, 245-247, 257, 259, 264, 270, 274, 291, 297, 337-340
Turbodrill, 161-162, 274
Turbosiren, 205, 243, 263
Turbulence, 31, 91, 109, 114, 181, 251
Turbulent, 8, 91, 209, 226, 229, 235, 250, 253
Twisted blades, 337

U

U.S. Patent 3,792,429, 180
U.S. Patent 5,583,827, 27, 44
U.S. Patent 5,626,200, 221
U.S. Patent 5,831,177, 243, 263, 269
U.S. Patent 5,969,638, 18
Ultrasonic, 164

V

Velocity, 5, 8, 14, 19, 32, 51-52, 65, 68, 71, 93, 143, 155, 163, 170-172, 177, 182, 185, 189, 191, 194-198, 214-215, 219-220, 234, 238-239, 246, 261
Vibration, 43, 162-163, 207-208, 235, 339
Viscosity, 8-9, 60, 145, 164-165, 167, 183-185, 209-211, 226, 257, 280, 287-289
Viscous drag, 183
Vortex, 91, 164, 208-209, 219-220, 222, 226, 239, 246-248, 250-251, 253

W

Water hammer, 5, 19, 41, 51, 178-179, 184, 248, 256-257
Wave equation, 6-7, 12-13, 18, 48, 66, 87, 154-155, 157, 170, 172, 204, 257, 284
Wave propagation, 1, 3-4, 13, 32, 47, 51, 58-60, 65, 68, 70, 77, 87, 92, 162, 171, 176, 210, 227-228, 259, 265-266, 287
Waveguide, 4-7, 15, 24, 26, 44, 47, 49, 51, 53, 55, 57-59, 61-67, 69-71, 73, 75-77, 79, 81, 83, 85-90, 92, 94, 96, 161-162, 172, 232, 249, 265, 268, 274, 276-277
White noise, 91, 250
Wind tunnel, 8-10, 12, 15-16, 19, 22-26, 28-29, 31-32, 34-40, 43-45, 51, 58, 60-61, 90-91, 127, 159, 175, 182, 206, 209-212, 214-215, 217, 221-243, 246, 248-255, 257-260, 263-264, 266, 269-270, 275, 291, 293-295, 337, 339
WLRS, 300-303, 309

About the Authors

Wilson C. Chin earned his Ph.D. from the Massachusetts Institute of Technology and his M.Sc. from the California Institute of Technology.

He is the author of eight books on drilling and cementing rheology, reservoir engineering, vibrations and wave propagation, well logging and formation evaluation, and MWD design. In addition, he has written over eighty papers in the same subject areas and won eighteen U.S. patents in modern well logging technology.

Prior to forming Stratamagnetic Software, LLC in 1999, Mr. Chin had been affiliated with Boeing, Pratt & Whitney Aircraft, Schlumberger, British Petroleum and Halliburton. Since then, he has been awarded four prestigious Small Business Innovation Research grants from the United States Department of Energy, and recently, a major DOE contract for model and software development for managed pressure drilling flow simulation and job planning.

Present and past corporate clients include Aldine Independent School District, Baker Atlas Oasis, BakerHughes Inteq, Brown & Root Energy Services, Calmena Energy Services, China National Offshore Oil Company (CNOOC), China National Petroleum Company (CNPC), China Oilfield Services Limited (COSL), Gas Research Institute, GE Oil & Gas, Geoservices, Gyrodata, Halliburton Carrollton Tools & Testing, Halliburton Duncan Technology Center, Halliburton Energy Services, Halliburton Sperry Sun, Harris County Education Foundation, Innovative Engineering Systems, Pan American Drilling Services, Pluspetrol Peru, Reed Elsevier, Research Partnership to Secure Energy for America (RPSEA), Schlumberger, Sondex, United States Department of Energy, and others. Mr. Chin is Honorary Professor at the University of Petroleum, Shandong, China, and also, at both the Research Institute of Petroleum Exploration and Development and the Drilling Research Institute, CNPC, in Beijing, China. During 2012, he served as Adjunct Professor, Drilling and Completions, at the University of Houston's new Petroleum Engineering Department.

Mr. Chin may be contacted through Stratamagnetic Software, LLC, Houston, Texas by email or phone at wilsonchin@aol.com and (832) 483-6899. Additional updates on company products and recent news are available at www.stratamagnetic.com.

Yinao Su   Limin Sheng   Lin Li   Hailong Bian   Rong Shi

Yinao Su, an Academician of the Chinese Academy of Engineering, is affiliated with China National Petroleum Corporation (CNPC) in Beijing where he directs its MWD program.   Professor Su has more than thirty years of research and development experience in oil and gas engineering, with major contributions to drilling mechanics, wellbore technology, trajectory control and downhole tool design.   He is an expert in control theory and leads a new research endeavor known as "Downhole Control Engineering." Su holds over thirty patents, has authored numerous books and more than two hundred papers.

Limin Sheng is Senior Technical Expert and Department Head in oil and gas drilling engineering at the CNPC Drilling Research Institute.   Professor Sheng has more than twenty-five years of experience in research and development focusing on MWD and downhole control engineering applications. Special interests include telemetry methodologies, automatic control, inertial navigation and downhole control. Sheng holds more than twenty patents and has published over two dozen papers.

Lin Li is Manager of the Downhole Control Engineering Research Institute, a laboratory for downhole information transmission at CNPC in Beijing. He holds joint positions as Senior Engineer, and Director, Continuous Wave MWD and Electromagnetic MWD Projects. Li is also a key contributor to CNPC's geosteering project efforts.

Hailong Bian earned his Doctorate from the University of Electronics Science and Technology in China. He works as a Postdoctoral Fellow and Engineer at the CNPC Downhole Control Engineering Research Institute. He is the lead technical focal point on CNPC's high-priority continuous wave MWD mud pulse telemetry project.

Rong Shi is an Engineer with the CNPC Downhole Control Engineering Research Institute.   Shi, a key technical contributor to the continuous wave telemetry project, specializes in mechanical design and data acquisition.

Also of Interest

Check out these other related titles from Scrivener Publishing

From the Same Author

Formation Testing: Pressure Transient and Formation Analysis, by Wilson C. Chin, Yanmin Zhou, Yongren Feng, Qiang Yu, and Lixin Zhao, ISBN 9781118831137. This is the only book available to the reservoir or petroleum engineer covering formation testing algorithms for wireline and LWD reservoir analysis that are developed for transient pressure, contamination modeling, permeability, and pore pressure prediction. *APRIL 2014.*

Electromagnetic Well Logging, by Wilson C. Chin, ISBN 9781118831038. Mathematically rigorous, computationally fast, and easy to use, this new approach to electromagnetic well logging does not bear the limitations of existing methods and gives the reservoir engineer a new dimension to MWD/LWD interpretation and tool design. *May 2014.*

Other Related Titles from Scrivener Publishing

Bioremediation of Petroleum and Petroleum Products, by James Speight and Karuna Arjoon, ISBN 9780470938492. With petroleum-related spills, explosions, and health issues in the headlines almost every day, the issue of remediation of petroleum and petroleum products is taking on increasing importance, for the survival of our environment, our planet, and our future. This book is the first of its kind to explore this difficult issue from an engineering and scientific point of view and offer solutions and reasonable courses of action. *NOW AVAILABLE!*

Sustainable Resource Development, by Gary Zatzman, ISBN 9781118290392. Taking a new, fresh look at how the energy industry and we, as a planet, are developing our energy resources, this book looks at what is right and wrong about energy resource development. *NOW AVAILABLE!*

An Introduction to Petroleum Technology, Economics, and Politics, by James Speight, ISBN 9781118012994. The perfect primer for anyone wishing to learn about the petroleum industry, for the layperson or the engineer. *NOW AVAILABLE!*

Ethics in Engineering, by James Speight and Russell Foote, ISBN 9780470626023. Covers the most thought-provoking ethical questions in engineering. *NOW AVAILABLE!*

Fundamentals of the Petrophysics of Oil and Gas Reservoirs, by Buryakovsky, Chilingar, Rieke, and Shin. ISBN 9781118344477. The most comprehensive book ever written on the basics of petrophysics for oil and gas reservoirs. *NOW AVAILABLE!*

Petroleum Accumulation Zones on Continental Margins, by Grigorenko, Chilingar, Sobolev, Andiyeva, and Zhukova. ISBN 9781118385074. Some of the best-known petroleum engineers in the world have come together to produce one of the first comprehensive publications on the detailed (zonal) forecast of offshore petroleum potential, a must-have for any petroleum engineer or engineering student. *NOW AVAILABLE!*

Mechanics of Fluid Flow, by Basniev, Dmitriev, and Chilingar, ISBN 9781118385067. The mechanics of fluid flow is one of the most important fundamental engineering disciplines explaining both natural phenomena and human-induced processes. A group of some of the best-known petroleum engineers in the world give a thorough understanding of this important discipline, central to the operations of the oil and gas industry. *NOW AVAILABLE!*

Zero-Waste Engineering, by Rafiqul Islam, ISBN 9780470626047. In this controversial new volume, the author explores the question of zero-waste engineering and how it can be done, efficiently and profitably. *NOW AVAILABLE!*

Formulas and Calculations for Drilling Engineers, by Robello Samuel, ISBN 9780470625996. The most comprehensive coverage of solutions for daily drilling problems ever published. *NOW AVAILABLE!*

Emergency Response Management for Offshore Oil Spills, by Nicholas P. Cheremisinoff, PhD, and Anton Davletshin, ISBN 9780470927120. The first book to examine the Deepwater Horizon disaster and offer processes for safety and environmental protection. *NOW AVAILABLE!*